Ergonomics *and* Psychology

Developments in Theory and Practice

Edited by

Olexiy Y. Chebykin
Gregory Z. Bedny
Waldemar Karwowski

T0136383

CRC Press
Taylor & Francis Group
Boca Raton London New York

CRC Press is an imprint of the
Taylor & Francis Group, an **informa** business

CRC Press
Taylor & Francis Group
6000 Broken Sound Parkway NW, Suite 300
Boca Raton, FL 33487-2742

First issued in paperback 2019

ISBN-13: 978-1-4200-6700-2 (hbk)
ISBN-13: 978-0-367-38737-2 (pbk)

Library of Congress Cataloging-in-Publication Data

Ergonomics and psychology : developments in theory and practice / editors,
   Olexiy Y. Chebykin, Gregory Z. Bedny, and Waldemar Karwowski.
      p. cm. -- (Ergonomics design and management : theory and applications ;
      v1)
   Includes bibliographical references and index.
   ISBN 1-4200-6700-1 (alk. paper)
   1. Psychology. 2. Human engineering. 3. Psychology, Applied. I. Chebykin,
Olexiy Y. II. Bednyi, G. Z. (Grigorii Zakharovich). III. Karwowski, Waldemar,
1953-

BF149.E73 2008
158.7--dc22   2007048937

Visit the Taylor & Francis Web site at
http://www.taylorandfrancis.com

and the CRC Press Web site at
http://www.crcpress.com

Ergonomics *and* Psychology

Developments in Theory and Practice

Ergonomics Design and Management: Theory and Applications

Series Editor

Waldemar Karwowski
Industrial Engineering and Management Systems
University of Central Florida (UCF) – Orlando, Florida

Published Titles

Trust Management in Virtual Organizations: A Human Factors Perspective
*Wiesław M. Grudzewski, Irena K. Hejduk, Anna Sankowska,
and Monika Wańtuchowicz*

Ergonomics and Psychology: Developments in Theory and Practice
Olexiy Y. Chebykin, Gregory Z. Bedny, and Waldemar Karwowski

Forthcoming Titles

Ergonomics in Developing Regions: Needs and Applications
Patricia A. Scott

Handbook of Human Factors in Consumer Product Design, 3 volume set
Neville A. Stanton and Waldemar Karwowski

Nanotechnology, Occupational and Environmental Health and Safety
Ash M. Genaidy and Waldemar Karwowski

Contents

Preface ix
Acknowledgments xiii
Editors xv
Contributors xvii

Section I Work Psychology and Ergonomics

1 Ecological Ergonomics 3
Marvin J. Dainoff

2 Integrating Cognitive and Digital Human Models for Virtual Product Design 29
Daniel W. Carruth and Vincent G. Duffy

3 Time Study during Vocational Training 41
Gregory Z. Bedny and Waldemar Karwowski

4 The Laws of Ergonomics Applied to Design and Testing of Workstations 71
V.F. Venda, V.K. Kalin, and A.Y. Trofimov

5 Day-to-Day Monitoring of an Operator's Functional State and Fitness-for-Work: A Psychophysiological and Engineering Approach 89
Oleksandr Burov

Section II Modular Processes in Mind and Brain

6 Identification of Mental Modules 111
Saul Sternberg

7 Identification of Neural Modules 135
Saul Sternberg

Section III   Psychophysiology of Work

8   **The New Interface of Brain, Mind, and Machine: Will the Emergent Whole Be Greater than the Sum of the Parts?**........... 167
     Chris Berka, Daniel J. Levendowski, Gene Davis, Vladimir T. Zivkovic, Milenko M. Cvetinovic, and Richard E. Olmstead

9   **The Interaction of Sleep and Memory** 189
     Jeffrey M. Ellenbogen

10   **Attention, Selection for Action, Error Processing, and Safety**.. 203
     Magdalena Fafrowicz and Tadeusz Marek

Section IV   Activity Theory and Ecological Psychology and Their Application

11   **Activity Theory: Comparative Analysis of Eastern and Western Approaches**.................................... 221
     Waldemar Karwowski, Gregory Z. Bedny, and Olexiy Y. Chebykin

12   **Discourse in Activity** 247
     Harry Daniels

13   **Movements of the Cane Prior to Locomotion Judgments: The Informer Fallacy and the Training Fallacy versus the Role of Exploration**.. 267
     Gregory Burton and Jennifer Cyr

Section V   Emotional Regulation of Activity and Education

14   **Emotional Intelligence: A Novel Approach to Operationalizing the Construct**.................................... 303
     E. L. Nosenko

15   **Emotional Regulation of the Learning Process**.......................... 325
     Olexiy Y. Chebykin and S.D. Maksymenko

16   **Emotional Resources of the Professional Trainer**...................... 341
     G.V. Lozhkin

Section VI   Personality

17   Good Judgment: The Intersection of Intelligence and Personality...357

Robert Hogan, Joyce Hogan, and Paul Barrett

18   Relational Self in Action: Relationships and Behavior 377

Susan E. Cross and Kari A. Terzino

Index...397

Preface

One of the main questions of psychology is concerned with the relationship between different fields and schools of psychology and their influence on theoretical and applied studies. The aim of this book is to give the reader representative ideas of the most successfully applied basic approaches in psychology, ergonomics, education, training, etc.

Such issues as interaction and mutual influences of the diverse schools of psychology on each other become particularly important during the globalization of the world economy and science. There is a fundamental need to understand what is positive and negative in different directions of psychology, and how we can overcome what is negative and utilize what is positive in our theoretical and applied studies.

The cognitive approach has been a dominant one in the United States for a long time now. Fundamental theoretical data and multiple techniques adopted for the study of mental rather than manual work have been developed, and new theoretical concepts of learning and instruction have been suggested utilizing this approach. However, no one scientific approach can be considered as a perfect one and as an absolute truth in psychology. Every approach has its positive and negative aspects. This is also true for cognitive psychology. As a result, new approaches in psychology and an attempt to utilize them in practice have emerged in the United States and other countries. To these approaches one can relate ecological psychology, situated cognition, action theory, activity theory, etc. Often, representatives of the new approaches start with overemphasizing negative aspects of other approaches and attempting to exaggerate advantages of their own approach. This negatively affects objective analysis of the advantages and disadvantages of different directions in psychology and the possibilities of utilizing them in practice. Therefore, there is a need to integrate different directions in psychology to further their successful application.

The purpose of this volume is to demonstrate that different scientific approaches are tightly interconnected and their uniqueness is relative. Areas of psychology that can successfully accommodate the achievements of other directions and modify them can later be significantly developed and efficiently applied in practice.

In this volume we present chapters by scientists who work in different directions of psychology and, particularly, in cognitive psychology, ecological psychology, and activity theory. Interaction and interdependence of these directions of science can be traced in the work of a variety of authors in this book. For example, one of the leading specialists in cognitive psychology, Saul Sternberg, who presents his research in this book, developed the additive factors method from the late 1970s through the early 1980s, facilitating the discovery of the existence and distinctiveness of different stages of the information processing. These stages are of a very short duration in time and can be measured in milliseconds. The results of his study were not only utilized in engineering psychology in the United States but also accumulated and were elaborated upon in activity theory. As a result, microstructural methods of the analysis of human cognitive actions have been developed in activity theory.

Here is another example: Cognitive processes in activity theory are always considered as "mental processes." However, such concepts as sensory memory, short-term memory, long-term memory, etc., were transferred into activity theory from cognitive psychology. At the same time, in general and systemic-structural activity, theory cognition is considered not only as a process, but also as a system of cognitive actions and operations. On the other hand, systemic-structural activity theory outlines different stages of activity analysis based on this data. Some directions of Western psychology such as German action theory, situated cognition, ecological psychology, etc., have been influenced by activity theory. For example, the works of Berenshtein on self-regulation of motor actions and movements was an important source for development of the ecological psychology.

Representatives of not only various scientific approaches, but also representatives of different countries contributed to this volume. Most of the authors are scientists from the United States and the Ukraine. The Ukraine is no longer just a republic of the former Soviet Union but is an independent country now. Disintegration of the Soviet Union was associated not only with the changes in the political climate in the Ukraine and other republics, but also with the drastic deterioration of the economy. The political turmoil in the region also negatively affected the cooperation of scientific communities of the former republics. All of this, in turn, negatively influenced Ukrainian science. However, we can now observe that the economy and science are moving forward in Ukraine. It was a pleasure to see how enthusiastically Ukrainian scientists took part in the development of this volume. We want to note that some contributors who live in the United States present data gathered in Ukraine (see, for example, the work of Susan E. Cross and Kari A. Terzino). The cooperative work of scientists from the United States, the Ukraine, and other countries will promote the further development of science.

This book is the collective achievement of scientists in diverse areas of psychology and ergonomics. Papers that are presented in this book allow the readers to develop their own vision and concept about psychological issues and apply them in ergonomics, education, and other important areas of human activity.

The book touches on major recent developments in psychology and their application. The major areas covered in this book are cognitive psychology and psychophysiology, activity theory and ecological psychology, emotion regulation, social psychology and personality, work psychology and ergonomics. The authors who have contributed to this volume are leading scientists in their respective fields.

Section I of the book, "Work Psychology and Ergonomics" provides a review on studies of human work from different perspectives. It has brought together experts in the field from diverse branches of ergonomics and psychology. This section lays the foundation for demonstrating the possibilities for future research by utilizing different approaches in psychology.

In Section II of the book, "Modular Processes in Mind and Brain," a method for the modular decomposition of mental and neural processes is discussed, together with examples of its application to behavioral data (Chapter 2.1) and neural data (Chapter 2.2). A special case of this approach, the method of additive factors (AFM), was first described by Sternberg in 1969, as a way to use reaction-time data to analyze mental processes that are organized as a sequence of processing stages. The AFM has been applied in numerous experiments in cognitive psychology, and has also triggered studies in activity theory. The more general process decomposition approach of Section II can be applied to a far wider range of measures and process structures.

Section III, "Psychophysiology of Work" presents some recent discoveries regarding the cognitive and neural processes that can take place during performance of different kinds of tasks. We believe that a solid grasp of theory in these basic branches of psychology provides a strong base for the development of the efficient methods in the study of human work.

Section IV, "Activity Theory and Ecological Psychology" includes three chapters: two of them are related to activity theory and one to ecological psychology. It is worth noting that activity theory influenced ecological psychology; hence these two related topics will logically appear in the same section.

Section V is titled "Emotion Regulation of Activity". The purpose of this section is to demonstrate theoretical and experimental justifications of emotions during the study of human work, learning, and training.

Section VI called "Personality" includes two chapters in which the authors consider such theoretical and applied aspects of psychology as personality and individual differences. These areas of applied psychology study human activity or behavior through the prism of personality.

The book can be of use for a broad spectrum of scientists, practitioners, and students who work in such areas as psychology, human factors and ergonomics, education, and other areas of research. The presented studies performed from various theoretical perspectives have not only their specific methods but also have some common principles.

Presenting the variety of research methods and approaches in one book allows the reader to achieve a broader perspective on how these approaches can benefit from each other.

Olexiy Y. Chebykin
Gregory Z. Bedny
Waldemar Karwowski

Acknowledgments

I want to thank the many people who helped me to edit this book. First, the wonderful volunteers of the U.S. Peace Corps at South Ukrainian State Pedagogical University, Samrong So and Jason Tolub, who helped with the English translation. I want to extend my gratitude to my colleagues at the Foreign Languages Department of the University of South Ukraine and especially to Marina Yakovleva for their assistance. And finally, I want to express my appreciation to my family, my wife Tamara and my son Dmitriy, for their love, patience, and support throughout my career.

Olexiy Y. Chebykin

I would like to express my great appreciation to all the contributors to this book who worked so diligently with the editors to make it happen.

My family was my greatest support, as has been the case in all my undertakings.

So, I'd like to thank my wife Inna and my daughter Marina for their help.

My warmest thanks also go to my life-long friend Mark Zeltser whose wealth of knowledge and command of both languages has been instrumental in this effort.

Gregory Z. Bedny

This book is an example of the successful international collaboration between psychologists and human factors professionals across two continents. My sincere thanks go to my two coeditors, Drs. Chebykin and Bedny. It has been a real pleasure to work with them from the inception of this book. I would also like to express my appreciation to Laura Abell, who served as editorial assistant on this challenging project.

Waldemar Karwowski

Editors

Olexiy Y. Chebykin holds a Ph.D. and an Sc.D. in psychology and specializes in the field of the emotional regulation of human activity. He is a full member and Academician-Secretary of the Academy of Pedagogical Science, president of South Ukrainian Pedagogical University after K. D. Ushinskiy, and a member of the International and American Association of Psychologists. Prof. Chebykin obtained significant results in the elucidation of the nature of stressed and poststressed states of students and the development of methods of emotional regulation in students during the educative process. He is an editor and a member of the editorial boards of many scientific journals. Prof. Chebykin holds the title of an Honored Worker of Science of the Ukraine.

Gregory Z. Bedny, Sc.D., Ph.D., presently works at Essex County College in Newark, New Jersey. He was awarded a Ph.D. in industrial/organizational psychology by Moscow Educational University and holds a postdoctorate degree (Sc.D.) in experimental psychology from the Institute of General and Educational Psychology, Russian National Academy of Pedagogical Sciences. Dr. Bedny is a member of the International Academy of Human Problems in Aviation and Astronautics (Russia) and was awarded an honorary doctor of science degree by the University of South Ukraine. He is a board-certified professional ergonomist. A member of the editorial boards of several international journals, he has authored over 100 scientific publications, including five monograph books and several textbooks. His latest are the monographs *The Russian Theory of Activity: Current Applications to Design and Learning* (coauthored with D. Meister) and published in 1997, and *A Systemic–Structural Theory of Activity: Application to Human Performance and Work Design* (coauthored with W. Karwowski) and published in 2006. He also served as a guest editor of theoretical issues in *Ergonomics Science*, Vol. 5 (4), 2004. His research focuses on experimental and industrial/organizational psychology as well as ergonomics. He developed systemic–structural activity theory, which he has applied to the design of a human–machine system and to human–computer interaction, safety, education and training, work motivation, efficiency of performance, etc. He can be reached at gbedny@optonline.net.

Waldemar Karwowski, Sc.D., Ph.D., CPE, P.E., is a professor and chairman of the Department of Industrial Engineering and Management Systems at the University of Central Florida, Orlando. Dr. Karwowski is a board certified professional ergonomist. He served as secretary-general (1997–2000) and president (2000–2003) of the International Ergonomics Association. Dr. Karwowski is the author or co-author of over 300 scientific publications. He has recently published the monograph *A Systemic–Structural Theory of Activ-*

ity: Application to Human Performance and Work Design, coauthored with Dr. Bedny. He is an editor of several scholarly books including the *International Encyclopedia of Ergonomics and Human Factors*. He holds an honorary doctor of science degree from several universities. He also is a member of the International Academy of Human Problems in Aviation and Astronautics (Russia). Dr. Karwowski serves as the editor of *Human Factors and Ergonomics in Manufacturing*, an international journal published by John Wiley & Sons (New York) and is the editor-in-chief of *Theoretical Issues in Ergonomics Science* (TIES). He can be reached at wkar@mail.ucf.edu.

Contributors

Paul Barrett
Hogan Assessment Systems
Tulsa, Oklahoma

Gregory Z. Bedny
Essex County College
Newark, New Jersey

Chris Berka
Advanced Brain Monitoring
Carlsbad, California

Oleksandr Burov
International Academy of Human
    Factors Challenges
Kiev, Ukraine

Gregory Burton
Psychology Department
Seton Hall University
South Orange, New Jersey

Daniel W. Carruth
Department of Cognitive Science
Center for Advanced Vehicular
    Systems
Mississippi State University
Starkville, Mississippi

Olexiy Y. Chebykin
Academy of Pedagogical Science
South Ukrainian Pedagogical
    University
Odessa, Ukraine

Susan E. Cross
Department of Psychology
Iowa State University
Ames, Iowa

Milenko M. Cvetinovic
Advanced Brain Monitoring, Inc.
Carlsbad, California

Jennifer Cyr
Rutgers University
New Brunswick, New Jersey

Marvin J. Dainoff
Department of Psychology
Miami University
Oxford, Ohio

Harry Daniels
Director of Centre for Sociocultural
    and Activity Theory Research
The University of Bath
Bath, United Kingdom

Gene Davis
Advanced Brain Monitoring
Carlsbad, California

Vincent G. Duffy
Department of Industrial Engineering
Center for Advanced Vehicular
    Systems
Mississippi State University
Starkville, Mississippi

Jeffrey M. Ellenbogen
Massachusetts General Hospital
    Sleep Medicine Program
Harvard Medical School
Boston, Massachusetts

Magdalena Fafrowicz
Institute of Applied Psychology
Jagiellonian University
Krakow, Poland

Joyce Hogan
Hogan Assessment Systems
Tulsa, Oklahoma

Robert Hogan
Hogan Assessment Systems
Tulsa, Oklahoma

V.K. Kalin
Department of Psychology
Tavric National University
Simferopol, Ukraine

Waldemar Karwowski
Department of Industrial
   Engineering and Management
   Systems
University of Central Florida
Orlando, Florida

Daniel J. Levendowski
Advanced Brain Monitoring, Inc.
Carlsbad, California

G.V. Lozhkin
National University of Physical
   Education and Sports
Kiev, Ukraine

S.D. Maksymenko
Academy of Pedagogical Science
G.S. Kostyuk Institute of Psychology
Kiev, Ukraine

Tadeusz Marek
Jagiellonian University
Krakow, Poland
and
Warsaw School of Psychology
Warsaw, Poland

E.L. Nosenko
Department of General and
   Educational Psychology
Dnipropetrovsk National University
Dnipropetrovsk, Ukraine

Richard E. Olmstead
VA Greater Los Angeles Healthcare
   System
Los Angeles, California

Saul Sternberg
Department of Psychology and
   the Institute for Research in
   Cognitive Science
University of Pennsylvania
Philadelphia, Pennsylvania

Kari A. Terzino
Social Psychology Program
Iowa State University
Ames, Iowa

A.Y. Trofimov
Department of General and Applied
   Psychology
Kiev State University
Kiev, Ukraine

V.F. Venda
The Venda Ergonomic Advantages,
   Inc.
Mississauga, Canada
and
Alupka, Crimea, Ukraine

Vladimir T. Zivkovic
Advanced Brain Monitoring, Inc.
Carlsbad, California

Section I

Work Psychology
and Ergonomics

1

Ecological Ergonomics

Marvin J. Dainoff

CONTENTS

1.1 The Problem: Ergonomics' Identity Crisis 3
1.2 The Problem with Ergonomics—Also Its Strength 6
1.3 An Example—The Case of Seated Work Posture 8
    1.3.1 Relationship between Working Conditions and Disorders 8
    1.3.2 Interdependencies in Seated Posture 9
1.4 Beyond Anthropometry—Implementing Fit with Multiple
    Degrees of Freedom 15
1.5 Necessity for an Ecological Approach 17
1.6 A Framework for Ecological Research and Practice 20
    1.6.1 Cognitive Work Analysis as a Tool for Integration 21
    1.6.2 Epistemological Issues 23
1.7 Conclusions 24
Acknowledgments 25
References 25

The premise of this chapter is that the field of human factors/ergonomics (HF/E) represents a unique intersection among professional and academic specialties with a particular focus on psychology and engineering. As such, HF/E is in a very advantageous position to both understand and improve a major societal issue of our time—the relationship between people and technology. However, the very multidisciplinary and cross-disciplinary attributes that are the strength of this field are also a source of considerable challenge and disputation. It is, therefore, particularly appropriate for this volume to offer the proposition that a synthesis of theories from the American ecological psychologist J.J. Gibson and the Russian motor control theorist N.A. Bernstein can serve as an integrating framework for HF/E.

1.1 The Problem: Ergonomics' Identity Crisis

It may be useful to examine three "official" definitions of HF/E. The first is from the International Ergonomics Association (IEA), the second from the Human Factors and Ergonomics Society (HFES), and the third from the Board of Certification in Professional Ergonomics (BCPE):

> Ergonomics (or human factors) is the scientific discipline concerned with the understanding of interactions among humans and other elements of a system, and the profession that applies theory, principles, data and methods to design in order to optimize human well-being and overall system performance.
>
> Ergonomists contribute to the design and evaluation of tasks, jobs, products, environments and systems in order to make them compatible with the needs, abilities and limitations of people. (International Ergonomics Association, 2000)
>
> The Society's mission is to promote the discovery and exchange of knowledge concerning the characteristics of human beings that are applicable to the design of systems and devices of all kinds.
>
> The Society furthers serious consideration of knowledge about the assignment of appropriate functions for humans and machines, whether people serve as operators, maintainers, or users in the system. And, it advocates systematic use of such knowledge to achieve compatibility in the design of interactive systems of people, machines, and environments to ensure their effectiveness, safety, and ease of performance. (Human Factors and Ergonomics Society, 2000)
>
> Ergonomics is the discipline that applies scientific data and principles about people to the design of equipment, products, tasks, devices, facilities, environments, and systems to meet the needs for human productivity, comfort, safety and health. (Board of Certification in Professional Ergonomics, 2002)

Although there seems to be consensus among those who consider themselves professional ergonomists (e.g., IEA BCPE, HFES) that the field can be basically categorized into physical, cognitive, and organizational components (see in particular Hendrick, 2000), this is not the view from outside the field. A common perception of the general public is that ergonomics is a *cause* of musculoskeletal disorders in the workplace. The official Web site of OSHA (http://www.osha-slc.gov/SLTC/ergonomics/) focuses on determining whether there are "ergonomic-related injuries" associated with certain jobs or tasks. Ergonomics is seen as a problem or hazard to be solved rather than a solution to problems of poor design/work organization. At the same time, a recent textbook on human–computer interaction presents a similarly limited view of the field:

> Ergonomics is a huge area which is distinct from HCI but sits alongside it. Its contribution to HCI is in determining constraints on the way we design systems and suggesting detailed and specific guidelines and

standards. Ergonomic factors are in general more established and better understood than cognition and are therefore used as the basis for
standardizing hardware design. (Dix, Finlay, Abowd, and Beale, 1998,
p. 115)

The constraints referred to here are primarily focused on issues such as display
and keyboard layout. Therefore, Dix et al. argue that the focus of ergonomics
is *physical* rather than cognitive. Accordingly, principles of usability, which
form the core of their text, do not seem to be the province of ergonomics.

To further confuse the issue, consider the following arguments made in a
recent special issue of the *International Journal of Human Computer Interaction*
assessing the scientific foundations of usability (Gillen and Bias, 2001):

a) A consensus of leaders in HCI is that traditional cognitive psychology has had minimal impact on the problems of interface design
   and usability. Hence, usability engineers do not find this research
   helpful in design.
b) A new professional field of "usability science" needs to be created.
c) This field is different from both HF/E and HCI.

According to Gillen and Bias (2001, p. 357), HF/E is "the study of how humans
accomplish work-related tasks in the context of human–machine system operation ..." Therefore, they argue that (i) much of HF, such as anthropometry and
biomechanics, is outside the purview of usability science, and (ii) given the
emphasis on work-related tasks and machines, the scope of HF/E is narrower
than the scope of usability science. At the same time, HCI is characterized
in terms of an "... exclusive focus on computing and often on the interactive
aspects of using computers to the exclusion of less interactive aspects such as
reading displays" (Gillen and Bias, 2001, p. 358). However, usability science is
broader, considering noncomputing as well as computing artifacts.

Gillen and Bias seem unaware of the logical inconsistency in including
noncomputing artifacts (tools and machines) within the purview of usability
science while rejecting anthropometry and biomechanics as being irrelevant
to usability. How can, for example, the usability of a hand-held power tool
be considered without reference to issues of arm/hand anthropometry, and
force moments around wrist, elbow, and shoulder joints?

Gillen and Bias argue that a new conceptual approach to usability science
needs to emerge. This approach, which has its origins in the functionalist
school of psychology, should include the following components: (a) goal orientation, (b) adaptation, (c) close links between cognition and action, and (d)
explicit consideration of physical, social, cultural, and task context.

Karwowski (2000) has raised a parallel set of arguments. He discusses the
"misconceptions about perceived limitations of current ergonomic theory,"
including "lack of objective measures of human performance, uncertainty
of the cause-effect paradigm explanation, or low predictive powers ... with
respect to ergonomic intervention effects." He further observes that "many of
the misconceptions ... by critics ... [who] ... have very limited if any under-

standing of the underlying phenomena of complexity and non-linear properties of many human-artifact systems" and that "it does not help that many ...
members of the ergonomic community do not recognize the importance of
these phenomena either" (p. 80).

Karwowski argues that what is required to deal with these concerns is a
new scientific discipline that focuses on compatibility relationships between
humans and artifacts. He calls this discipline *symvatology* (reasoning about
compatibility). The study of compatibility relationships is essential in order
to improve system–human fit.

1.2   The Problem with Ergonomics—Also Its Strength

Ergonomics exists at the interstices of many different scientific and engineering disciplines. These include industrial and manufacturing engineering, work physiology and biomechanics, medicine, anthropometry, cognitive
psychology, computer science, statistics, and management. The strength of
ergonomics is directly derived from the overlapping of disciplinary boundaries. MacArthur Fellow and immunologist David Root-Bernstein has argued
persuasively that this is generally true in scientific creativity. He quotes the
conclusion from a conference of Nobel Prize Laureates:

> The most dramatic progress was to be made by those who mastered the
> tools and ideas of a range of previously weakly related disciplines, and
> based their work on the emerging new mathematics. (Root-Bernstein,
> 1989, p.384)

Following directly from Root-Bernstein's generalization of this conclusion,
the field of ergonomics could greatly benefit from a conceptual (not necessarily mathematical) framework within which ideas and findings from contributing disciplines could be expressed and discussed. The purpose of this
chapter is to provide such a framework, based on the seminal contributions
of J.J. Gibson and N.A. Bernstein.

> Two books were written at the end of the 1940s, which together should
> have changed the face of psychology. Both of these books offered natu-
> ralistic approaches to basic psychological processes, which, prior to that
> time, had been treated in highly abstract ways, or torn out of their func-
> tional contexts. (Reed and Bril, 1996, p. 242)

As Reed and Bril eloquently remind us, the first of these books was J.J. Gibson's *Perception of the Visual World* (1950); the second, which was suppressed
for political reasons, was N.A. Bernstein's *On Dexterity*. The volume was
finally published (Bernstein, 1996), and it remained for those who followed

and elaborated the Gibsonian position to incorporate Bernstein's insights into the ecological position (see for example, Turvey, Shaw, and Mace, 1978).

Gibson's ecological approach to psychology, as outlined in his last book (Gibson, 1979), provides part of the foundational basis for an ecological ergonomic framework. Gibson rejected the prevailing view of psychology as the study of an individual organism in which environmental context is simplified or ignored. (The argument is also at the heart of the critique by Gillen and Bias, discussed earlier.) Instead, he argued that the individual and environment are so tightly and reciprocally coupled that they cannot be studied independently of one another. Thus, to understand the simple case of a person walking across a field, a detailed physical analysis of the terrain is required—including characterization of the projected optical flow patterns across the retina associated with movement in a given direction. Because this approach takes into account the interacting aspects of both person and environment, Gibson called it an "ecological" approach. Gibson's arguments are similar to the *systems* versus *psychological* distinction made by Meister (1989), and, in fact, Gibson himself saw similarities between his view and systems theory (Vicente, 1997).

Finally, consider an alternative definition of HF/E (consistent with the "official" definitions described earlier) as the *fit* between people and the elements of the physical environment with which they interact (Dainoff and Dainoff, 1986, p. 1). As such, ergonomics is inherently relationship oriented in that absolute dimensions and physical characteristics of objects in the work environment must be defined relative to the relevant characteristics of the user. The ecological perspective provides a principled approach to conceptualizing such relationships.

Whereas Gibson argued for the importance of viewing perception within the functional context of behavior in the environment, Bernstein presented a parallel argument for viewing actions within their functional/adaptive contexts. What he called "dexterity" is a capacity of solving motor control problems under dynamically changing parameters (Reed and Bril, 1996). Hence, movement science cannot study movements as abstract patterns without taking into account functional demands of task constraints, environmental constraints, and (changing) constraints within the individuals themselves (Newell, 1996).

The application of these basic concepts to the practical problems of HF/E will be further developed later in this chapter. However, at this point it is of interest to point out that the ecological framework has already had a considerable impact on the field of ergonomics (Flach, Hancock, Caird, and Vicente, 1995; Hancock, Flach, Caird, and Vicente, 1995), particularly in its direct influence on cognitive work analysis (see Rasmussen, Pejtersen, and Goodstein, 1994; Vicente, 1999; Vicente, 2002). Ironically, given the foundational work on perception and action, the bulk of this influence has been in the areas of cognitive and organizational rather than physical ergonomics. However, there are some exceptions (Dainoff and Mark, 2001; Dainoff and Wagman, 2004; Dainoff, Mark, and Gardner, 1999; Mark, Dainoff, Moritz, and Vogele,

1991; Mark et al., 1997; Wagman and Taylor, 2004). We will now explore these exceptions, during the course of which we hope to argue that the strong differentiation among physical–cognitive–organizational components itself has been problematic and a disservice to the field.

1.3   An Example—The Case of Seated Work Posture

Consider the case of a person sitting in a chair working at a computer. If one were to pick a prototypic work task for the turn of the millennium, this would probably be it.

1.3.1   Relationship between Working Conditions and Disorders

Since the introduction of the computer terminal into the office environment in early 1970s, there has been a growing body of evidence describing musculoskeletal and visual problems associated with prolonged periods of work at such terminals (Dainoff, 1980, 1982, 2000; National Research Council, 2001). As reports of such disorders began to accumulate, so did the availability of new kinds of office equipment (e.g., adjustable chairs and furniture, lighting fixtures, glare control treatments). Eventually, there arose an international awareness of the need to provide guidance and direction to both users and manufacturers of computer terminals and associated office equipment. This was manifest in a series of governmental and nongovernmental standards and regulations, e.g., ISO 9241, EC Directive 90/270/EEC, ANSI/HFS 100. In the United States, concerns over what came to be called "ergonomic disorders" in the office became part of a broader concern with "work-related musculoskeletal disorders" which culminated in the promulgation of an ergonomic regulation through the U.S. Occupational Safety and Health Administration (OSHA). The political reaction to this regulation and its subsequent overturn is well known to the ergonomic community.

What can we learn from this controversy? For purposes of this chapter, our focus will be narrowed so as to address two fundamental scientific questions: (a) were the reported problems amongst office workers sufficiently attributable to working conditions that they could, in fact, be classified as "work-related (medical) disorders," and (b) were the "interventions" (e.g., providing better-designed furniture) effective in reducing such disorders?

During the debate over the OSHA regulation, Congress directed the National Research Council to conduct a thorough analysis of the issue, including an extensive review of the published literature. With respect to the two questions just posed, their conclusions were as follows:

> We can conclude with confidence that there is a relationship between exposure to many workplace factors and an increased risk of musculo-skeletal disorders. ... there is some evidence that using ergonomic principles to modify chairs, workstations, and keyboards can be effective in reducing the prevalence of upper extremity symptoms; in the office setting, results concerning the effects of these interventions on physical findings are mixed. (National Research Council, 2001, pp. 362–363)

Reviewing much of the same literature with respect to interventions, Westgaard and Winkel (1997) argued that *none* of the intervention studies they review are definitive in the sense of meeting standards of scientific rigor associated with clinical trials (e.g., random assignment to experimental and control conditions, double-blind assessment). A parallel assessment by Karsh et al. (2001) came to much the same conclusion.

Thus, in summary, there is some agreement that the disorders observed in office workers should be taken seriously; however, the effectiveness of using equipment designed according to ergonomic principles to alleviate these disorders has not been demonstrated.

The argument to be put forward here is that the reason the effectiveness of interventions has not been demonstrated is that they have not been studied properly. Specifically, what has been lacking is an ecological perspective. This argument has been developed in more detail elsewhere (Dainoff, 2006), but will be summarized here.

What might constitute an ergonomic intervention that *should* be effective? Much office work with VDTs tends be characterized by prolonged periods of static posture coupled with high visual demands. In such postures, movement occurs in only a few muscle groups such as wrists and fingers. To the extent that poorly designed workstations require the operator to take up working postures that are inefficient ("awkward"), the onset of fatigue—with consequent discomfort or pain—will be relatively rapid. Conversely, the application of ergonomic principles to workplace design will delay the onset of fatigue, resulting in improved work efficiency and increased feelings of well-being. In the longer term, ergonomic improvements should reduce the incidence of those musculoskeletal disorders that might be linked to poor working postures (Dainoff, 2000; Kroemer and Kroemer, 2001).

1.3.2 Interdependencies in Seated Posture

What might these improvements consist of? The core of an office ergonomic intervention has traditionally been the ergonomic chair. Initially (in the early 1970s), the key components differentiating an ergonomic chair from a traditional task chair was the capability of height adjustability of the seat, along with a rounded front edge and perhaps a bit of support in the lumbar region of the back. The resulting working posture constrains the angle between thighs and trunk to around 90° (see Figure 1.1). However, this "cubist" posture was criticized by two leading researchers, E. Grandjean and A. Mandal, as not

FIGURE 1.1

Early version of ergonomic chair constrains trunk and thighs to 90° posture.

providing sufficient support for the lumbar spine. Grandjean et al. (1983) proposed that both the backrest and the seat should be angled such that the user could work in a backward-leaning posture. Mandal (1981), however, argued that a forward angle of the seat pan would allow the lumbar spine to maintain a healthy posture while keeping the trunk upright. Figure 1.2 illustrates both postures (see Figure 1.2).

Dainoff and Mark (1987) reasoned that both arguments were valid in the sense that different task demands required alternative working postures. So, for example, the backward-leaning posture might be preferable if the task was predominantly screen intensive. However, this posture substantially increases the viewing distance to the working surface. If the user is required to read paper documents, which typically have smaller font sizes, the increased viewing distance is likely to make the backward posture unsatisfactory. Hence, a forward-leaning posture, with the trunk upright and a reduced viewing distance, is preferable for tasks involving paper documents.

Note that although the initial focus on improving chair design was directed at the goal of supporting the lumbar spine, it became clear that other task-related goals needed to be accommodated (e.g., maintaining appropriate viewing distance). Consequently, the ergonomic chair increased in complexity, requiring controls for back and seat angle adjustment along with the original height adjustment mechanism. Present-day chairs also have controls for arm rest height and angle, seat pan depth, and tension control.

Accordingly, as ergonomists began collaborating with other interested industry and user representatives to develop ergonomic standards and guidelines (e.g., the American National Standard [ANSI/HFS 100] in 1988), it became clear that individual elements of the workplace (such as the chair) could not be considered in isolation, but as components in an integrated workplace system (Human Factors Society, 1988). Consider, for example, the dilemma faced by standards writers in dealing with the seemingly simple problem of specifying the range of height adjustability of an ergonomic chair. These specifications can have major financial implications to the extent that furniture manufacturers will modify designs on the assumption that customers will demand products that meet such specifications.

FIGURE 1.2A
Forward-leaning posture. Seat pan angled forward with backrest upright. Distance to copy-holder is 660 mm.

FIGURE 1.2B
Backward-leaning posture. Distance to copyholder is 890 mm.

The underlying goal of the specification is to allow a comfortable (not awkward) seated working posture while taking into account the variability in body dimensions (anthropometry) in the working populations. This goal is typically operationalized by attempting to accommodate the middle 90% of the user population. In the case of the ANSI/HFS 100 (Human Factors Society, 1988), the target population was the United States *civilian* workforce. (Insofar as work on ANSI/HFS 100 preceded efforts on ISO 4921, this is where the discussion will focus.) However, in the current revision to that Standard (BRS /HFES 100), a recent and extensive anthropometric survey of U.S. Army personnel (Gordon et al., 1989) was considered the best approximation to the U.S. civilian population (Human Factors and Ergonomics Society, 2002).

How does seat height adjustability relate to comfortable seated work posture? The simplest principle is that the height adjustment should allow the feet to be firmly support by the floor while the lower legs are vertical. The relevant anthropometric dimension is called *popliteal height*—the distance from the floor to the popliteal crease just behind the knee. Thus, if the seat height can be set to the popliteal height of a user, comfortable seat posture can be achieved. From the army data, the popliteal height of a small (5th percentile) female is 351 mm, and that of a large (95th percentile) male is 477 mm (Gordon et al., 1989). Hence, if the seat height can be made adjustable over this range, most of the target population can sit in a comfortable posture.

However, users do not just sit in chairs; they must also accomplish work. In particular, for the problem under consideration, they must operate a keyboard. Hence, a second postural criterion is required—the user's hands and arms must also be in a comfortable (not awkward) posture while operating the keyboard. This posture can be achieved when the user's elbow is at the level of the keyboard. This allows the upper arms to be vertical and the lower arms, hands, and wrist to be horizontal. The key anthropometric dimension to be accommodated is called *elbow rest height*—the vertical distance from the surface of the seat to the tip of the elbow when the upper arms are vertical. Thus, if the keyboard height is set to the user's elbow rest height, a comfortable working posture can be achieved.

Unfortunately, nature has given us a problem. Those individuals who have small (or large) elbow rest heights are not the same people who have small (or large) popliteal heights. In fact, in the army data for females, the correlation between these two dimensions is slightly negative ($r = -0.217$). Figure 1.3 depicts the problem; displaying a scatter plot of elbow rest height against popliteal height for army females. Indicated by arrows are the locations of the 5th percentile popliteal height and the 5th and 95th percentile elbow rest height values.

The practical implication of this relationship is that if we select a group of females who are all at the 5th percentile of popliteal height (351 mm) and set their chair heights to this value, we will find that the comfortable working height for their keyboards might be anywhere within the range 176 to 264 mm above seat height. (These are the 5th to 95th percentile values of elbow rest height for females.) Moreover, if these individuals are required to work

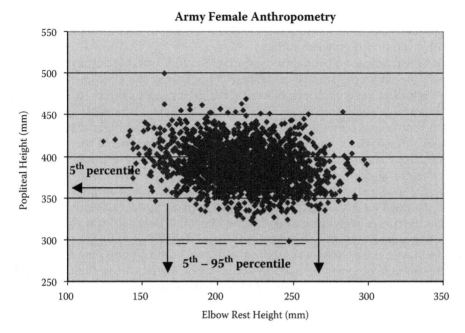

FIGURE 1.3
For those women who are at the 5th percentile of popliteal height, corresponding values of elbow-rest height cover a large range. U.S. Army female data. (From Gordon, C. C., Churchill, T., Clauser, T. E., Bradtmiller, B., McConville, J. T., Tebbetts, I., Walker, R. A. (1989). Anthropometric survey of U.S. Army personnel: Summary statistics interim report (No. Technical Report NATICK/TR-89-027). Natick, MA: U.S. Army Natick Research, Development, and Engineering Center.)

at a typical fixed-height work surface of 710 mm (28 in.), we see that *none* will be able to achieve a comfortable working posture; a 351 mm seat height plus the maximum elbow rest height of 300 mm yields an elbow position of 651 mm, which is 60 mm below the work surface height. Consequently, our goal of accommodating the small (5th percentile) female with a traditional fixed-height workstation with respect to two postural criteria (feet on the floor and elbow at the work surface) cannot be achieved. It is interesting that the 1988 ANSI/HFES Standard (Human Factors Society, 1988) approached this problem by using the *upper* range of chair height adjustability as the height that would allow the elbow to be at the proper working height, and then assumed a footrest would be needed. However, in the revision of this standard (BRS/HFES 100), it was accepted that adjustability of the keyboard support surface as well as the chair height would be necessary (Human Factors and Ergonomics Society, 2002).

Now a second complication emerges. For some perspectives, it is useful to review typical office furniture in the precomputer era For people who used typewriters (such as secretaries), the typical desk would have a work surface at a height of about 760 mm (30 in.) with a depression or well into which the typewriter would fit. Thus, the working height of the keyboard would be lower than the surface of desk—about 710 mm (28 in.) Those who were not typewriter users (managers, professionals) would have a simple flat desk with a 760-mm-high work surface.

As the computer terminals moved into the office workplace, they were initially placed on the surface of the desk. However, it soon became clear that the resulting keyboard heights were too high for most people—particularly compared to the lower typewriter keyboard heights. A cost-effective solution that became widely applied was to provide a traditional fixed-height work surface with a keyboard tray affixed to the underside of the work surface. This brought the keyboard support surface down to the 710 mm (28 in.) height typical for typewriters and became the prototype of fixed-height computer workstations referenced in the 1988 version of ANSI/HFES 100. As the need for adjustability in keyboard support surfaces became evident, manufacturers began providing height adjustable keyboard trays (as well as more expensive desks with adjustable work surfaces).

However, the problems associated with keyboard trays become evident as soon as they begin to be used in actual work. By requiring the user to sit away from the desk, the viewing and reach distances to the work surface are increased. In effect, the keyboard tray renders a large portion of the work surface unreachable without bending forward (which is, in itself, an awkward posture). The situation is worsened if the user chooses to utilize the backward-leaning posture described previously.

In practice, many keyboard trays seen in actual work environments are used for under the desk storage rather than for their intended purposes. Several years ago, the author conducted an ergonomic walk-through of a local government facility that provided such trays for its computer user. He observed that less than 10% of the trays actually were supporting keyboards. Instead, the users were simply tolerating the awkward postures associated with placing keyboards on standard height desks. Female operators mostly used the trays as places to store their purses.

To recapitulate, it can be seen that the apparently simple issue of designing the appropriate range of seat height adjustment in fact necessitates consideration of a number of interdependencies among ergonomic design criteria. These criteria should be in service of providing comfortable and efficient working postures for a large segment (90%) of the target population. We have already seen that the interaction between the goals of providing foot/leg support and the goals of getting the elbows/arms to an appropriate working height for keying necessitates adjustability of keyboard support surface height as well as chair height. However, as we have seen, satisfying the former goal may result in the failure to satisfy an additional criterion: that of ensuring that frequently used work objects are within range of easy reach.

This is only the beginning. Other design criteria include the necessity for the head/eyes to be in proper relationship with the monitor height, and the need for the lower back to be supported by being in contact with a back rest. The anthropometric dimensions which are most important in satisfying these requirements are *seated eye height* (vertical distance from seat surface to corner of eye) and *buttock-popliteal length* (horizontal distance from the buttocks to the popliteal crease.) Finally, there are a host of relationships related to effective hand/wrist posture with respect to input devices such as keyboards, mice, trackballs, etc. These are elaborated in BRS/HFES 100 (Human Factors and Ergonomics Society, 2002).

The preceding discussion is concerned with a computerized office environment. It is of some interest to briefly consider a different seated workplace: the cockpit of a high-performance military aircraft. This situation presents a set of very stringent requirements in which failure may be a matter of life or death. The pilot must be able to operate manual controls with both hands, must be able to reach foot peddles, must operate a control lever between his or her legs, must be able to see over the nose of the aircraft as well as the display panel, all while in a restraining harness during which high *g* forces are present. Finally, in extreme situations, the entire workspace (i.e., ejection seat) is blown out of the cockpit after which (hopefully) an attached parachute opens up.

In approaching the general question of ergonomic fit for both environments, it is first necessary to determine the critical task-related postural requirements (e.g., operating the keyboard and reaching the control stick). The next step is to jointly consider the equipment constraints together with anthropometric variability among the relevant anthropometric dimensions in the target population. For the office environment, this entails designing office furniture with ranges of adjustability that accommodate the target population and are also commercially feasible. BRS/HFES 100 provides such design specifications along with a description of the anthropometric basis and justification for these specifications (Human Factors and Ergonomics Society, 2002, pp. 72–94); see also (Dainoff, 2006). In a military aircraft, adjustability is limited by operational requirements, so that the target population is restricted by military selection criteria. In the U.S. military, a recent development has been the need to accommodate female pilots in aircraft formerly designed for male pilots. Zehner (2001) provides an in-depth discussion of fit testing for a military aircraft (see also Human Factors and Ergonomics Society, 2004) for a general discussion of applications of anthropometry to design, particularly when multivariate solutions are required.

1.4 Beyond Anthropometry—Implementing Fit with Multiple Degrees of Freedom

Let us assume that the engineering design problems have been solved for the case of the computerized office. Furniture is available with sufficient adjustability to accommodate, in theory, a large percentage of the user population. What is required to ensure that this adjustability is utilized as designed? This is the question of the effectiveness of ergonomic interventions discussed earlier in reference to literature reviews by Westgaard and Winkel (1997).

To illustrate, consider an intervention carried out by the author as part of a multinational, multidisciplinary investigation: the so-called MEPS study (Dainoff et al., 2005a, 2005b). (The term MEPS is an acronym for Musculoskeletal, Ergonomic, and Psychosocial Stress.) An extensive characterization of the musculoskeletal, visual, postural, and psychosocial attributes of the workplace of a target group of data entry employees was carried out, including symptom checklists and physical/optometric examination. Then, for the U.S. component of the study, an ergonomic intervention was implemented that included advanced ergonomic chairs; a motorized adjustable workstation; a keyboard that was divided into three components, each of which was fully adjustable; and an adjustable copyholder. Participants were also provided eyeglasses with an appropriate correction for the viewing distance of the monitor. The assessment battery was repeated after 1 month's use of the equipment and again after 11 months' use.

The operators' primary task was to enter alphanumerical data from paper tax forms into a database. They worked under an incentive pay system in which errors were severely penalized.

Consider the task of the operator faced with the new ergonomic equipment. Whereas their previous chairs had a simple height adjustment, the new chair had adjustable seat heights and angles, adjustable backrests, and adjustable arm supports. They could work in backward-leaning, upright, or upright with seat angle inclined forward postures. They could use the motorized control to place the level of their work surface and monitor at any height they wished; they could even stand while working if they liked. The keyboard was split into three segments: the two alphabetic components and the numeric component. Each component was independently adjustable so as to allow wrist angles to achieve a neutral working posture. Finally, an angle-adjustable copy holder was provided so as to allow the paper documents to be in approximately the same vertical plane as the monitor.

The operators in this situation have multiple degrees of freedom of adjustability of workstation components available. Exploration of these degrees of freedom has the potential for allowing the operator to achieve one or several comfortable/efficient work postures. However, for this potential to be realized, the operator must be aware of the concepts underlying healthy work posture, and how the various equipment adjustments can be coordinated to achieve this goal. Accordingly, a critical component of the overall intervention was a formal operator training program followed by a period of on-site coaching. The result, for this group of operators, was a substantial reduction of musculoskeletal signs and symptoms as well as a greatly improved satisfaction with their work environment (Dainoff et al., 2005).

A similar conclusion was reached by Amick et al. (2003) with respect to the combination of adjustability plus training. In their case, the researchers were able to utilize a control group as well as a formal productivity analysis along with assessment of musculoskeletal symptoms. The results indicated not only symptom reduction but a financial cost/benefit ratio of 1:25 in the first year.

1.5 Necessity for an Ecological Approach

Unfortunately, the epidemiologically oriented literature on effectiveness of ergonomic interventions seems to ignore the interdependence of multiple components as well as the critical role of operator education. Ergonomic components tend to be treated as black boxes that are either present or absent in a given investigation. The ergonomic research literature tends, as well, to reflect this reductionist approach. A recent extensive compilation of research data on working postures (Delleman, Haslegrave, and Chaffin, 2004) contains separate chapters on anthropometry, reach, hand and neck, vision, trunk, sitting, pelvis, leg and foot, force, performance measures, and digital human models. The single chapter devoted to multiple factor models concentrates on descriptive systems for risk assessment such as RULA and OCRA (Colombini and Occhipinti, 2004). Thus, any functional integration of this material is left to the reader.

What follows is a conceptual framework within which such functional integration can occur. This ecological framework based on the combined work of J.J. Gibson and N. Bernstein. Although it will be explicitly applied to the office workplace problem described above, it is broadly applicable to all areas of HF/E.

The first component of the ecological framework is the concept of *affordance*. Affordances are attributes of the environment of an individual (or "actor") defined with respect to the action capabilities of that individual (Dainoff and Mark, 2001; Gibson, 1979; Stoffregan and Michaels, 2004).* Insofar as the fundamental definition of ergonomics can be construed in terms of the fit between individual and environment (Dainoff and Dainoff, 1986; see above), the concept of affordance, as developed within the theoretical framework of ecological psychology, provides a systematic approach to an understanding and critical analysis of person–environment complementarity. Thus, components of the ergonomic chair described in the previous section are affordances for alternating between different seated work postures, but only for a particular set of users. The chair is not usable as designed for a 2-year-old child who is too small, an extremely obese adult who is too large, or a person

---

* There is a large and often contentious literature on affordances, both within and outside the field of ecological psychology. See particularly Stoffregen and Michaels. Norman (1988) introduced the term *affordance* to a wider audience—albeit with a different theoretical perspective, and Helander included it as part of his theoretical overview of the field of ergonomics in his presidential address to the International Ergonomics Association. In this paper, an attempt has been made to present an approach to affordances that minimizes theoretical differences, but maximizes practical utility for ergonomic design and problem solving.

with muscle impairment who is unable to adjust the controls. The chair is also not functionally usable by a person who may be of appropriate size and strength but does not understand either how to adjust the controls or why such adjustments might be useful.

The concept of affordance is particularly relevant to ergonomic aspects of design, because it requires the designer to explicitly take into account how physical objects relate to the action capabilities of users. Action capabilities, in turn, are determined by certain classes of *constraints*: personal, environmental, and task (Newell, 1996). Personal constraints refer to individual variability, including body dimensions (anthropometry), biodynamics (body strength, mass, and flexibility), and relevant psychological factors (perceptual, cognitive, and motivational). Environmental constraints include both size and shape of objects and surfaces as well as their physical properties relevant to the action.

The second component of the ecological framework for ergonomics is the *perception–action cycle*. Any integrated behavior pattern (task) can be decomposed into a series of steps in which perceived information about the possibility of action is followed by the action itself, which in turn reveals new information about potential actions, etc. For example, information is extracted from the home page of a Web site indicating the presence of a clickable button. Hand and finger muscles move the mouse to the location of the button and click. The effect of the click is act on the environment—a new Web page appears. The new page has additional information, some of which is extracted (perceived) and the cycle continues.

The perception–action cycle is the theoretical conception that links Gibson's concept of affordance with Bernstein's concept of skill as the coordination of multiple degrees of freedom. It is the information about affordances that, when perceived, initiates the perception–action cycle by revealing the capabilities for action within the environment. At the same time, the existence of multiple affordances within the system requires a degree of coordination through selection and timing of appropriate task-relevant actions associated with appropriate affordances. When this coordination can be achieved under changing environmental constraints, the user has achieved what Bernstein called *dexterity* (Bernstein, 1996). Consequently, we argue that the perception–action cycle should be the fundamental unit of analysis of work systems, and, therefore, a basic tool for ergonomics.

Let us consider in detail the example of a user/actor seated at a fully adjustable computer workstation while writing an article for a book chapter. The adjustability of the height and angle of the seat, angle of the backrest, height and angle of the keyboard support, height and angle of the monitor, as well as the scalable font size of the word processing program are all affordances that allow this particular actor to achieve a comfortable working posture. Note that each of these components refers to physical properties of the workstation defined in terms of corresponding attributes of the actor's anthropometry, motor control ability, and understanding of operation and coordination of the control mechanisms.

The term *comfortable* will be used as a surrogate for a complex set of postural, biomechanical, and behavioral criteria. An epidemiological approach might define *comfort* as those postural configurations which do *not* result in musculoskeletal disorders when maintained for reasonable durations of work tasks (National Research Council, 2001). At a more molecular level, biomechanical criteria might include postural orientations located within the centers of joint ranges of motions, and which are economical in terms of minimizing the effort required to either maintain position or perform action (see, e.g., Newell, 1961, 1996). Alternatively, comfort can also have a more immediately emotional/aesthetic component (Helander, 2003).

To elucidate further, consider the microstructure of the task. The actor is composing a portion of her article. This entails periods of rapid keyboard operation while entering text, interspersed with periods of reflection and pondering. An efficient working posture is afforded by the adjustability mechanisms in that the seat and backrest can be adjusted so that the trunk can be inclined backward with the feet flat on the floor and the lower back supported. The keyboard support is adjusted so that the hands are flat and forearms parallel to the floor. The head is erect, and the monitor is within a field of view 30° below the horizontal. The backward-inclined working posture places the eyes over 100 cm from the display screen, but this is compensated for by increasing the font size of the displayed characters (refer again to Figure 1.2).

Now, the chapter is written. The actor has a new task, to print a paper copy of the cited references and to verify that all of items in the list are properly cross-referenced in the body of the written text displayed on the computer screen. The paper document is placed in a document holder adjacent to the monitor. Thus, a new set of task and environmental constraints comes into play. The printed characters on the paper document are smaller; moreover, the actor now needs to be able to use a pen or pencil to check off the references on the paper as they are verified. If the actor were unaware of the adjustability affordances of the workstation system, the task would have to be accomplished by bending forward at the waist so as to read the text and extending her arm so as to place her hand in position to mark the page. This new postural configuration, which is at the extreme of her reach envelope, would not be acceptable (comfortable) for more than a brief period of time.

However, we will assume that the actor possesses the understanding that allows the adjustability mechanisms to become actual (rather than potential) affordances for her. Having placed the copy in the document holder, she initially perceives that the size of the text and the distance of the document do not afford a comfortable working posture. The action resulting from this perception is to move her hands to the area under the seat where the control surfaces of the chair are located. This action reveals new information: namely, the tactile structure of the controls. The tactile information leads to further action: the seat angle adjust mechanism is unlocked while the trunk moves to an upright position. The upright position is sensed as proprioceptive information and the seat angle adjust mechanism is released, locking the seat angle into position slightly below the horizontal.

The location of the backrest is now perceived as no longer supporting the trunk. A second set of perception–action cycles takes place in which the backrest is moved to a more upright position. The change in inclination of the seat lowers its average height above the floor. The actor perceives that her legs are no longer vertical and executes a third perception–action cycle, raising the height of the seat. At this point, the chair configuration affords a comfortable work posture, but the chair is still too far from the copy holder. The actor perceives yet another affordance; the keyboard tray can be slid under the work surface allowing the chair to be moved into its final optimal position. Finally, the armrest height can be adjusted so as to afford support for the elbow while the actor marks the paper document.

In this extended description, we hope to elucidate the intricate relationships among affordances, perception–action cycles, and coordination of multiple degrees of freedom. The adjustment mechanisms are affordances to the extent that information specifying them is available to the actor and that they correspond to the actor's action capabilities. If such specifying information is *not* perceived by the actors, the perception–action cycles required by the task constraints may be very different from what was anticipated by the designer. The author has the experience (shared by colleagues) of visiting a workplace supplied with new ergonomic chairs whose seat height control mechanisms were still covered with tape and the seat height was at its lowest setting—the way it was shipped from the dealer! Thus, for almost all of the operators, such "ergonomic" chairs resulted in poorer, less comfortable work postures. If this workplace had been included in an epidemiologic survey of the effects of ergonomic interventions, the conclusion might have been that, for this case, ergonomic chairs were provided but the outcome was negative in that working posture was worse than before. To the extent that ergonomic interventions are simply described in terms of equipment provided, it is not surprising that the overall evidence for effectiveness of such interventions in the office environment is reported as less than compelling (National Research Council, 2000). On the other hand, systematic application of an ecological perspective requires designers and implementers to explicitly consider interacting and interdependent system components. We next consider what such a systematic application might entail and the broader implications for ergonomics in general.

1.6 A Framework for Ecological Research and Practice

Earlier, we referred to two critics who were so disenchanted with the current state of HF/E that they argued for the creation of new disciplines. Karwowski (2000) called for the establishment of *symvatology*, a discipline that focuses on

TABLE 1.1

Example of Simplified Means-Ends Abstraction Hierarchy for
Height-Adjustable Seated Worktable

Functional purpose/goal	Person–workplace fit
Values and priorities	Posture–workstation fit
Purpose-related functions	Clearance under work surface for 90% of user population
Physical function	Adjustability of work surface height
Physical form	Work surface height should be adjusted between 520 and 760 mm

Note: See Vicente, K. J. (1999). *Cognitive Work Analysis: Toward Safe, Productive, and Healthy Computer-Based Work.* Lawrence Erlbaum Associates, Mahwah, NJ.

compatibility relationships between humans and artifacts. Gillen and Bias (2001) argued for a functionalistic science of usability with the following characteristics: (a) goal orientation, (b) adaptation, (c) close link between cognition and action, and (d) explicit consideration of physical, social, cultural, and task context. It is argued here that both of these sets of criteria can be explicitly addressed within the current theoretical and experimental framework of ecological psychology.

1.6.1 Cognitive Work Analysis as a Tool for Integration

At the core of this integration is *cognitive work analysis (CWA)*. (This description of CWA will follow that of Vicente (1999).) Although the majority of CWA efforts have been devoted to analysis of complex sociotechnical systems, such as power plant controls, and to the design of ecological interfaces for such systems, we have argued that this framework can be effectively applied to problems within the purview of physical ergonomics, such as workstation design (Dainoff, 1998, 2006; Dainoff et al., 1999; Xu et al., 1999). The starting place for CWA is *work domain analysis,* a process that characterizes the landscape of the work domain in terms of a hierarchy of affordances (see also Vicente and Rasmussen, 1990). This is conceptualized in terms of a *means-end abstraction hierarchy,* in that the affordances at any one level of the hierarchy serve as a means of attaining the goals or ends of the next higher level.*

Table 1.1 illustrates a simplified example (see Dainoff, 1998, 2006; Dainoff et al., 1999; and Xu et al., 1999 for a more extended treatment). At the lowest end of the hierarchy, *physical form* refers to the representation of relevant components of the physical structure of the elements comprising the system which is, in this case, a height-adjustable worktable. The actual range of adjustability in physical units (centimeters or inches) is represented at the physical form level. Moving upward, *physical function* constitutes the ends

---

* Work domain analysis has a second orthogonal component called *part-whole decomposition.* For simplicity, we will not discuss this part of the analysis.

for which the physical-form level is the means. In this case, physical function refers to the capability of, or requirement for, height adjustability. Continuing further upward, a *purpose-related function* called *clearance*, which refers to the area under the work surface, is an end for which height adjustability is a means. Clearance itself is subsumed under the broader concept of *posture–workstation fit*, which is a *values and priorities* characterization. Finally, the overall system *functional purpose or goal* (person–workplace fit) is found at the highest level of the hierarchy.

It should be noted that the hierarchical means–end structure provides an explicit means of incorporating the goal-directed and contextual criteria required by Gillen and Bias (2001). In particular, this example dealt with an adjustable workstation. In an expanded version of this analysis, we might find organizational goals at the values and priorities level of the hierarchy that value good working posture as means of increasing productivity and decreasing musculoskeletal symptoms. On the other hand, a different set of financially based values and priorities might reflect an unwillingness to allocate funds for operator training. Thus, organizational constraints become an essential part of the analysis rather than an afterthought to explain away poor results. At the same time, expanding the analysis at the purpose-related function level to include reach and support would introduce the functional interdependencies described earlier.

Work domain analysis is only the first stage of CWA. It is followed by *control task analysis*, which characterizes user actions on the work domain (Vicente, 1999). In effect, control task analysis is a formal representation that instantiates the perception–action cycles required to achieve the desired goals. The previous illustration of actions involved in writing a book chapter constitutes, in effect, a control task analysis.

Three other stages are necessary to fully complete a CWA. Whereas the control task analysis describes the perception–action cycles that are *possible* on the work domain, a *strategies* analysis examines those which might be *most effective* (Vicente, 1999). This level includes a consideration of relationships that might be automated. For multiple degrees of freedom of adjustability in chairs and workstations, automation might entail computer chips that "remember" desired locations. This component of CWA particularly embodies the adaptive criteria of Gillen and Bias (2001).

The *social organization and cooperation* stage of CWA reflects the reality that all work systems operate within a social context. Effective perception–action cycles might entail multiple actors interacting in multiple ways. The final stage of analysis, *worker competencies*, incorporates individual differences relevant to the previous stages of analysis (Vicente, 1999). The last two stages are where CWA can be transformed from a simple examination of an individual workstation into the analytic basis for the needs-assessment of a full-fledged ergonomic program. It is also at the worker competencies stage that the full realization of Karwowski's (2000) compatibility relationships is explored.

This very brief overview has had the purpose of arguing that CWA provides the conceptual structure within which both physical ergonomics and the ecological approach can be combined and overcome some of the practical concerns facing implementation of ergonomic solutions in the area of work-station design.

At the same time, CWA has already demonstrated its effectiveness in the realm of cognitive ergonomics with respect to the development of ecological interface design. Recent overviews may be found in Vicente, 2002, and Burns and Hajdukiewicz, 2004. Of particular interest is recent work for the Australian Defense and Technology Organisation by Naikar and her colleagues, in which CWA was utilized as a basis for awarding a major contract for a new military aircraft and, subsequently, to convince military authorities to reorganize the crew structure for that aircraft (Crone, Sanderson, and Naikar, 2003; Naikar and Pearce, 2003).

1.6.2 Epistemological Issues

Although CWA provides an important tool for integration, it alone is not sufficient. The CWA framework does not *replace* the domain-specific research findings from the multiple professional and academic disciplines—described earlier—that intersect with the field of HF/E. In Rasmussen's words, this framework should serve as a "cross-disciplinary market place" (1994, p. xv). However, there are important epistemological issues that arise when research findings from a wide variety of research traditions must be integrated. For example, the terms *comfort* and *fatigue* are higher-order abstractions that have been instantiated in a variety of contexts, ranging from cultural to biomechanical and physiological, in the ergonomics literature. (See Xiao and Vicente, 2000, and Vicente, 2000 for an extended general discussion of these issues, and Dainoff, 2006, for an application to workplace ergonomics.)

As a particular example, a leading researcher in seating research has asked if we should "forget about ergonomics in chair design? Focus on aesthetics and comfort!" (Helander, 2003, p. 1306). The core of his argument is that ergonomic chairs are overdesigned in terms of the amount of adjustability; users are unable to differentiate, on the basis of reduction in discomfort, ergonomic chairs with different ergonomic features. On the other hand, the same chairs could be differentiated on the basis of satisfaction and comfort. Thus, Helander concludes: "Discomfort is based on poor biomechanics and fatigue. Comfort is based on aesthetics and plushness of chair design and a sense of relaxation and relief" (Helander, 2003, p. 1315).

There are at least two major concerns with this conclusion. First, the methodology utilized in these studies is essentially that of consumer research. Users were given a selection of chairs and a series of ratings scales on a questionnaire. There is no indication that users were able to perceive the affordances of the chair controls, or the resulting perception–action cycles that might result from their employment. Simply put, these users did not receive the training that is essential for successful ergonomic intervention (Amick et al., 2003). Hence, it is not surprising that preference ratings would be more highly loaded on aspects of the seat cushion rather than the control mechanisms. However, this conclusion, although important for chair marketers, is hardly a basis for a scientific discussion of design of ergonomic seating.

The second concern with Helander's conclusion is the epistemological confusion inherent in the categorical linkages {biomechanics->discomfort} versus {aesthetics->comfort}. The problems with these linkages are readily apparent to any serious athlete who has searched for the "sweet spot" in a tennis racquet, or a carpenter who has appreciated the "feel" of his or her favorite hammer (see Dainoff and Wagman, 2004; Wagman and Carello, 2001; Wagman and Taylor, 2004 for an example of a systematic approach to this question). Any serious consideration of aesthetics must take into account the emotions associated with ease and lack of effort in successfully accomplishing perception–action cycles.

These kinds of concerns are not unique to ergonomics. Galison (1997) has described the ways in which the development of laboratory apparatus in microphysics transformed the social/organizational structure of the field from individual investigators working alone or in small groups (e.g., Wilson's cloud chamber) with total control and understanding of their apparatus, to industrial-style organizations requiring collaboration among many professionals (e.g., CERN, Fermilab). Galison emphasized the need for researchers in multidisciplinary relationships to develop what he calls *trading zones*. Derived from the field of linguistics, trading zones refer to simplified languages (creoles, pidgins) that arise when adjacent cultures require a mutually understandable means of communication to transact business. Such trading zones emerged among the various disciplines and professions involved in high-energy particle physics and enabled the highly complex, and highly successful, physical/organizational structures that have resulted. The need for such trading zones in HF/E is just as important.

1.7 Conclusions

The field of HF/E exists at the nexus of technology and human capabilities. Professionals in this field should be leaders in the quest for human-centered approaches to technology. However, as we have seen, there are certain barriers to this leadership. Despite "official" statements to the contrary by professional organizations, there seems to exist a conceptual divide between physical/biomechanical and cognitive approaches.

What is urgently required are, in Galison's (1997) terms, trading zones, emerging concepts that allow for communication across subdisciplines. The argument presented here is that the theoretical legacies of J.J. Gibson (1979) and N.A. Bernstein (1996), along with the CWA framework derived, in part, from this work (Rasmussen et al., 1994; Vicente, 1999), can provide the conceptual framework for such trading zones. The concepts of affordance and perception–action cycle, instantiated within CWA as means–ends abstraction hierarchies and control task analysis, allow for systematic incorporation of research findings from across the wide breadth of ergonomics and related disciplines. Such research does not have to have been explicitly conducted within the ecological framework to be utilized.

This view offers the promise of allowing understanding of complex interdependencies while maintaining focus on functionalistic aspects of context and goal-oriented adaptation as argued by critics such as Karwowski (2000) and Gillen and Bias (2001).

Acknowledgments

I would like to dedicate this paper to the memory of my uncle, Aaron Alexander, PhD, microbiologist, researcher, and educator, who greatly influenced my own scientific path. I also want to express gratitude to my long-time colleague, Leonard Mark, and my even longer-time colleague and partner, Marilyn Hecht Dainoff, for their intellectual stimulation. They have made major contributions to this work.

References

Amick, B. C., III, Robertson, M., DeRango, K., Bazzani, L., Moore, A., and Rooney, T. (2003). The impact of a highly adjustable chair and office ergonomics training on musculoskeletal symptoms: Two month post intervention findings. Paper presented at the 47th Annual Meeting of the Human Factors and Ergonomics Society.

Bernstein, N. A. (1996). On dexterity and its development. In Latash, M. and Turvey, M. T. (Eds.), *Dexterity and Its Development*. (pp. 1–244). Hillsdale, NJ: Lawrence Earlbaum.

Board of Certification in Professional Ergonomics. (2002). Mission statement. http://bcpe.org/aboutus/detail.asp?RecordID=18

Burns, C. M., and Hajdukiewicz, J. R. (2004). *Ecological Interface Design*. Boca Raton, FL.: CRC Press.

Colombini, D. and Occhipinti, E. (2004). Multiple factor models. In Delleman, N. J., Haslegrave C. M., and Chaffin, D. B. (Eds.), *Working Postures and Movements*. Boca Raton, FL: CRC Press.

Crone, D. J., Sanderson, P. M., and Naikar, N. (2003). Using cognitive work analysis to develop a capability for the evaluation of future systems. Paper presented at the Human Factors and Ergonomics Society 47th Annual Meeting, Denver.

Dainoff, M. J. (1980). Occupational stress factors in video display terminal (VDT) operation: A review of empirical research. Cincinnati, Ohio: U.S. Dept. of Health and Human Services, Centers for Disease Control, Public Health Service, National Institute for Occupational Safety and Health, Division of Biomedical and Behavioral Science.

Dainoff, M. J. (1982). Occupational stress factors in visual display terminal (VDT) operation: A review of empirical research. *Behaviour and Information Technology*, 1(2), 141–176.

Dainoff, M. J. (1998). Ergonomics of seating and chairs. In Karwowski, W. and Marras, W. S. (Eds.), *The Occupational Ergonomics Handbook*. Boca Raton, FL: CRC Press.

Dainoff, M. J. (2000). Safety and health effects of video display terminals. In Harris, R. L. (Ed.), *Patty's Industrial Hygiene*, 5th ed. New York: Wiley.

Dainoff, M. J. (2006). Ergonomic industrial interventions: chairs and furniture. In W. S. Marras and W. Karwowski (Eds.), *Interventions, Controls, and Applications in Occupational Ergonomics*. Boca Raton, CRC Press.

Dainoff, M. J. and Dainoff, M. H. (1986). *People and Productivity: A Manager's Guide to Ergonomics in the Electronic Office*. Toronto; London: Holt Rinehart and Winston of Canada.

Dainoff, M. J. and Mark, L. S. (Eds.) (1987). *Task and the Adjustment of Ergonomic Chairs*. Amsterdam: Elsevier (North-Holland).

Dainoff, M. J. and Mark, L. S. (2001). Affordances. In Karwowski, W. (Ed.), *Encyclopedia of Ergonomics*. London: Taylor and Francis.

Dainoff, M. J. and Wagman, J. B. (2004). Implications of dynamic touch for human factors/ergonomics; contributions from ecological psychology. Paper presented at the 48th Annual Meeting of the Human Factors and Ergonomics Society, New Orleans, LA.

Dainoff, M. J., Mark, L. S., and Gardner, D. L. (1999). Scaling problems in the design of work spaces for human use. In Hancock, P. A. (Ed.), *Human Performance and Ergonomics*. (pp. 265–290): Academic Press, San Diego, CA.

Dainoff, M. J., Åaras, A., Horgen, G., Konarska, M., Larsen, S., Thoresen, M., and Cohen, B.G.F, (2005a). The effect of an ergonomic intervention on musculoskeletal, psychosocial and visual strain of VDT entry work: organization and methodology of the international study. *International Journal of Occupational Safety and Ergonomics*, 11(1), 9–23.

Dainoff, M. J., Cohen, B. G. F., and Dainoff, M. H. (2005b). The effect of an ergonomic intervention of musculoskeletel, psychosocial, and visual strain of VDT data entry work: the United States part of the international MEPS study. *International Journal of Occupational Safety and Ergonomics*, 11(1), 49–63.

Delleman, N. J., Haslegrave, C. M., and Chaffin, D., B. (Eds.). (2004). *Working Postures and Movements*. Boca Raton, FL: CRC Press.

Dix, A., Finlay, J., Abowd, G., and Beale, R. (1998). *Human Computer Interaction*, 2nd ed. Hemel Hempstead, Hertfordshire: Prentice Hall Europe.

Flach, J., Hancock, P. A., Caird, J., and Vicente, K. J. (1995). *Global Perspectives on the Ecology of Human-Machine Systems*. Hillsdale, NJ: Lawrence Erlbaum Associates.

Galison, P. (1997). *Image and logic*. Chicago: University of Chicago Press.

Gibson, J. J. (1950). *The Perception of the Visual World*. Boston, MA: Houghton-Mifflin.

Gibson, J. J. (1979). *The Ecological Approach to Visual Perception*. Boston, MA: Houghton Mifflin.

Gillen, D. J. and Bias, R. G. (2001). Usability science. I: Foundations. *International Journal of Human-Computer Interaction*, 13, 351–372.

Gordon, C. C., Churchill, T., Clauser, T. E., Bradtmiller, B., McConville, J. T., and Tebbetts, I., and Walker, R. A. (1989). Anthropometric survey of U.S. Army personnel: Summary statistics interim report (Technical Report NATICK/TR-89-027). Natick, MA: U.S. Army Natick Research, Development, and Engineering Center.

Grandjean, E., Hünting, W., and Pidermann, M. (1983). VDT workstation design; preferred settings and their effects. *Human Factors*, 25, 161–173.

Hancock, P. A., Flach, J., Caird, J., and Vicente, K. J. (1995). *Local Applications of the Ecological Approach to Human-Machine System*. Hillsdale, NJ: Lawrence Erlbaum Associates.

Helander, M. G. (2003). Forget about ergonomics in chair design? Focus on aesthetics and comfort! *Ergonomics*, 46(13/14), 1306–1307.

Hendrick, H. W. (2000). The technology of ergonomics. *Theoretical Issues in Ergonomics Science*, 1, 22–23.

Human Factors and Ergonomics Society. (2000). What is HFES? http://www.hfes. org/About/Menu.html

Human Factors and Ergonomics Society. (2002). BRS/HFES 100 Human factors engineering of computer workstations. Santa Monica, California: Human Factors and Ergonomics Society.

Human Factors and Ergonomics Society. (2004). Guidelines for using anthropometric data in product design. Santa Monica, California: Human Factors and Ergonomics Society.

Human Factors Society. (1988). ANSI/HFS 100–1988 American national standard or human factors engineering of visual display terminal workstations. Santa Monica, California: Human Factors Society.

International Ergonomics Association. (2000). What is ergonomics? International Ergonomics Association. http://www.iea.cc/ergonomics/

Karsh, B.-T., Moro, F. B. P., and Smith, M. J. (2001). The efficacy of workplace ergonomic interventions to control musculoskeletal disorders: A critical analysis of the peer-reviewed literature. *Theoretical Issues in Ergonomics Science*, 2, 23–96.

Karwowski, W. (2000). Symvatology: The science of an artifact-human compatibility. *Theoretical Issues in Ergonomics Science*, 1, 76–91.

Mandal, A. C. (1981). The seated man (homo sedans). The seated work position. Theory and practice. *Applied Ergonomics*, 12.1, 19–26.

Mark, L. S., Dainoff, M. J., Moritz, R., and Vogele, D. (1991). An ecological framework for ergonomic research and design. In Hoffman, R. R. and Palermo, D. S. (Eds.), *Cognition and the Symbolic Processes: Applied and Ecological Perspectives* (pp. 477–505), Hillsdale, NJ: Lawrence Erlbaum Associates.

Mark, L. S., Nemeth, K., Gardner, D., Dainoff, M. J., Paasche, J., Duffy, M., and Grandt, K. (1997). Postural dynamics and the preferred critical boundary for visually guided reaching. *Journal of Experimental Psychology: Human Perception and Performance*, 23(5), 1365–1379.

Meister, D. (1989). *Conceptual Aspects of Human Factors*. Baltimore, MD: Johns Hopkins University Press.

Naikar, N. and Pearce, B. (2003). Analysing activity for future systems. Paper presented at the Human Factors and Ergonomics Society 47th Annual Meeting, Denver, CO.

National Research Council. (2000). *How People Learn: Brain, Mind, Experience, and School*. Washington, DC: National Academy Press.

National Research Council. (2001). *Musculoskeletal Disorders and the Workplace*. Washington, DC: National Academy Press.

Newell, K. M. (1996). Change in movement and skill: Learning, retention, and transfer. In Latash, M. and Turvey, M. T. (Eds.), *Dexterity and its Development*. Mahwah, NJ: Lawrence Earlbaum Associates.

Norman, D. A. (1988). *The Psychology of Everyday Things*. New York: Basic Books.

Nubar, Y. and Contini, R. (1961). A minimal principle in biomechanics. *Bulletin of Mathematical Biophysics*, 23, 377–391.

Rasmussen, J., Pejtersen, A. M., and Goodstein, L. P. (1994). *Cognitive Systems Engineering*. New York: Wiley.

Reed, E. S. and Bril, B. (1996). The primacy of action in development. In Latash, M. and Turvey, M. T. (Eds.), *Dexterity and its Development*. Mahwah, NJ: Lawrence Erlbaum Associates.

Root-Bernstein, R. S. (1989). *Discovering*. Cambridge, MA: Harvard University Press.

Stoffregen, T. A. (2004). Breadth and limits of the affordance concept. *Ecological Psychology*, 16(1), 79–85.

Turvey, M. T., Shaw, R. E., and Mace, W. (1978). Issues in the theory of action: Degrees of freedom, coordinative structures, and coordination. In Requin, J. (Ed.), *Attention and Performance* (Vol. VII, pp. 557–595). Hillsdale, NJ: Lawrence Earlbaum.

Vicente, K. J. (1997). Heeding the legacy of Meister, Brunswik, and Gibson: Toward a broader view of human factors research. *Human Factors*, 39(2), 323–328.

Vicente, K. J. (1999). *Cognitive Work Analysis: Toward Safe, Productive, and Healthy Computer-based Work*: Lawrence Erlbaum Associates, Mahwah, NJ.

Vicente, K. J. (2000). Towards Jeffersonian research programmes in ergonomics science. *Theoretical Issues in Ergonomics Science*, 1, 93–112.

Vicente, K. J. (2002). Ecological interface design: Progress and challenges. *Human Factors*, 44(1), 62–78.

Vicente, K. J. and Rasmussen, J. (1990). The ecology of human machine systems ii: Mediating direct perception in complex work domains. *Ecological Psychology*, 2, 207–249.

Wagman, J. B. and Carello, C. (2001). Affordances and inertial constraints on tool use. *Ecological Psychology*, 13, 173–195.

Wagman, J. B. and Taylor, K. R. (2004). Chosen striking location and the user-tool-environment system. *Journal of Experimental Psychology: Applied*, 10, 267–280.

Westgaard, R. H. and Winkel, J. (1997). Ergonomic intervention research for improved musculoskeletal health: A critical review. *International Journal of Industrial Ergonomics*, 20(6), 463–500.

Xiao, Y. and Vicente, K. J. (2000). A framework for epistemological analysis in empirical (laboratory and field) studies. *Human Factors*, 42(1), 87–101.

Xu, W., Dainoff, M. J., and Mark, L. S. (1999). Facilitate complex search tasks in hypertext by externalizing functional properties of a work domain. *International Journal of Human-Computer Interaction*, 11(3), 201–229.

Zehner, G. F. (2001). Prediction of anthropometric accommodation in aircraft cockpits (afrl-he-wp-tr-2001-0137). Wright-Patterson Air Force Base, OH: United States Air Force Research Laboratory.

2

Integrating Cognitive and Digital Human Models for Virtual Product Design

Daniel W. Carruth and Vincent G. Duffy

CONTENTS

2.1 Introduction 29
2.2 Digital Human Models 30
2.3 HUMOSIM...32
    2.3.1 Simulating the Mind...33
        2.3.1.1 How It Works...34
        2.3.1.2 The Motor System in Detail 35
    2.3.2 Integrating HUMOSIM into a Cognitive Architecture..............35
        2.3.2.1 Issues..36
    2.3.3 A Specific Task...37
2.4 Summary..38
Acknowledgments 38
References 38

2.1 Introduction

In the field of ergonomic analysis, a number of computer-based design tools and methodologies are being developed to provide tools that assist designers in improving product and workspace design (Duffy, 2007a). Designers create virtual versions of their products and the environments in which the products will be used. Digital human models (DHMs), or avatars, can be placed in the environment, and various ergonomic data can be acquired based on the DHM. These data allow the designer to analyze the quality of designs with virtual prototypes, avoiding the cost of constructing physical prototypes and the involvement of actual human participants. However, many of the digital human models included in these software tools have significant limitations, particularly in the simulation of posture and motion.

A potential solution is to improve the software simulation of the human so that the software can accurately predict posture and movement during task performance. To fully generate the data provided by motion capture, it is not enough to be able to simulate the postures and movement. This would still require the designer to specify the motions the DHM should perform within the environment. Although this would reduce time and improve results, it should be possible to place the DHM into the environment with knowledge of the task and expect it to perform the task acting on its own, as a human participant would. In addition, such a model might provide tools for the analysis of cognitive issues relevant to the product and the environment. A cognitive model for the task that included components such as memory, learning, and planning, as well as the capability to request that the avatar simulate specific movements, could provide such a tool.

We believe that the necessary components for creating a true digital human exist and are reaching a level of maturity that allows us to begin to attempt to integrate the currently separate models. Two models must be integrated to create a DHM capable of moving and acting within a virtual environment: a model of human motion and a model of human task performance. The human motion simulation component would need to be able to accurately predict the full range of postures and movements required to complete the task of interest. The motion component would be directed by a cognitive model capable of simulating cognitive, perceptual, and motor request aspects of task performance. We will examine the two fields of research investigating these types of models and discuss how they might be integrated to form an integrated digital human model. The goal of our current research is to integrate cognitive models of human performance with digital human models in comprehensive simulations generating both cognitive and ergonomic data for human–machine interactions in completely virtual environments.

2.2   Digital Human Models

Digital human models are software models of the physical human in the form of an avatar, or virtual human. These models are intended to assist designers in testing the physical design of workspaces at an early stage.

Current digital human models provide designers with tools for analyzing the physical design of workspaces in a virtual environment. These tools are limited in a number of significant ways. First, the models of human motion for predicting posture and movement are limited in capability. DHMs are basically virtual manikins that must be carefully positioned in a reasonable posture by the designer. Some models currently in use allow designers to utilize inverse kinematics (IK) algorithms. These algorithms may have limited validity in predicting reasonable postures (Chaffin, 2007). Either the designer or the IK system may select erroneous postures that do not accurately represent likely postures in the environment. Given a vehicle interior, a designer must directly arrange the avatar in the seat in a plausible way with little assistance from the DHM software. Any dynamic motion, such as reaching from the steering wheel to the center console, must be provided by the designer, potentially through the use of motion capture or by direct manipulation of the avatar.

Second, DHMs simulate only the physical attributes and capabilities of a human. The DHM is an empty shell directed by the designer and the software toolset. It is not currently possible for a designer to place a DHM into a vehicle interior and simulate the physical motion of the DHM during the driving task. For example, if a designer is interested in the ability of a driver to access a particular control during different driving scenarios, the designer must either place the avatar into each posture of interest directly or access some existing posture or motion database. It would be beneficial if a designer could create a model of task performance that would allow the DHM to simulate human performance of the task within the virtual environment of interest.

Some recent work has demonstrated a methodology for interactive virtual design assessment for seated reach from a wheelchair during ATM design (Kang et al., 2006). Optimization tools provide some opportunities to make design more efficient, considering the many potential outcomes (Duffy et al., 2005). Validation of such methods (Wu et al., 2005) and the impact of using physical prototypes and force feedback on postures in virtual design (Duffy, 2005) are among the questions that need to be considered for their future effective use.

Figure 2.1 represents an example of an integrated cognitive and physical test bed for cognitive and physical ergonomic analysis and design originally described by Duffy (2007b).

The upper-left figure represents a driving simulator scene, whereas the upper-right figure represents the driver in the simulator in CAVS at Missis-

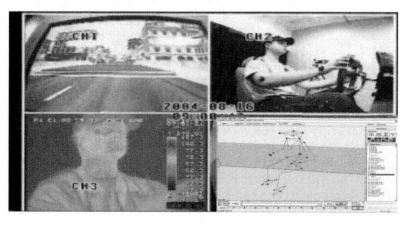

FIGURE 2.1
An example of an integrated cognitive and physical test bed for cognitive and physical ergonomic analysis and design.

sippi State. The lower-left figure represents the thermographic images that have shown relationships with mental workload (Or and Duffy, 2007), and the lower-right figure represents the markers from the motion capture system (Du et al., 2004).

2.3 HUMOSIM

Chaffin (2007) and the HUMOSIM group at the University of Michigan recognize that the inability of DHMs to accurately model the posture and motion of humans is a significant issue. The group is focused on developing models that allow the prediction of human motion for ergonomic analysis (Chaffin, 2007). Within the HUMOSIM group, there are two major efforts: the development of a large database of motion capture data and the development of models that can predict posture and motion based on the database.

In the process of assembling a database of movements, the HUMOSIM group has captured data from over 200 subjects to collect almost 100,000 motion data sets (Chaffin, 2007; HUMOSIM Group, 2003). The database includes motions for reaching toward a target, transferring objects, reaching and grasping objects, looking toward points, exerting force on handles, and pushing buttons. The continuously expanding database provides real-motion capture data to provide a foundation for the development of models capable of predicting motion.

There are two methods for predicting motion under development within the HUMOSIM group. The primary method, functional regression analysis, is spearheaded by Faraway (1997, 2001). This regression technique can predict joint angles during the motion of a hand toward a target in terms of inputs such as the stature, age, and gender of a subject. The technique was used to compare the relative importance of stature, age, and gender in seated reaching motions (Chaffin, Faraway, Zhang, and Woolley, 2000).

The method works by recognizing that the change over time of individual components of motion, such as joint angles, can be described as functions. Each motion can be described using a set of functions describing different components of the motion. Only a small number of functions are necessary because human motion is generally smooth (Faraway, 2001). Using these functions, models capable of predicting the trajectories of specific body markers can be created. Faraway (2001) describes a method referred to as the "Stretch Pivot" method, which is able to predict motion of an entire multisegmented body with specific start and end locations. The new method was devised to reduce error in the predicted location of the endpoint that occurred using the functional regression model of joint angles and trajectory predictions. The "Stretch Pivot" method ensures that the hand will arrive at a specific location. In Faraway (2001), the reported median errors were less than 7 cm.

The second method relies more directly on the motion database. The motion modification method, developed by Park, Chaffin, and Martin (2004), searches the motion database for "root motions" that can be analyzed and modified slightly to create a desired motion. Both methods are potential tools for creating dynamic human models capable of predicting valid postures and motions for ergonomic analyses.

2.3.1 Simulating the Mind

HUMOSIM provides a potential method for modeling realistic human postures and motions and improving the usability of DHMs in existing software packages. However, HUMOSIM still requires that the designer guide the DHM model through the task by identifying each motion required for the task. Ideally, the designer should only need to place a DHM into a virtual environment and provide the DHM with knowledge about the task for the DHM to perform it in a manner representative of human performance. To do this, the DHM must include a model that accurately represents how a human thinks about and performs the task.

Cognitive architectures are frameworks for constructing models of human performance focusing primarily on cognitive aspects such as memory, learning, planning, etc. The architecture itself provides an infrastructure for building executable programs that can perform the same activities as a human. The infrastructure provides scaffolding for building models of specific tasks that conform to a general theory for simulating human cognition. Models built within cognitive architectures can span a broad range of tasks requiring complex and dynamic cognition (Anderson and Lebiere, 1998). A number of different architectures exist, including ACT-R (Anderson, Bothell, Byrne, Douglass, Lebiere, and Qin, 2004), APEX (Freed, Shafto, and Remington, 1998), C/I (Kintsch, 1998), COGNET/iGEN (Zachary, Ryder and Hicinbothom, 1998), EPIC (Kieras and Meyer, 1997), QN-MHP (Liu et al., 2006), and SOAR (Newell, 1990). Each architecture has different underlying theories, structures, strengths, and weaknesses.

Anderson's ACT-R has some particular characteristics that make it suitable for the task at hand. First, ACT-R allows researchers to collect quantitative measures, such as reaction time, that can be directly compared to quantitative data from human subjects. Second, ACT-R's architecture is modular (Anderson et al., 2004). The core modules implemented in the architecture include perceptual modules, motor modules, a declarative memory module, and a goal module. A central production system coordinates activity between the modules. The central production system recognizes patterns stored in buffers associated with the modules and can request changes via the buffers (Anderson et al., 2004). Third, ACT-R has been applied to a wide range of tasks with varying levels of cognitive complexity in areas such as perception, attention, memory skills, learning, problem solving, decision making, language processing, and intelligent agents.

2.3.1.1 *How It Works*

The ACT-R cognitive architecture has evolved over 20 years to its current state (Anderson and Lebiere, 1998). Briefly, we will discuss the basic operation of the ACT-R architecture. ACT-R has two basic types of knowledge, declarative and procedural. Declarative knowledge includes knowledge about what the world is like. For example, the knowledge that "an apple is on the table" or "five times two is ten" is declarative. This knowledge is stored in chunks (Anderson and Lebiere, 1998). These chunks are collections of elements representing the knowledge. Procedural knowledge is knowledge about how to do something and is described as being outside our awareness. Procedural knowledge is stored in production rules in the form of a condition and an action. The condition tests the contents of the modular buffers, and the action specifies the action to take if the conditions are met. The actions are requests to modify the buffers or request an action from the modules. The process of matching rules to the current conditions and performing the specified actions is accomplished by the central production system.

As previously mentioned, the ACT-R architecture has four types of modules: the perceptual modules, the motor modules, the declarative memory module, and the goal module. Each module has one or more buffers associated with it that holds information based on the processing of the module. The perceptual modules include modules for visual and auditory perception. The visual system makes available to the central production system information about where items in the visual field are and what items are currently visible. The motor modules control movement of the hands and are primarily used to model typing and the handling of a mouse. Within the motor system, there is also a vocal module for generating speech.

The declarative memory module handles retrieval of declarative knowledge from memory. Only one chunk of memory is available to the central production system during a cycle. The contents of the single chunk are governed by a complex memory process that is described in detail by Anderson et al. (2004). The goal module maintains a chunk that contains information about the current task. As the central production system applies rules, the slots of the goal chunk may be updated to track progress in solving a problem.

Cognition is performed by ACT-R through a cycle in which the contents of the buffers of the modules are determined by the internal mechanisms of each module and the state of the external world, the production rules are matched against the contents of the buffers, a production rule is selected, its actions are applied, and the contents of the buffers are updated for the next cycle. By assuming that a cycle takes approximately 50 ms to complete, the central production system provides timing information for atomic levels of cognition. This, together with timing information calculated within each individual module, allows ACT-R to report detailed predictions for reaction times during performance of the task.

2.3.1.2 The Motor System in Detail

The design of ACT-R's motor module is of particular interest as it will provide an interface between the cognitive architecture and HUMOSIM. The existing motor system is largely based on the Kieras and Meyer (1997) EPIC Manual Motor Processor. Through the motor module, ACT-R's central production system issues motor commands as production rules are fired. While the motor module carries out the requested action, the central production system is able to continue firing other productions, allowing cognition to continue while the motor action is carried out (Anderson and Lebiere, 1998). To request an action, the central production system specifies a number of parameters representing an action to the motor module's buffer. The parameters depend on the action but can include specifying the hand and finger to move, the key to press, etc. For example, the following sample command would request a simple punch movement with the index finger of the right hand.

+manual>

ISA punch

Hand right

Finger index

When an action is requested, the motor module must first prepare the movement. As it prepares the movement, the motor module will not accept any other requests. During this period, the motor module creates a list of features representing the actual movement. The length of the preparation period increases as the difference between the features of the current and the previous movement increases.

After the movement is prepared, it takes 50 ms to initiate the movement, which frees the system to prepare another movement. After the movement is initiated, it is executed, and the motor buffer's state is set to "free," allowing another movement to be executed. The duration of the execution of the movement depends on the characteristics of the movement, with simple keystrokes taking a fixed time while the time taken to move a hand or finger is calculated based on Fitts's law (Fitts, 1954).

2.3.2 Integrating HUMOSIM into a Cognitive Architecture

Many of the cognitive architectures may provide a framework for specifying models of task performance, but, to create the integrated DHM, the architecture must be able to request that HUMOSIM simulate the movements required to actually perform the task. ACT-R's motor module provides a modular system that can be extended to allow a cognitive model to request complex motions to be simulated by the HUMOSIM tools.

To attach HUMOSIM to ACT-R's motor module, there must be a way to specify movements in ACT-R that can be translated to HUMOSIM, an appropriate extension to the calculation of the preparation period, and HUMOSIM must provide timing feedback to ACT-R.

ACT-R's current system for issuing motor commands can be extended to include the available range of movement styles supported by HUMOSIM. For example, a reaching motion would include parameters for the coordinates to reach to and the hand to reach with. In the following example, the issued command would request a simulation of a seated reaching motion towards a target 30° to the right and 60° from the up vector and 24 in. from the *y*-axis. Alternatively, the coordinates could be specified as an *x*, *y*, and *z* position with the origin located at the bottom of the seat or the head.

+manual>

ISA seated-reach

Theta-Up 60

Theta-Y 30

Distance 24

Once the motor command has been issued by the central production system, the preparation period will need to be calculated. The feature system utilized by ACT-R's current motor module can be extended to the larger range of motions that HUMOSIM can simulate. ACT-R's feature system uses a hierarchy of movements. If the motion style changes, all of the features of the motion must be prepared again. If the motion style remains the same, but the body part changes, the features at or below the body part will have to be prepared again. The issued command would be translated into a motion request for HUMOSIM and submitted to the HUMOSIM system for simulation. Once HUMOSIM has completed the simulation of the motion, the motor module will be free and the time taken to perform the motion will need to be reported.

2.3.2.1 Issues

HUMOSIM can be integrated into the ACT-R cognitive architecture in some form, extending or mimicking the existing motor module system. However, there are a number of potential issues involved in doing so. The first issue is a theoretical one. Cognitive architectures, including ACT-R, typically claim some theoretical underpinnings that guide the implementation of the architecture and therefore constrain the cognitive models built within the architecture. HUMOSIM is an empirically evaluated statistical methodology that is informed by real-motion capture data and previous work on biomechanical theories of motion, but is not itself based on those theories (Faraway, 2001). There may be some question about the validity of integrating a system with fewer theoretical underpinnings into ACT-R.

There is also an issue regarding feedback from HUMOSIM to ACT-R. ACT-R is perceptually aware of what is provided to it through its visual system. In an HCI domain, the relative importance of ACT-R being aware of the world outside the computer monitor is limited. In a workspace or vehicle interior, ACT-R will need to be aware of how its body movements may or may not be interfering with its view of the environment. It will be necessary to inform ACT-R's visual system if the body movements affect the position of the head or if the view is blocked by a body part such as an arm.

HUMOSIM is currently unaware of obstacles, including its own body parts (Faraway, 2001). For example, a seated reaching motion toward the floor might pass through the knees of the simulated body. This issue could significantly limit the application of an integrated ACT-R/HUMOSIM system. However, this is an area of current research within the HUMOSIM group (HUMOSIM Group, 2003). Additionally, neither the current motor system in ACT-R nor HUMOSIM provide tactile feedback to ACT-R. Adding the capability to detect obstacles and report impacts may be useful for certain tasks.

Speed of processing is also a concern. Within the ACT-R cognitive architecture, models can be created that will run in real time or faster, and a priority goal of the HUMOSIM group is to provide real-time simulations (Chaffin, 2004). A successful integration of the two systems will need to maintain real-time capability. A real-time model may be more easily integrated into existing systems built for interaction with human subjects. For example, a real-time model could be more easily integrated with the JACK software.

2.3.3 A Specific Task

The goal in our current research is to create a dynamic digital human model capable of performing a complex, dynamic task from start to finish with little or no designer intervention. Our laboratory is specifically interested in developing a model that can be used to test the impact of novel interfaces in a vehicle interior on driving performance. The addition of new and potentially complex interfaces such as navigation systems may have a serious impact on driving performance. Factors affecting the impact of the system might include location, design, and cognitive workload associated with using the interface. The potential cost of developing and constructing physical prototypes of each design is significant. Currently, rules of thumb for interface design, static models of reach in vehicle interiors, standards in vehicle design, and some intuition would be used to reduce the number of potential designs to one or two that could be constructed and tested. However, this process may curtail the development of unique new interfaces that may be very beneficial. By developing an intelligent DHM capable of driving a simulated vehicle and interacting with a simulated interior, many more designs could be thoroughly tested for both cognitive and ergonomic impact.

To create such a model of driver performance, we need a cognitive model of driver performance capable of performing basic driving tasks as well as simple secondary tasks and a physical model of reaching motions within the vehicle interior. Within the ACT-R community, a model of human driving performance has been developed and used in investigations of the effects of cell phone use on driving performance (Salvucci, Boer, and Liu, 2001; Salvucci, 2001; Salvucci and Macuga, 2001). In the HUMOSIM group, seated reaching motions have been a significant area of research (Reed, Parkinson, and Chaffin, 2003; Zhang and Chaffin, 2000). By building on this prior research, an integrated cognitive and ergonomic model of driver performance can be created. In addition to our effort, the HUMOSIM Group is working to integrate the QN-MHP cognitive architecture (Liu et al., 2006) into the HUMOSIM framework.

2.4 Summary

Knowing the potential for integrating existing models as well as the limitations regarding where these integrated models are applicable will enable other researchers to build in this work in other multidisciplinary research environments. The work described should facilitate new developments in the psychological sciences and their application to engineering design. Models in the cognitive sciences that can be integrated into such virtual design test beds will help highlight achievements in this field in the new millennium.

Acknowledgments

The authors would like to thank Prof. Larry Brown and the Department of Industrial Engineering as well as Prof. Don Trotter and Zach Rowland from the Center for Advanced Vehicular Systems at Mississippi State University for their support throughout the early development of the project. As well, the authors would like to thank Prof. Don B. Chaffin and the University of Michigan for their willingness to allow access to the HUMOSIM software for research and educational purposes at Mississippi State. Finally, we appreciate the cooperation of Prof. Dario Salvucci at Drexel University for his willingness to allow the driver model to be integrated for the purposes of this project.

References

Anderson, J. R., Bothell, D., Byrne, M., Douglass, S., Lebiere, C., and Qin, Y. (2004). An Integrated Theory of the Mind. *Psychological Review*, 111(4), 1036–1060.

Anderson, J. and Lebiere, C. (1998). *The Atomic Components of Thought*. Mahwah, NJ: Erlbaum.

Chaffin, D. B. (2007). Human motion simulation for vehicle and workplace design, *Human Factors and Ergonomics in Manufacturing, 17*(5), 475–484.

Chaffin, D. B., Faraway, J. J., Zhang, X., and Woolley, C. (2000). Stature, age, and gender effects on reach motion postures, *Human Factors, 42*(3), 408–420.

Du, C. J., Williams, S. N., Duffy, V. G., Yu, Q., McGinley, J., and Carruth, D. (2005). The use of computer-aided ergonomic analysis tools in the assessment of an automotive lifting task, *HCII 2005, 9th Conference on Human-Computer Interaction International,* Las Vegas, July, 2005, CD-ROM.

Duffy, V.G. (2005). Impact of a force feedback on a virtual interactive design assessment, *Human Aspects of Adveancd Manufacturing and Hybrid Agility (HAAMAHA) 2005 Conference Proceedings,* San Diego, July 18–21, CD-ROM.

Duffy, V.G. (Ed.) (2007a). *Digital Human Modeling,* LNCS 4561. Springer-Verlag: Germany.

Duffy, V.G. (2007b). Modified virtual build methodology for computer-aided ergonomics and safety, *Human Factors and Ergonomics in Manufacturing, 17*(5), 413–422.

Duffy, V. G., Jin, M., Eksioglu, B., Yu, Q., Kang, J., and Du, C. J. (2005). Virtual design optimization tool for improved interaction with hybrid automation, *HCII 2005, 9th Conference on Human-Computer Interaction International,* Las Vegas, July, 2005, CD-ROM.

Faraway, J. J. (1997). Regression analysis for a functional response, *Technometrics, 39*(3), 254–261.

Faraway, J. J. (2001). Statistical modeling of reaching motions using functional regression with endpoint constraints, (Tech. Rep. No. 384). Michigan, United States: University of Michigan, Department of Statistics.

Fitts, P. M. (1954). The information capacity of the human motor system in controlling the amplitude of movement. *Journal of Experimental Psychology, 47,* 381–391.

Freed, M., Shafto, M., and Remington, R. (1998). Using simulation to evaluate designs: The APEX approach. In Chatty, S. and Dewan, P. (Eds.), *Engineering for Human-Computer Interaction.* Kluwer Academic.

HUMOSIM Group. (2005). *From motion modeling to improved ergonomics in designed systems.* Retrieved January 2, 2008, from http://www.engin.umich.edu/dept/ioe/HUMOSIM/papers/History_4-05.pdf.

Kang, L., Duffy, V.G., and Li, Z. (2006). Universal accessibility assessments through virtual interactive design, *International Journal of Human Factors Modeling and Simulation, 1*(1), 52–68.

Kintsch, W. (1998). *Comprehension: A Paradigm for Cognition.* New York: Cambridge University Press.

Kieras, D. E. and Meyer, D. E. (1997). An overview of the EPIC architecture for cognition and performance with application to human-computer interaction. *Human-Computer Interaction, 12,* 391–438.

Liu, Y., Feyen, R., and Tsimhoni, O. (2006). Queueing Network-Model Human Processor (QN-MHP): a computational architecture for multitask performance in human-machine systems, *ACM Transactions on Computer-Human Interaction, 13*(1), 37–70.

Newell, A. (1990). *Unified Theories of Cognition.* Cambridge, MA: Cambridge University Press.

Or, C.K.L. and Duffy, V.G. (2007). Development of a facial skin temperature-based methodology for non-intrusive mental workload measurement, *Occupational Ergonomics, 7*(2), 83-94.

Park, W., Chaffin, D. B., and Martin, B. J. (2004). Toward memory-based human motion simulation: development and validation of a motion modification algorithm. *IEEE Transactions on Systems, Man and Cybernetics—Part A: Systems and Humans, 34*(3), 376–386.

Reed, M. P., Parkinson, M. B., and Chaffin, D. B. (2003). A new approach to modeling driver reach. Technical Paper 2003-01-0587. SAE International, Warrendale, PA.

Salvucci, D. D. (2001). Predicting the effects of in-car interfaces on driver behavior using a cognitive architecture. In *Human Factors in Computing Systems: CHI 2001 Conference Proceedings*, pp. 120–127. New York: ACM Press.

Salvucci, D. D. and Macuga, K. L. (2001). Predicting the effects of cellular-phone dialing on driver performance. In *Proceedings of the Fourth International Conference on Cognitive Modeling*, pp. 25–32. Mahwah, NJ: Lawrence Erlbaum Associates.

Salvucci, D. D., Boer, E. R., and Liu, A. (2001). Toward an integrated model of driver behavior in a cognitive architecture. *Transportation Research Record*, No. 1779.

Wu, L. T., Duffy, V. G., Kang, J., and Carruth, D. (2005). Validation of the virtual build methodology for automotive manufacturing, *HCII 2005, 9th Conference on Human-Computer Interaction International*, Las Vegas, July 2005, CD-ROM.

Zachary, W., Ryder, J., and Hicinbothom, J. (1998). Cognitive task analysis and modeling of decision making in complex environments. In Cannon-Bower, J. and Salas. E. (Eds.), *Making Decisions Under Stress*, Washington, DC: APA.

Zhang, X. and Chaffin, D. B. (2000). A three-dimensional dynamic posture prediction model for simulating in-vehicle seated reaching movements: development and validation. *Ergonomics, 43*(9), 1314–1331.

3

Time Study during Vocational Training

Gregory Z. Bedny and Waldemar Karwowski

CONTENTS

3.1 Introduction 41
3.2 Knowledge and Skills Classification 43
3.3 Acquisition of Professional Knowledge and Skills 45
3.4 Pace of Performance..48
3.5 Functional Analysis of a Pace Formation Process...................................53
3.6 Influence of the Time Standard on Vocational School Students'
    Performance..57
3.7 Time Study and Work Motivation during Vocational Training............63
3.8 Conclusion..68
References 69

3.1 Introduction

Over the years Ukraine evolved a well-developed system of vocational education, consisting of various vocational schools. To graduate from such a school one had to put in 1 to 2 years of study. Vocational school students had to attend lectures several days per week. They also applied their skills in vocational school workshops (apprentice training). At the end of the semester or at the final stage of their education, students underwent on-the-job training directly at the factory. After completion of vocational school, they received a certificate and started to work in industry. The vocational system was responsible for preparing a qualified workforce for industry, where workers could not be educated through on-the-job training. On-the-job training (3 to 6 months) was used only for preparing workers for jobs requiring the lowest qualifications.

The vocational school system encountered problems when students completed the program, started to work in industry, and were found to be unable to perform their production operations (tasks) in the required period of time. Evaluation of labor productivity or "output per hour" is the major purpose of time studies in industry. The time study was devised by Frederick Taylor in the United States. It was also an important area of research and practical application in Ukraine. The objectives of the time study in industry are the elimination of unnecessary effort, the development of more efficient job performance methods, improvement of productivity, and the development of principles of job estimation. The purpose of the time study is to determine a time standard for a qualified worker performing a particular task working at a normal pace. Time standards (standardized task completion time) can be used for different purposes. They can be used for scheduling work, evaluation of plant productivity and cost estimation, development of a wage incentive plan, etc.

The major purpose of time studies in the vocational system is to prepare young workers to perform their job according to time standard requirements that exist in industry. Therefore, special job training is required for this purpose. It was discovered that the major emphasis in vocational training had been on the quality of task performance. Time requirements were not considered an important aspect of the training process. For example, according to the existing program, the instructor can use a time standard for task performance only when a student performs production operations at the final stage of practice and presumably possesses the required skills and knowledge. This is based on the assumption that introducing a time standard early in the training can decrease the quality of work because the student does not yet have the required knowledge and skills. Therefore, during almost the entire training period students concentrate their efforts on quality of work and do not know how to organize their work according to the time requirements. Preliminary observation of students in the workshop showed that, in most cases, the students did not know how much time they should spend on a particular job. The emphasis was only on the quality of the output. A stopwatch study discovered that students spend a great deal of time unproductively and did not organize their time in an optimal way. Students had poor motivation, and instructors expended a great deal of effort on disciplining them. After being trained without time limits, students could not increase the pace of task performance under production conditions (Bedny, 1981).

Systemic–structural activity theory outlines the following basic approaches in the study of activity: parametric and systemic approach. Parametric analysis entails the study of distinct components of activity. For example, the cognitive approach belongs to parametric analysis. The systemic approach considers activity or behavior as a system. The systemic approach includes a morphological and functional analysis of the activity. The major units of morphological analysis are cognitive and motor actions and operations. Functional analysis considers activity as a goal-directed self-regulative system. In functional analysis, the major units of analysis are functional mechanisms or function blocks. Learning in this case is regarded as a self-regulation process (Bedny and Karwowski, 2006). In this work, we conduct our study mostly from the functional analysis perspective. According to functional approach, learning is considered an acquisition of diverse strategies for activity performance and transformation from less efficient to more efficient work practices. The term *strategy* is treated as a dynamic plan for goal achievement that is responsive to external contingencies, as well as to the internal state of the subject. Strategies are the result of a goal-directed self-regulative process. The goal is considered to be one of the most important mechanisms of self-regulation. The method of goal assignment can affect the blue-collar workers' training process. The transition to faster performance requires reconstructing strategies of performance and, therefore, development of different skills (Bedny, 1981). It was hypothesized that introducing to the student a goal that includes time standard requirements could increase performance speed without decreasing work quality. The purpose of this chapter is to describe how time standards influence students' performance during vocational training and how students can be trained to perform the task at the required level of pace.

3.2 Knowledge and Skills Classification

Training can be considered to be the management of students' strategy of activity performance, during which they acquire professional knowledge, skills, values, attitudes, and so forth.

Knowledge is the information encoded in our memory in the form of images, concepts, or propositions. To know something means to have an image of an object or a notion of it, or be able to state a proposition about it. The ability of a subject to describe verbally the major features of an object demonstrates that the subject has a notion or a concept of an object. The ability of a subject to give a definition of an object or make a statement about its features is an example of a proposition. A subject conceiving a picture of an object or being able to define it as a sound is an example of imaging. A combination of these elements and their structural organization in our memory constitutes our knowledge. Knowledge may reflect externally observable events or unobservable processes. According to activity theory one can distinguish between knowledge about objects (physical and mental) and knowledge about actions or operations with objects (Landa, 1976; Novikov, 1986; Platonov, 1970). In cognitive psychology knowledge about objects is defined as *declarative knowledge*, and knowledge about actions and operations with them is defined as *procedural knowledge* (Anderson, 1985). Procedural knowledge is particularly important in vocational training. Novikov (1986) introduced the following classification of declarative and procedural knowledge in vocational education.

The first class of declarative knowledge is:

1. Directly sensed knowledge about objects used during task performance

2. Phenomenological (descriptive) knowledge about technical objects (devices, equipment, computer, etc.)

3. Knowledge about functioning technical objects and their construction
4. Knowledge about the scientific basis and principles of organization and operation of technical objects.

The second class of procedural knowledge includes:

1. Directly sensed knowledge about performance of actions, based on perceiving and imagining of actions with objects
2. Phenomenological (descriptive) knowledge about actions with technical objects
3. Knowledge about general rules (algorithms) of actions with specific objects under different conditions
4. Knowledge about the scientific bases and general principles of performing technological processes and organizing labor.

Spontaneous knowledge that is obtained directly from sensory-perceptual experience very often is implicit and cannot be verbalized. In contrast, other categories of knowledge are explicit and can be verbalized. A subject can have in memory a vast amount of declarative knowledge and limited procedural knowledge. As a result, he or she will not be able to perform mental or physical actions and operations based on declarative knowledge. In this situation a subject would have a problem applying his or her knowledge in practice. If knowledge is structurally organized, a subject can apply it more efficiently. Discrepancy between declarative and procedural knowledge results often in a situation in which a subject applies his or her knowledge intuitively.

The level of generalization of declarative and procedural knowledge can be different. Very often declarative knowledge is at a higher level of generalization than procedural knowledge. For example, in vocational schools theoretical classes provide a higher level of general declarative knowledge than procedural knowledge obtained during practical training. This discrepancy provides a better opportunity for further professional growth. Generalization and transfer of knowledge requires development of conscious and unconscious strategies of extracting critical or essential features of objects or phenomena and neglecting those that are not essential. It is important to find out the regularities of the variations of these features from situation to situation. It is possible to develop algorithms of identification that help a subject to acquire diverse strategies to extract essential features. However, not all problems are algorithmic in nature. In such cases it is recommended that heuristic principles of analysis be used (Landa, 1976).

One of the main purposes of teaching is to develop the student's capability to think. Thinking is not simply knowledge, but what one does with that knowledge. Thinking can be considered as a structurally organized system of mental operations carried out on knowledge and objects to solve problems and make decisions. One can have a large repertoire of images, concepts, and propositions (knowledge) but only a small repertoire of mental actions; one cannot then know which mental actions to apply to which knowledge. This means that students' capability to think is not sufficiently developed.

The other important units of professional experience are habits and skills. In activity theory they have different meanings. In Russian they are designated as *naviky* and *umenia*. Habits (*naviky*) are considered to be cognitive, and motor actions are consciously automated during learning and training. Sometimes, the high level of automation causes transfer of cognitive and motor actions into cognitive or motor operations. Habits are associated with care, quickness, and economy, with lower level of attention, and a high level of efficiency. Habits can be integrated into motor or mental patterns. This pattern or structure creates a first level of cognitive or motor skills (*umenia*). There are more complicated or second-level skills. These kinds of skills constitute an individual ability to organize knowledge and the first-level skills into a system, efficiently use them to perform a particular class of tasks or solve particular classes of problems.

We have already considered above how conscious actions can transfer into habits during training. When these habits are simple, they can be used as building blocks or operations to form more complicated actions (Leont'ev, 1978). However, there are cognitive and motor operations that almost never become conscious; the other group of operations at the first step of training is unconscious. At the following step of special training, these operations become available at the conscious level, and, therefore, after their structural organization the operations are transferred to conscious cognitive or motor actions. For example, a gymnast's unconscious motor operations can be transferred into consciously regulated actions. This makes the gymnast's skills more flexible and consciously regulated. In some instances transformation of unconscious operations into consciously regulated actions can be undesirable. Transformation from an unconscious level of regulation into a conscious level can cause deautomation of skills (Bedny and Meister, 1997).

During the skills acquisition process a student can transfer his or her attention from controlling actions to the goal and results of actions. In activity theory, depending on the content of actions associated with corresponding skills, they are classified as sensory, perceptual, mnemonic, thinking, imaginative, or motor (Gil'bukh, 1979). The other system of classification distinguishes sensory (visual, auditory, etc.), mental (calculation, reading, solving practical problems, etc.), and motor (connected with performance of motor actions) skills (Platonov, 1970).

3.3 Acquisition of Professional Knowledge and Skills

The training program must provide an opportunity to practice actively the skills required on the job. Repeated performance of the same activity with evaluation of obtained result leads to the changes in activity structure. These changes can be described as follows:

1. *Changes in methods of performing actions:* The individual operations are integrated. Unnecessary action components (unnecessary because enhanced skills render them unnecessary) are progressively eliminated. Performance speed is increased, and actions that have been performed sequentially before can now be performed simultaneously.

2. *Changes in methods of regulating actions:* The reference point for regulating actions is extracted very quickly. For example, during acquisition of sensory-perceptual skills, students develop an ability to extract required acoustical features from the environment. An external contour of regulation is replaced by an internal one. For instance, visual evaluation of motor actions is replaced by kinesthetic evaluation. Levels of action regulation are changed, or leading levels of action regulation associated with our consciousness and meaningful interpretation of actions is gradually replaced by lower level of action regulation. As a result, individual attention is transferred from perception of his or her own actions to the actions' output. Actions are evaluated not only consciously, but also intuitively.

3. *Changes in strategies of activity performance:* Separate actions are integrated into a holistic structure of activity. The individual prepares for (anticipates) subsequent actions while performing preliminary ones. Transfer from one action to another is performed automatically. The number of actions that can be performed simultaneously increases.

4. *Changes in emotionally-motivational regulation of activity:* Energetic components of activity associated with the emotional-motivational state of the subject change during skill acquisition. Subjectively, this state is experienced as changes in effort of activity performance, reduced emotional tension, changes in evaluation of significance of different components of activity, and changes in motivation of activity performance in general. Emotional-motivational regulation of activity influences cognitive strategies of activity performance.

Changes in the skill structure during practice can be explained by mechanisms of activity self-regulation (Bedny and Karwowski, 2006). Activity acquisition can be considered a self-regulative process when a subject transfers from less efficient strategies of performance to the more efficient strategies of performance. Strategies of performance include conscious and unconscious components of activity. The more complicated the skills, the more intermittent strategies are required during the skill acquisition process. Systemic-structural activity theory (SSAT) demonstrates that learning can be viewed as an active regulative process, and strategies of performance can be described based on analysis of self-regulation mechanisms. Venda and Ribal'chenko (1983) also discovered that for the more complex skills the students utilize a greater number of intermittent strategies. However, the obtained data were explained from a different theoretical perspective.

During the skills acquisition process, students can transfer their attention from controlling their actions to the goal and results of these actions.

Increasing efficiency of the training process can reduce the number of intermittent strategies of performance. The more efficient the training process, the more important the role of second-level skills. Acquisition of this kind of skill enables a student to utilize different strategies of performance for the same activity. The skills become flexible and adaptive to the ever-changing situation. According to Bernshtein (1966), skill acquisition is a complicated constructive process. During each new trial, a student develops a new structure of the skill adapted to the particular trial. Therefore, repetition of trials cannot be considered the same activity performance. Activity is reconstructed during each trial, and this process can be considered "repetition without repetition" (Bernshtein, 1966). Changes in strategies of performance during the acquisition process can be discovered by utilizing a variety of indices. The most important of them are analysis of errors and of performance time, as well as skill acquisition curves. The shape of the learning curve is affected by the nature of the task, the idiosyncratic features of the trainee, and the methods of training. Averaging curves permits one to discover the general features of the skill acquisition process. At the same time, averaging curves can result in missing some individual features of skill acquisition of different trainees. Therefore, the individual curve and the averaging curve should be used in combination. It is also useful to develop several learning curves, one describing acquisition of task performance as a whole, and others describing acquisition of separate elements of the tasks (Bedny and Zelenin, 1988). Several learning curves permit one to describe the acquisition process of different subskills and how this is related to overall performance. Figure 3.1 demonstrates two

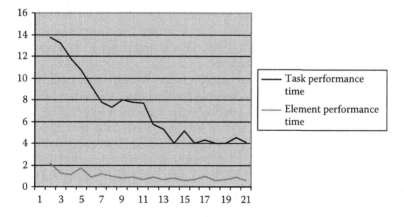

FIGURE 3.1
Learning curves that depict "turning a part on the lathe" acquisition process (on the horizontal axe is plotted the trial number and the vertical axe shows performance time in minutes).

curves, where the top curve depicts the acquisition of a complete production operation, "turning a part on the lathe," and the bottom curve reflects the acquisition of one of these production operation elements, namely, "fixing a part into a three-jaw check." From this figure one can see that students are relatively skilled at performing the operational element "fixing a part into three-jaw check," and the performance time is reduced owing to the turning process.

The other important method of strategy analysis involves description of activity time structure during different stages of skill acquisition. This is particularly relevant when we study strategies of activity during task performance and conduct of the time study (Bedny and Meister, 1997).

3.4   Pace of Performance

There is no precise definition of work pace. Barnes (1980) defines *work pace* as the speed of operator's motions. However, this definition is unsatisfactory because it ignores cognitive components of activity and logical organization of cognitive and behavioral actions. The pace can be considered the speed of performance of different components of activity that are structurally organized in time. Hence, pace of performance can be defined as the operator's ability to sustain a specific speed (below maximum) of holistic activity structure that unfolds during task performance. This pace should be sustained during the shift and subjectively evaluated by the operator as an optimal pace. It has been discovered that the slowest workers' pace of performance can be slower than the fastest blue-collar workers' pace by a factor of 2 (Barnes, 1980). Therefore, in a large group of workers who perform the same task by using the same method, the fastest operator would produce approximately twice as much as the slowest operator. Our study shows that the performance time ratio in vocational school can vary on average from one to four (Bedny and Zelenin, 1986). This is apparent when students work without time standard requirements.

Pace evaluation is difficult to perform. One widely used method of pace evaluation in industry is based on subjective judgment. This method is called *rating* (Barnes, 1980). Rating is a process during which a specialist compares the pace of the operator's work with the observer's own concept of normal or standard pace. The latest can be understood as an average worker's pace that can be maintained during a shift without excessive mental and physical effort, assuming that quality of work would be within the assigned standard. An average person walking on a level grade at 3 mi (4.8 km) per hour along a straight road is used to represent a normal walking pace. This criterion has been supported by physiological studies. It is a traditional type of activity that is also easy to compare with subjective feelings and psychophysiological measurements. Physiological studies demonstrate that energy expenditure per unit of covered distance is minimal if the speed of walking is between 4 and 5 km/h (Frolov, 1976).

The other method, also based on physiological evaluation of pace of performance, is an experimental one. In cases when a practitioner evaluates medium and heavy physical tasks, physiological evaluation of pace of performance is possible. Oxygen consumption in calories per minute and heart rate in beats per minute can be utilized. It is more difficult to evaluate the pace of performance when cognitive components of activity dominate during task performance.

Expenditure of energy at 4.17 kcal/min is equivalent to a pulse rate of 100 beats/min. Analysis in publications—(Lehmann, 1962) and Rozenblat (1966)—demonstrate that a pulse rate of 100 beats/min, or 4.17 kcal/min, should be used as the benchmark for the boundary between acceptable and unacceptable strenuousness of work. It corresponds to the boundary between low and heavy physical work intensity, according to Rosenblat's classification. In work conditions when the pulse rate increases beyond this standard, additional standard break time is recommended.

There are several different rating scales for evaluation of work pace. These kinds of scales are developed on the basis of psychophysical methods. For example, there is a scale where standard or normal pace is defined as 100. If the actual pace of performance is less than normal, it receives a number less than 100. If the actual pace is higher than the standard pace, a number above 100 is assigned. The last number that is assigned to the real pace of performance should be "0" or "5" (70; 75; 80, etc.). Pace evaluation is produced for separate elements of the task. However, duration of separate elements should be no more than 30 s (Barnes, 1980). After the evaluation of the pace of the element performance and the real measurement of the performance time, the standardized performance time for each element of the task is determined according to the following formula.

$$S = T \times P$$

where S is the standardized time for a particular element of the task; T is the time obtained during chronometrical measurement; and P is the coefficient of pace performance (it defines the relationship between the evaluated by expert pace of performance and standardized pace of performance). For example, real time of task element performance is 0.30 min; pace of performance is 90. Therefore, $S = 0.30 \times 90/100 = 0.27$ min.

In vocational schools, when physical components of work dominate, energy expenditure for 14- to 17-year-old students should not exceed 3.7 kcal per minute for boys and 3.2 kcal per minute for girls (Kosilov, 1975). This means that the standard or normal pace assigned for teenage students based on subjective judgment should be equivalent to 70 units instead of 100 units assigned for adult workers. Such psychological methods as analysis of error rate, subjective evaluation of pace by students, and observation of external symptoms of fatigue can be used for evaluation of students' pace.

One of the most important aspects of pace study in vocational training is the study of the pace formation process. We will consider it now. Chebisheva (1969) conducted the following laboratory experiment. Subjects had to sort wooden sticks with different colors while matching the pace of the metronome strokes. At the beginning of the experiment metronome strokes were set to a slow pace. Gradually, the pace of the strokes was increased. Hence, the students should sort the sticks at a different pace. The following levels of pace performance were discovered during this study: (1) very low pace, which was evaluated as an uncomfortable pace, (2) optimal pace, (3) intensive pace, (4) difficult-to-achieve pace, and (5) unachievable pace.

It was discovered that the transition from very slow pace to optimal pace reduces the amount of errors. This pace is associated with a more positive emotional state of subjects during task performance. However, further increase of pace causes an increase in the error rate. Intensive pace, which exceeds the optimal level, is associated with emotional tension and greater difficulty. The difficult-to-achievable pace is considered excessive and can be sustained only for a very short period of time. It causes an unacceptable error rate, and the work is interrupted. It was discovered that the optimal pace activates subjects and motivates them to seek more efficient task performance strategies. Gradual increase in pace is possible. After attainment of optimal pace, it is possible to perform at a higher pace. Performance of tasks at a pace that insignificantly exceeds the optimal pace stimulates better performance. Therefore, the concept of optimal pace during training can be changed accordingly. In the United States, training with gradual increasing of speed of the task performance is known as *above real-time training*. This method has been applied to air force pilot training (Miller, Stanney, D. Guckenbereg, and E. Guckenbereg, 1997).

The pace of experienced workers is relatively stable. In vocational training pace of performance is changed during skill acquisition. It was discovered that transition to the higher level of performance cannot be reduced to the increase in the speed of performance. The ability to perform a task at a higher pace is accompanied by changes in the structure of activity (Bedny, 1981). This is, to a great extent, a new kind of skill. Special training is required to train students to perform work at the required pace.

The concept of professional pace in ergonomics is important not only for training of blue-collar workers in manufacturing but also for the study of operator work in semiautomatic and automatic systems. The operator working with an automatic system under emergency conditions functions as a monitor entering the control loop to override the automatic system and enters new data when required. However, the concept of pace of performance has not been studied in this field. Usually, this problem is reduced to the study of reaction time. Very often, analysis of the operator's temporal parameters in ergonomics is associated with the Hick–Hyman law for evaluation of the speed of information processing, and Fitts' law when manual control is more important. For example, Fitts' law states that when a movement's amplitude

(A) and a target width (W) were manipulated, the time of the movement performance can be determined by the following equation:

$$M(T) = a + blog_2 (2A/W),$$

where *a* and *b* are constants and $log_2 (2A/W)$ is called *the index of difficulty*.

The index of difficulty integrates two characteristics: amplitude and precision. One specific aspect of Fitts' experiment was that the subject moved a metal stick with the maximum speed between two targets. Trying to transfer this result to the work environment, one can assume that each operator's action is at the maximum pace and that each action does not influence the previous or the subsequent action. However, it is important to know not only the speed of isolated reactions, but also how much time is needed for performance of the total task, particularly when it is performed under emergency conditions. The task is not a sum of independent reactions, but rather a system of logically organized actions integrated according to a set goal. An operator never performs the task with the speed that is equivalent to the speed of the isolated reactions. For instance, it has been discovered that when a subject has to hit four targets instead of two, the pace of performance slows down (Bedny and Karwowski, 2006a).

Nojivin (1974) determined that the speed of performance of even a simple reaction significantly changes depending on the instructions utilized by the experimenter. In general, our studies demonstrate that in operator performance one can distinguish between very high, high, and average levels of pace. A very high pace of activity is slightly slower than the operator's reaction time to various stimuli. This pace is possible only in those cases when an operator reacts to the isolated signals using discrete actions in highly predictable situations. A high pace is that in which an operator performs a sequence of logically organized mental and physical actions in response to the appearance of different signals. For motor activity it is essentially the same as that reported in system MTM-1. The pace in MTM-1 system is considered the optimal pace for mass production (Barnes, 1980). However, our analysis demonstrates that this pace is high for the operator and is suitable for his or her work only in time-restricted conditions in emergencies. The pace of the operator's mental activity should be determined on the basis of an analysis of cognitive strategies of task performance. An average pace is that in which an operator performs a task at his or her own timescale when there are no time constraints (Bedny and Karwowski, 2006a).

Task performance time is an important characteristic of a man–machine system. An operator very often performs various tasks under time-restricted conditions. In such circumstances the notion of reserve time is important. *Reserve time* is defined as the surplus of time over the minimum that is required for the operator to perform a required task. Reserve time is defined according to the following formula:

$$T_{res} = T - T_0,$$

where T is the time that cannot be exceeded and T_0 is the minimum time for the task performance.

To determine T_0, one has to define a quick but still reliable pace of performance that is lower then reaction time (high level of operator's pace according to our classification). From the self-regulation point of view, it is important to distinguish between the objectively existing reserve time and the operator's subjective evaluation of this time. In many cases they are not the same (Bedny and Meister, 1997). An operator often roughly evaluates his or her reserve time by making statements such as, "I have a little time" or "I have no time," etc. Such statements may change activity strategies during task performance. Discrepancy between objective reserve time and subjective evaluation of that time may lead to an inadequate evaluation of the situation, which in turn leads to inadequate strategies of task performance in accidental conditions. The subjective perception of reserve time influences cognitive components of activity and the emotional-motivational state of the operator. Psychic tension can emerge even when objectively there is plenty of time for the task performance. Subjective perception of reserve time is an important component of a dynamic model of the situation. In general, the relationship between objective and subjective reserve time is an example of the application of functional analysis of activity (activity is considered a self-regulative system) to studying temporal parameters of task performance. The emotional-motivational state of the operator under time-restricted conditions is an important aspect of functional analysis of activity. Nayenko (1976) distinguished between what he called *operational tension* and *emotional tension*. Operational tension is determined by a combination of task difficulty and lack of available time for the task performance. Emotional tension is determined by the personal significance of an activity for the performer.

Let us consider an example. A major catastrophic shipping accident, involving a collision between a freighter and a passenger ship, took place in the Black Sea in 1984. The captain himself could not explain the reasons for his refusal to heed the warning of his mate, who alerted him to the danger of a passenger ship cutting across his course when it was still 5 mi away. Although the captain saw the ship clearly and had been warned by his mate, he failed to take any emergency action. As a result, 400 people perished. This situation cannot be explained in a traditional way as a violation of safety rules. There are a number of different factors. However, the major one according to the functional analysis is the fact that the dynamic model of the situation was inadequately developed with respect to mapping from the spatial to the temporal domain (Bedny and Zelenin, 1988). In other words the functional mechanism of self-regulation of activity that is responsible for creation of such a model did not balance space and time components of the task. The distance of 5 mi between the ships could have been subjectively perceived by the captain as substantial, though analysis of the vessel's movement revealed that the ship's momentum could be halted only over a distance of as much as 2 mi. This situation required very quick transformation of the spatial characteristics of the task into temporal ones. The captain was not sufficiently alert at the time to make a decision quickly enough to avoid the catastrophe.

3.5 Functional Analysis of a Pace Formation Process

Study of the speed of performance of isolated actions is not adequate for an analysis of the pace formation process. It is important to understand how a subject can change pace of performance of his or her holistic activity during performance of a logically organized sequence of actions. Hence, the goal in front of us was to study a pace formation process when subjects have to perform a sequence of logically organized actions for attainment of the task goal. We had an opportunity to measure not only the task performance time but also the performance time of the separate actions. In this study we utilized functional analysis of activity, in which activity is considered a self-regulative system (Bedny and Karwowski, 2006a). To conduct this study, a special apparatus was designed (Bedny, 1985). There were horizontal and vertical panels for the subject on one side of the device. On the other side of the apparatus there was the experimenter's panel. At the subjects' horizontal panel there was a start position for the index finger—button 1. A middle button 2 and two right-edge buttons 3 and 4 (one red and the other, green) were placed on this panel, as shown in Figure 3.2.

The distance between the start position 1 and the middle button 2 was 12 cm. The distance between the middle button 2 and two edge buttons 3 and 4 was also 12 cm. At the subject's vertical panel there were one digital and four different colored bulbs. In this experiment we utilized only green and red bulbs (the same colors as the subject's two edge bottoms). There were also two stopwatches on the experimenter's panel.

After one bulb turns on, a subject has to move his or her index finger and press the middle button and then move the finger to one edge button and press it according to instructions. Therefore, this apparatus can measure performance time of the first and the second actions in combination with their corresponding cognitive components. Summation of performance time of these actions provides information about the whole task's performance time.

Three groups of subjects were selected. Each group performed a different task. Each subject performed the same task 50 times during one day. The first group performed the following task. After red bulb turns on, a subject moves his or her index finger from the start position and

FIGURE 3.2
Description of apparatus. 1—start position; 2—intermittent button; 3—red button; 4—green button.

Gregory Z. Bedny and Waldemar Karwowski

TABLE 3.1

Average Time of Task Performance(s)

Group Number	First Day			Second Day			Third Day		
	First Action	Second Action	Sum	First Action	Second Action	Sum	First Action	Second Action	Sum
1	0.50	0.36	0.86	0.48	0.33	0.84	0.47	0.31	0.78
2	0.61	0.42	1.03	0.49	0.39	0.88	0.46	0.36	0.82
3	0.52	0.41	0.93	0.38	0.36	0.74	0.39	0.35	0.74

presses the middle button and then moves the finger to the red edge button and presses it. If the green bulb turns on, a subject presses the green button. A subject should perform this task in 0.9 s. The second group performed the same task with the following differences. If red bulb turns on, they should press green button, and if the green bulb turns on, they should press the red button. If the subject from the second group made a mistake, he or she received an electric shock. The third group performed the same task as the second group. However, their performance time was 0.8 s. Each task includes the following actions. When a subject perceives which bulb is turned on, he decides which edge button must be pressed, and then the subject performs two motor actions. The first motor action includes movement of an index finger from the start position and pressing the middle button. The second motor action includes movement of index finger to the red edge or green edge button and pressing it. After each trial, subjects were informed about their task performance time.

In this experiment we did not measure reaction time to an isolated stimulus when a subject reacts with a single action, as is done in traditional reaction time measurement procedures. Subjects had to perform a logically organized sequence of actions. The second difference is the requirement that subjects not react with maximum speed as they do during reaction time measurements; they had to perform the task according to the time standard requirements.

Table 3.1 presents average data for all three groups.

Experimental data for the first group that performed the simplest task demonstrate that this group achieved the time standard requirements the first day (0.86 s). In the following 2 days, the task performance time reduced negligibly. The difference between performance time on the first and third days is statistically insignificant. This means that subjects achieved their time standard requirements on the very first day of the experiment. Hence, the pace of performance developed on the first day for this group of subjects. The discussion with the subjects and the observation of their work in combination with analysis of performance time of separate actions and of the whole task uncovered strategies utilized by the subjects during the task performance. Subjects selected the following strategy. If their performance time was more than the time standard requirements, they increased the speed of performance significantly. After a number of trials, they selected one subjectively preferable result. This result was considered a subjective standard of success. This subjective standard was less than the time standard assigned for task performance and was considered more suitable on the basis of the participants' subjective feelings. The pace of performance that corresponded to the subjective standard of success was considered optimal, and the task performance time that is slightly less than the selected standard is considered a successful result. Sometimes, subjects slightly corrected their subjective standard of a successful result. Therefore, the purpose of the strategy was not simply to increase the speed of performance but also to stabilize the result in relation to the subjective standard. Considering that the subjective standard was less than the required time standard, this strategy guaranteed that the required time standard would be achieved and participants did not spend more time than required.

An activity goal that is accepted or formulated by the subject does not always determine the exact result of activity. This can be explained by different factors. For example, the goal very often does not have all necessary information about the required results of activity in itself. Moreover, the mental representation of a desired result often can be developed only during the performance process. Different subjects can formulate a different mental representation of a desired result when they have the same goal. Hence, the other important mechanism of self-regulation that influences motivation is "a subjective standard of a successful result." For example, a non-specific goal accepted by the subject, such as "to react with the maximum speed," can be a source of diverse possible results depending on a subject's "subjective standard of successful results." A subject with a lower subjective standard of what constitutes a successful result will be satisfied with the slower reaction. In contrast, a subject who is more ambitious and therefore has a higher subjective standard of success attempts to react more quickly. Hence, the same goal can produce different results depending on other functional mechanisms of activity self-regulation. The previously described example where we were studying the performance time of two sequential actions demonstrated that performance time was less than the time determined by the goal formulated by the instructor. During the sequence of trials, subjects developed their own understanding of the successful result. The duration of performance accepted by the subjects as a successful result was less than the objectively given criterion. This criterion was arrived at gradually and could be corrected by subjects during the trials.

The other interesting aspect of the strategy of a task performance was the fact that the first motor action from start position to the middle button 2 required more time than the following movement over the same distance to the green or red button, and this difference was statistically significant (see Table 3.1). This can be explained by the fact that the decision to press the green or red button was made during the execution of the first action. The data demonstrate that a study of isolated reactions is not sufficient for the analysis of the performance time of the holistic task. Subjects develop their own strategy that is not completely predetermined by instructions that are given to them.

The experiment with the second group demonstrated that the complication of the task resulted in changes in task performance strategies and an increase in task performance time (see Table 3.1). The performance time for the first day of the experiment was 1.03 s and exceeded the required time standard (0.9 s). On the second day three subjects performed the task according to the time standard requirement. The average performance time matched the required time standard. On the third day the pace of performance slightly increased. Therefore, the time standard requirement was achieved mainly on the second day. The difference between task performance time on the first and the third days was 0.21 s and was statistically significant. Comparison of time performance on the third day in the first and the second groups demonstrated that the pace of task performance was approximately the same. However, the dynamic of the pace of performance was different. In the first group the required pace was achieved on the first day. In the second group, where the task was more complicated, this pace was achieved only on the second day. This can be explained if we consider strategies of task performance. One can see that in the second experiment, on the first day the difference between performance time for the first and the second action was 0.2 s, and on the second and third days it was 0.1 s. For the first group the difference between the first and the second actions was 0.14 s, 0.15 s, and 0.16 s, respectively. Therefore, the performance strategy during the three days did not change for the first group of subjects. Cognitive functions dominate during performance of the first action, and the second action is performed mostly automatically. In the second group the subjects gradually changed the strategy of task performance after receiving an electric shock.

The second action became more significant for the subject. The subjects made a decision about selection of the second action during the performance of the first action, and this decision was now double-checked before starting the second action. There were other changes in strategies of the task performance. According to the instructions, the second group had to use the bulb color and the button color for decision making. However, subjects in the second group gradually abandoned these distinguishing features. They started using the spatial position of the red and green buttons. Spatial position is a more complicated distinguishing feature than color, but the interference in color features during decision making made this feature more difficult to use during decision making. As a result, all subjects gradually started to ignore color and use only the spatial position of the buttons during decision making. This strategy eliminated the interference of the color in the second group of experiments. This means that subjectively relevant task conditions or mental representation of the task changed in an effort to increase the reliability of task performance. Not all components of performance strategies were changed consciously. For example, when we asked subjects in these two groups which action required more time, seven said that the second action required more time than the first one.

In the third experiment, in which subjects were faced with a more rigorous time standard, there were further changes in task performance strategies. The second action became more consciously controlled. Differences in time performance of the two actions kept decreasing. Two subjects even changed their strategies in such a way that the performance time of the second action became greater than the performance time of the first action. This can be explained by the fact that the increasing time standard requirements (0.8 s instead 0.9 s) and the differing bulb and button colors made the second action a more significant component in the task performance. As a result, the second action became more cognitively controlled.

This study demonstrates that externally given instructions do not exactly predetermine the strategies of task performance. Subjects develop their own understanding of the goal, develop a mental model of the situation and subjective standard of success, evaluate significantly different elements of the task, etc. As a result, complicated dynamic strategies of activity performance can be developed. All of this is the result of a complex process of activity self-regulation (Bedny and Karwowski, 2004, 2006a). Hence, the pace of task performance is actively developed during this self-regulative process. Information about the temporal parameters of activity that is presented to the subject becomes critically important. The duration of pace acquisition depends on the efficiency of the mechanisms of self-regulation. The more complicated the task, the more stages of strategy transformation are needed. As a result, the duration of the pace formation process is increased. The mental model of the situation continually changes depending on the stage of the learning process. Allocation of attention between different elements of activity changes as well. Some actions become more automatic, and others more cognitively controlled. Elements of activity that become automatic very often are performed simultaneously. As a result, the pace of activity performance can be increased without increasing the speed of separate actions. This in turn influences the pace formation process. Functional analysis of activity is the foundation of the self-regulative concept of learning. Learning is considered a process of strategy transformation during the activity performance. The more complicated the skills involved, the greater the number of intermittent strategies that are utilized by the students. The establishment of the pace of performance can be explained only through the analysis of the strategies utilized by the learners during their skill acquisition process. In the following section we will consider some principles of the pace formation process during professional training.

3.6 Influence of the Time Standard on Vocational School Students' Performance

The transition to the higher level of the task performance leads to the reconstruction of professional skills. Therefore, working under time-restricted conditions requires special training. However, there is an opinion that introducing time standard requirements during the training process can decrease quality of students' performance because the students do not yet have the required knowledge and skills. Official instructions distributed by the State Committee of Vocational Education of Ukraine recommended not informing students about the time constraints of their task at the stage when students did not possess the necessary skills. The instructors should pay attention only to the quality of work. Observation and recording of the time taken by trainees to perform their work in vocational workshop showed that trainees could not organize their work on time. Students spent significant amounts of time counterproductively. After being trained without time limits, students could not organize their work on time and increase the pace of task performance under production conditions. Trainees spent time unproductively, especially during the setup and auxiliary components of work. It was hypothesized that introducing students to the performance of the task according to established time standards could increase performance quantity without decreasing the quality of work. Therefore, an attempt was made to find out how the goal that includes not only qualitative but also quantitative requirements would influence students' motivation, discipline, fatigue, feeling of time, etc. This study was conducted over a long period. As an example several field experiments will now be discussed in brief.

The first experiment was performed with students who performed bench and assembly work in a vocational school workshop (Bedny, 1981); male (12 in each group) 16- to 18-year-olds were selected. They were first-year students. The experimental group had a lower average grade than the control group. In the former Soviet Union, numbers were utilized for grade assignments instead of letters (A equals 5; B, 4; C, 3; D, 2; F, 1). Both groups received the same task and the same instructions. The experimental group had to perform the task within 2 h (time standard 2 h). The control group worked as usual without time standard requirements. Observations of student work were made. Students did not know that they were involved in experimental study. For them it was regular work. Experiments with each group were performed separately. The average grade in their practice for the experimental group was 3.3 and for the control group was 3.8. This was done intentionally to understand how the time standard influences weaker students. The average time of task performance for the experimental group was 105 min, and it was 127 min for the control group. The difference in performance time was statistically significant. The variation in performance time was lower in the experimental group. The quality of work in the experimental group was also better despite the fact that less successful students were in that group at the beginning of the study. The average grade in the experimental group was 3.8 versus 3.4 in the control group. This data were not statistically significant. Nevertheless if we take into consideration the fact that the students in the experimental group had lower average grades prior to this study, this is a positive result. We divided the whole task into separate performance stages. Each stage was evaluated according to the qualitative criteria of performance. It was discovered that in the experimental group the quality of work had a tendency to be better, particularly during the final stage of the task performance. The quality of work in the control group was the same at all stages of task performance. Therefore, the dynamics of the quality of performance were better in the experimental group. Observation of student behavior showed that in the experimental group the students demonstrated better concentration on task performance. They were not criticized by the instructor. In contrast, the behavior of students in the control group was often criticized by the instructor. Students in the experimental group showed more interest in their work than those in the control group. In general, students of the experimental group performed work faster with better quality. The incorporation of the time standard into the training process also reduced variability of the work performance time.

Let us consider another example. The experiment was conducted with 10 first-year female students. The task included milling special bolts. The experiment was conducted with each student individually over 2 d, and the study took about 3 weeks. The trainees did not know that they were taking part in an experiment. For them this was just a regular workshop. In a preliminary study with a different group of trainees, it was discovered that the work under consideration was repetitive and monotonous. In spite of the fact that students became familiar with the task, they usually reduced their productivity on the second day of performing this task when they worked as usual without time standard requirements. It was suggested that presenting to the trainees a goal with precise quantitative and qualitative requirements could increase the work productivity and reduce monotony.

On the first day, the trainees worked under regular conditions without being asked to produce a particular number of pieces (the time standard was absent). On the second day, the trainees performed the same task, but had to produce 30 parts per day. Each part contained 3 facets and therefore each student has to mill 90 facets per shift. However, they were milling 110 facets on average. This is an evidence that the students introduced their own subjective standard of success that exceeded objective requirements. The result of this study demonstrated that the introduction of time standard requirements sharply increased productivity. The average number of milling facets per student on the first day (without time standard requirements) was 52. On the second day, when students were introduced to the time standard, they were milling 110 facets on average. The difference in productivity was statistically significant.

We also recorded the number of facets students were milling every 30 min. Figure 3.3 demonstrates the dynamics of productivity over the whole shift (vertical axis—number of facets students were milling; horizontal axis—current time). Curve 1 shows the dynamics of student productivity during the first day, when students worked with no time standard, and curve 2 demonstrates the dynamics of productivity for the second day, when students performed the same task but with the introduced time standard. As can be seen, productivity rose faster on the second day and was maintained at a higher

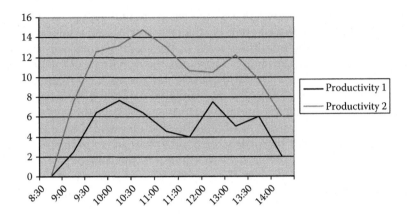

FIGURE 3.3
Dynamics of productivity during shift. 1—work without time standard; 2—work with time standard.

level for a longer period of time. We have conducted multiple similar studies that also proved our main conclusion that preset goals for quality and quantity standards have a positive effect on task performance during training.

Craft classes used to be conducted in public schools in Ukraine. The purpose of these classes was to develop general knowledge and the skills to perform certain kinds of work. The acquired knowledge and skills did not correspond to any professional standards. They were just perceived to be useful in the students' life. These classes also aimed to develop a positive attitude to possible future work. Two 45-min craft classes were held once a week. We decided to find out how introducing a time standard in the middle-school workshop would influence quality and quantity of student performance. These students were younger than vocational school students. We conducted our study with fifth- to seventh-grade students. The result of the experiment with these students was similar. Time standard requirements led not only to an increase in the quantity of work but also in the quality of performance. However, the teacher should explain how students can organize their work under time restriction conditions. Therefore, time standard requirements always have a positive impact on student performance independent of age.

An awareness of time during task performance is an important professional skill. This skill includes an ability to plan the temporal aspects of work. A longer training period is required to develop these professional skills. Inculcating these skills is very difficult and to a significant degree is an unconscious process for workers in industry. We have developed a special training program that helps develop these skills more quickly (Bedny, 1981). In our study we have demonstrated that even young students can develop such skills by means of specially designed training.

We will now briefly describe an experiment with sixth-grade middle-school students (11 to 12 years old). We have developed special templates, or standardized samples, of typical elements of possible bench work for sixth-grade students. These elements of work can be encountered in different combinations when students perform their tasks in a school workshop. Typical elements of work were made of metal sheet. Each standardized sample for a typical element of work has several different-sized examples, and a time for their performance is also assigned. Students had to perform various kinds of tasks that involved production of different kinds of hand-made metal goods. They also received drawings and written instructions on how to perform these tasks. Students broke down the received assignment into standardized elements of work and compared them to existing templates or standardized samples and times of their performance. After that, students determined the performance time for all standardized elements of work in their particular task and then calculated the performance time for the whole task.

As an example, we describe one such experiment. Students had to make a metal hook used for latching a door. This entails bench work that includes different bench operations and requires special training. It was a sufficiently complex task for the sixth-grade students. It included such bench-work operations as laying out, drilling, filing, sawing a slot, and cutting chamfers. In the first step, students determine the duration of all future operations and the total time for performance of the task based on their opinion. In the second step, students determine the duration of performance of separate operations by utilizing the standardized typical elements of work described earlier. For this purpose they selected standardized samples for typical elements of work that were similar to the operations required to be performed. On the basis of a comparison of the performance time for the selected typical elements, students determined the performance time of the required operations. During this assignment, students introduced some corrections to the estimated required time for the separate operations. After determining the duration of all operations, students calculated the duration of the total task performance. After this theoretical calculation, students made the metal part described earlier. Figure 3.4 presents the results obtained from one group of students.

The bold line demonstrates the predicted time for different operations and the total time of task performance when students simply used their past experience. The gray line demonstrates the calculated time when students utilized samples of the standardized typical elements of work. The white line demonstrates the time actually spent on the performance of separate operations and the total time for task performance. Numbers 1 to 5 on the horizontal axis indicate how many operations were performed. Number 6 on the same axis denotes total time for task performance.

Time in minutes is represented on the vertical axis. From this figure, one can see that students' prediction time, when they did not use special templates for the typical elements of possible bench work, significantly deviated from the actual performance time (compare the top white line with the bottom bold line in Figure 3.4). At the same time, when students utilize templates of typical elements of work, their prediction of performance time of separate operations and the whole task was much more accurate (compare

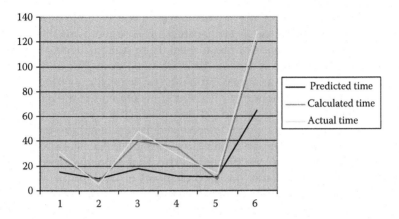

FIGURE 3.4
Student's actual, predicted, and calculated time task performance. 1—laying out; 2—drilling; 3—filing; 4—sawing of slot; 5—cutting of chamfers; 6—total time.

the top white line with the middle gray line). This curve closely replicates the shape of the actual time line. When students attempted to predict performance time for separate operations on the basis of just their experience, the following facts have been discovered: For operations 1, 3, and 4, students' predictions were much lower than the required times. Only for operations 2 and 5, which were of short duration, were the predictions more accurate. In general, students had a tendency to underestimate the time required for more complicated operations. Students also had a tendency to assign similar times for all operations. There was not much variation in predicted time for different operations. For example, for the complicated operations 1, 3, and 4, students predicted time that was not significantly more than the times for operations 2 and 5, which are simpler and required much less time. The predictions were totally different when students utilized standardized samples of work. This time, the predictions were close to the actual performance times. Similar results were obtained with other groups of students when they performed different tasks.

The study demonstrated that students developed an ability to predict with higher precision the performance time for various tasks when they utilized standardized samples of typical elements of possible work with time standard. It is important to mention that experienced workers also utilize similar strategies to predict their performance time without even realizing it; workers store such standardized samples of work and associated times in their memory. These experiments demonstrate that it is possible to develop a special training program directed toward the development of the skill "feeling for task performance time." Usually, such skills are formed unconsciously and require extensive experience.

Similar methods of training can be used in vocational schools. The study demonstrates that introducing time standards had a significant impact on trainee performance. However, it was discovered that this occurs only when time standards are introduced to first-year students. If students have worked without time standards for a long time and time requirements are introduced, for example, at the end of the second year of the training program, introducing time standard requirements was not effective. This can be explained by the fact that not just cognitive but also social and motivational factors become important in the training process. Time standard requirements after their acceptance by a group of students become an important component of the group goal and source of intersubjective process of communication and comparison between students. An individual, on the basis of time requirements accepted by the group, evaluates his or her own performance. Time requirements are not the only factors that facilitate social communication based on comparison; social factors are also important mechanisms of acceptance or rejection of goals with time standard requirements. Students, as members of social groups, compare their temporal components of task performance with socially accepted goals that include time standard requirements. Maintenance of established time requirements becomes an important motivational factor and source of self-actualization. When time standards are introduced at an earlier stage of professional training, students accept these norms and they perform according to socially accepted norms for the group. Introduction of time standards has this important effect even when students work independently. In described experiments students perform bench-type work, where each student is responsible for his or her own task. However, when students receive time standard requirements, and if these requirements are accepted by students as an important aspect of their goal, this goal is perceived by students as a socially accepted standard for all members of the group. This process permits the goal for a group of students to be converted into a personal goal that plays an important role in self-assessment. The goal becomes more significant for students. As a result, their motivation to achieve the goal increases. Students develop their own subjective standard of successful performance. This standard often exceeds the stated requirements (goal), and students' productivity increases even more than was expected. In this situation, students continually transform the solution of practical problems into self-assessment evaluation. Their confidence in their ability to work efficiently improves. In the training process, this is particularly important because students are more sensitive to their own achievements than are experts.

3.7 Time Study and Work Motivation during Vocational Training

The relationship between motivation and performance during vocational training is especially important. We will now examine how goals that include time standard requirements influence student motivation and performance. From a functional analysis perspective, motivation is considered to be an important mechanism of a goal-directed self-regulative process. Functional model of self-regulation includes different functional mechanisms or function blocks such as set, goal, dynamic mental model, assessment of task difficulty, assessment of sense of task, formation of the motivational level, subjective standards of successful results, etc. Interaction of these mechanisms determines strategies of task performance during the training process. The model of self-regulation gives an opportunity to define stages of self-regulation (Bedny and Karwowski, 2006b). The first stage is called a preconscious motivational stage that determines motivational tendencies. This stage is not associated with a conscious goal. An unconscious set is important at this stage of motivation. An unconscious set can be transferred into a conscious goal, and vice versa. The second stage of motivation (goal-related motivational stage) is the formation of or acceptance of the conscious goal. This stage of motivation can be developed in two ways: by bypassing the preconscious stage of motivation or through the transformation of an unconscious set into a conscious goal. The preconscious stage of motivation not only precedes other motivational stages, but also functions in parallel. When the current task is interrupted and the attention shifts to a new goal, the former goal does not disappear, but is transformed into a preconscious set. As a result, a goal-related motivational stage is transformed into a preconscious motivational stage. The availability of the already-formed goal in the form of a set allows the individual, if necessary, to return to the interrupted task.

The third motivational stage is related to the evaluation of the difficulty and significance of the task. This motivational stage is called *task-evaluative aspect of motivation*. Evaluation of difficulty and significance of the task or of the situation involves two important interacting functional mechanisms (function blocks). One is called "assessment of task difficulty" and another one is "assessment of the sense of task." These two function blocks work interdependently. For example, there is an objective characteristic of the task called *task complexity*. Subjectively, this characteristic is evaluated as the task difficulty. The more difficult the task is perceived to be, the more significant it becomes for the subject. Evaluation of the task significance is performed by another important motivational mechanism of activity self-regulation that is called *assessment of the sense of the task*. There is a complicated relationship between such self-regulative mechanisms as evaluation of the task difficulty, significance, and motivation. For example, if attainment of the task goal is evaluated as very difficult, and significance of the task is very low, the person will not be motivated to perform the task. If difficulty of the task is perceived as very low and its significance is very high, the motivation can be moderate. If attainment of the task goal is evaluated to have a low level of difficulty and significance of the task is also evaluated as low, this will result in a very low level of motivation. This kind of work is usually perceived as monotonous and boring. Hence, an increase of the task difficulty does not always increase the level of motivation, as stated by Lee et al. (1989), and the concept of self-efficacy (Bandura, 1987) is also not sufficient to explain work motivation. For instance, if the subject evaluates the goal as very difficult because of low self-efficacy and the task has very high level of significance, the subject will still be motivated to perform the task. Motivation cannot be explained by a single functional mechanism but can be understood as a dynamic system that depends on a complex relationship between different mechanisms of self-regulation.

The fourth stage of motivation is associated with its executive aspect of task performance and is associated with goal attainment. This stage is called *executive- or process-related stage of motivation*. The last, or fifth, stage of motivation is related to evaluation of activity result and involves other functional mechanisms of self-regulation. The function block "subjective standard of successful result" is very important for the analysis of this stage of motivation. All stages of motivation can be in agreement or in conflict. For example, positive or negative evaluation of student work has meaning for the student only if the goal attainment is significant for him or her (goal-related motivational stage). Motivation to achieve a goal may be unrelated to the executive- or process-related stage of motivation. In such cases, a positive motivation to achieve the goal may be combined with a negative motivation for the task performance.

Learning activity has not only an individual-psychological character but also a sociopsychological character. This raises the question of how motivation is changed depending on sociopsychological factors. Not only cognitive but also emotional-motivational aspects of student activity change during social interaction. In the course of learning, groups formulate their own norms and activity goals. Each student within the group evaluates the goal of the task depending on how others evaluate the same goal. Therefore, a goal-related motivational stage in a group can be different, depending on the group dynamics. Norms and goals set by the instructor during the learning process can contradict the socially accepted goals and norms of the group. As a result of social learning, the objective norms and goals given to the students can motivate them in various ways.

It has been discovered that for students in vocational school who had worked without time standards for a long time and had almost finished the training course, introducing the time standard was ineffective. Students ignored the time standard requirements. This can be explained by the fact that students had already developed their own performance standards as a function of peer comparison. Students develop their own informal standards and social norms within the group. These standards and norms can be developed consciously or unconsciously. Because there were no objective time standards for a long time, students did not accept time standard requirements. Members of a group evaluate themselves and others according to their own norms and standards. Students disregard any external requirements associated with the temporal parameters of their work owing to informally developed norms.

As an example we will now briefly describe how time standards influence motivation and thereby change productivity dynamics. Two groups of younger workers were trained to perform turning operations on cylinder bushings. This work was repetitive, and students felt monotony. This kind of monotonous work is very often encountered in mass manufacturing. Our suggestion was that a precisely formulated goal to perform the required number of pieces of product would have motivational influences on trainees' performance and increase their productivity. Two groups of male students performed this repetitive monotonous task for three days. The experiment was conducted with each student individually. There were six students in each group. This was a field study, and the trainees did not know that they were involved in an experiment.

The task was simple, and a trainee could acquire the necessary skills during the first day of training. The control group was not informed about the standard production rate. The trainees in the experimental group were informed that they needed to produce 60 bushings per day. Preliminary study showed that this requirement was not difficult for the trainees. Considering that the task was not at all complicated but, rather, monotonous for students, productivity depended primarily on motivational factors. In the control group where students worked without the time standard requirements, the students produced on average 32.2 bushings while maintaining the required quality level. While on the first day students produced 36.1 pieces per shift, on the second day they produced 33.8 bushings, and on the third day, they produced 26.8 bushings on average per shift. However, these changes were not statistically significant. Therefore, we can talk only about the decreasing productivity tendency.

The second group of students were given the time standard requirements. The average performance for the experimental group over 3 days was 82.3 pieces per shift—almost double that of the first control group—and this difference was statistically significant. On the first day the trainees produced 76.5, on the second day, 83.1; and on the third day, 87.1 bushings per shift. These changes were statistically significant. Therefore, time standards not only increase productivity of work but also change the dynamics of productivity. The difference in the dynamics of productivity is particularly interesting for our study. The obtained result can be explained by the conflict between goal-related and process-related stages of motivation. On one hand, subjects accepted the given goal (goal-related stage of motivation), which was a part of trainees' duties and responsibilities related to their job requirements. On the other hand, the performance itself induced a negative motivational state. If a negative motivational state during work performance outbalances the goal-related stage of motivation, the trainee may quit the job altogether. This contradiction explains why productivity decreases from day to day in the control group.

We will now explain why productivity increased in the experimental group where time standard requirements were introduced. Analysis and observation demonstrate that intermittent productivity is periodically compared with the goal of the task. Therefore, introduction of the time standard helps students to properly distribute their energy during the shift.

However, it is not the only reason for the changes in productivity. Studies demonstrated that during task performance psychic saturation occurs, and this saturation increased from day to day in the control group. Trainees in the experimental group did not know in advance that they would attempt to increase their productivity. The formation of a more difficult goal, namely, "to produce more than is required on the first day and even more on the following day" emerged during task performance. The functional mechanisms responsible for evaluation of significance of the task and task difficulty are particularly important in this situation. To maintain the process-related stage of motivation, students increased their subjective standard of success, which is another mechanism of self-regulation. Voluntarily-increasing productivity up to a particular level during 3 days emerged as a dynamic criterion of success. Students did not know its value in advance. The level of aspiration, feeling of tiredness, etc., is very important in the formation of the subjective standard of success. This standard is subjectively associated with commands to self such as, "It is enough for me," "I can stop now," etc.

On the second day students realized that the assignment was not difficult for them. Moreover, students acquired the necessary skills on the first day of the job. This led to a decrease of subjective evaluation of task difficulty on the second day. A decrease in perceived difficulty results in a decrease in task significance. This in turn negatively affects motivation and other functional mechanisms of self-regulation. To maintain a high level of motivation on the second day, students increased the difficulty of their goal and subjective standard of successful result again by striving to produce a bigger quantity. However, even this higher result is achieved relatively easily on the second day and the subjective evaluation of task difficulty decreases again owing to further mastering of skills, which in turn leads to a reduction of significance and motivation. As a result, students "raise the bar" again, and the cycle repeats itself. Because of this dynamic, we can infer that subjects are aware that they are capable of better performance. This increases the self-esteem of subjects, because they have proved that they can perform a more difficult job. Students voluntarily developed goals to produce more the next day than produced the previous day. Therefore, different stages of motivation (specifically, the goal formation stage of motivation and process-related and evaluative stages of motivation) interact during students' work activity.

Analysis of different mechanisms of self-regulation and stages of motivation help explain why productivity and its dynamics changed when time standards were introduced into the training process. A subjective standard of successful result can vary within the same range in a probabilistic manner, but in some other situations a subjective standard of successful performance can be developed in advance and in a precise, quantitative manner.

Fatigue is also an important aspect of time study during vocational training. Introduction of time standards during vocational training increases trainee productivity. Therefore, instructors have to know whether students' fatigue is increased under these circumstances. To answer this question, experiments with different groups of students were conducted. Fatigue of students under two conditions were measured. One group of students performed the job without time standard requirements, whereas another group performed the job under time-restricted conditions. These types of experiments were performed several times with other groups that performed a number of various tasks. It was discovered that in spite of the fact that students produce more under time-restricted conditions, fatigue reduction and increase in their work capacity were observed. This shows that students better organized their work over time and had better motivation under time-restricted conditions (Bedny, 1981).

The other research study investigated temporal parameters of the students' skills. On the basis of this study, a method of determining time standards for vocational training was developed (Bedny, 1981). In general, it was found that the introduction of time standards during vocational training not only increases trainee productivity (quality and quantity of work), but also has a positive effect on the dynamics of productivity, increases students' work motivation, improves discipline, and reduces fatigue. Introduction of time standards is also helpful for developing a feeling for time during task performance.

3.8   Conclusion

Activity is a goal-directed self-regulative system. From the standpoint of self-regulation, learning can be considered a self-regulative process when a learner acquires diverse actions and strategies of activity performance that lead to acceptance of the formulated goals by a learner. Therefore, such stages of a learner's activity as goal interpretation, goal acceptance, and goal formation are important aspects of the learning process. Hence, the method of assigning goals can affect the learning process in general and training of blue-collar workers in particular. From the standpoint of self-regulation, a trainees' goal that contains objective data about time requirements for task completion has an important cognitive and motivational function in trainees' activity. However, a learner's goal as a mechanism of self-regulation does not function independently. We have demonstrated the importance of such mechanisms as subjectively relevant task conditions (dynamic model of situation or task), formation of motivational level, formation of performance program (executive strategy), subjective standard of successful result, etc. All these mechanisms interact during the process of activity self-regulation. The self-regulation allows one to develop more effective strategies of task performance.

A comparison of the time standard as a component of the trainees' goals with actual output increases the efficiency of the functional mechanisms of trainees' activity self-regulation.

Not only cognitive but also social and motivational factors become essential in the training process. Accepted by a group of students, the time standard requirement becomes an important component of the group goal and the source of intersubjective process of communication and comparison between students. An individual evaluates his or her own performance on the basis of the time requirement accepted by the group. Social factors are also important mechanisms of acceptance or rejection of the goal with time standard requirements. Students, being members of a social group, compare their temporal components of task performance with a socially accepted goal that includes a time standard requirement. Studies also demonstrate that time requirements are important motivational factors during the training process. However, it was discovered that students accept time requirements during their following training process only when these requirements are introduced at an early stage of professional training. Time standards play an important role even when students work independently. When students receive time standard requirements, if these requirements are accepted by students as an important aspect of their goal, this goal is perceived as a socially accepted standard for all members of the group. This process permits a goal for a group of students to become a personal goal that plays an important role in self-assessment. The goal becomes more significant for students. As a result, their motivation to achieve the goal increases. Students develop their own subjective standard of successful performance. This standard often exceeds the stated requirements (goal) and students' productivity increases even more than expected. Students continually transform the solution of practical problems into self-assessment. This increases the confidence of students regarding their efficiency. During the training process, this is particularly important because a student is more sensitive to his or her own achievements than an expert is. The presented data show that activity is a complicated self-regulated system. This system includes different mechanisms of self-regulation. In any particular situation, some of these mechanisms play a more important role than others. Consideration of learning and training as a self-regulative process is called *functional analysis of activity* (Bedny and Karwowski, 2006).

In general, this study demonstrates that introducing time requirements into the training process facilitates more efficient activity self-regulation that is based on cognitive and motivational parameters.

References

Anderson, J. R. (1985). *Cognitive Psychology and Its Application*, 2nd ed. New York: Freeman.

Barness, P. M. (1980). *Motion and Time Study Design and Measurement of Work*. New York, John Wiley and Sons.

Bedny, G. Z. (1981). *Psychological Aspects of Time Study during Vocational Training*. Moscow: Higher Education.

Bedny, G. Z. (1985). Some mechanisms of self-regulation in formation of pace of activity. *Questions of Psychology,* 1, 74–80.

Bedny, G. Z. and Zelenin, M. P. (1986). *Formation of Professional Pace and Time Study during Vocational Training.* Moscow: State Committee of Vocational Training.

Bedny, G. Z. and Zelenin, M. P. (1988). *Ergonomic Analysis of Work Activity and the Problem of Safety in Merchant Marine Transportation,* Moscow: Merchant marine publishers.

Bedny, G. Z. and Meister, D. (1997). *The Russian Theory of Activity: Current Applications to Design and Learning.* Mahwah, NJ: Lawrence Erlbaum Associates.

Bedny, G. Z. and Karwowski, W. (2004). *A functional model of human orienting activity.* In Bedny, G. Z. (Guest Editor), Special Issue: Activity Theory. *Theoretical Issues in Ergonomic Science.* 4, 255–275.

Bedny, G. Z. and Karwowski, W. (2006a). *Systemic Structural Theory of Activity: Application to human performance and work design.* Taylor and Francis Company.

Bedny, G. Z. and Karwowski, W. (2006b). The self-regulation concept of work motivation, *Theoretical Issues in Ergonomic Science,* pp. 413 – 436.

Bernstein, N. A. (1966). *The Physiology of Movement and Activity.* Moscow: Medical Publishers.

Chebisheva, V. V. (1969). *Psychology of Professional Training.* Moscow: Education.

Frolov, N. I. (1976). *Physiology of Movement.* Leningrad: Science Publishers.

Gil'bukh, U. Z. (1979). *Simulator Devices in Vocational Training.* Kiev: Higher Education.

Landa. L. M. (1976). *Instructional Regulation and Control: Cybernetics, Algorithmization and Heuristic in Education.* Englewood Cliffs, NJ: Educational Technology Publication (English translation).

Lee, T. W., Locke, E. A., and Latham, G. P. (1989). Goal setting theory and job performance. In Pervin, A. (Ed.), *Goal Concept in Personality and Social Psychology,* pp. 291–326.

Lehmann, G. (1962). *Practical Work Physiology.* Stuttgart: George Theme Verlag.

Leont'ev, A. N. (1978). *Activity, Consciousness and Personality.* Englewood Glifts. Prentice Hall.

Miller, L., Stanney, K., Guckenbereg, D., and Guckenbereg, E. (1997). Above real-time training. *Ergonomic in Design.* V. 5, # 3, pp. 21–24.

Nayenko, I. I. (1976). *Psychic Tension.* Moscow: Moscow University Publisher.

Nojivin, U. S. (1974). On psychological self-regulation of sensory — motor actions. In Shadrikov, V. D. (Ed.), *Engineering and Work Psychology,* vol. 1., pp. 206–210, Yaroslav: Yaroslav University.

Novikov, A. M. (1986). *Process and Method of Formation of Vocational Skills.* Moscow: Higher Education.

Platonov, K. K. (1970). *Questions of Work Psychology.* Moscow: Medicine Publisher.

Kosilov, S. A. (1975). *Work Physiology.* Moscow: Medical Publishers.

Rozenblat, V. V. (1966). *The Problem of Fatigue.* Moscow: Medicine Publisher.

Venda, V. F. and Ribal'chenko, M. V. (1983). Transformational theory of learning and questions of planning engineering—psychological experiment. In Venda, V. F. and Vavilov, V. A. (Eds.), *Theory and Experiment in Analysis of Operator's Work.* Moscow: Science Publisher, pp. 147–157.

4

The Laws of Ergonomics Applied to Design and Testing of Workstations

V.F. Venda, V.K. Kalin, and A.Y. Trofimov

CONTENTS

Abstract...71
4.1   Introduction 72
4.2   Disabilities caused by MSD...73
4.3   Laws of Ergonomics..74
4.4   Experimental Study of Optimal Angle of Printed Circuit Board
      (PCB) 77
      4.4.1   Method...77
              4.4.1.1   Objective..77
      4.4.2   Equipment...77
      4.4.3   Participants 78
      4.4.4   Procedure 78
4.5   Invention of Assembly Workstations with Indirect Observation
      of Operations and Negative Tilt of Work Surface 81
4.6   Laboratory Testing of the New Workstations 82
4.7   Industrial Comparative Testing of Traditional and New
      Assembly Workstations 83
4.8   Conclusion..86
References 86

Abstract

Electronic assembly workstations are designed so workers can directly observe the manual operations they are performing. Sustained work posture at such workstations causes static contractions of the neck and shoulder muscles, leading to various musculoskeletal diseases (MSDs) (Venda and Hendrick, 1994). Three laws of ergonomics are presented and used as a theoretical basis for designing and testing new electronic assembly workstations making possible the indirect observation of operations and giving a negative tilt to the work surface. Laboratory and industrial test results were conducted at the university laboratories and electronic assembly plants in parallel in Canada and Ukraine. The tests proved that the new workstations significantly improve work posture and productivity, lowering muscle strain as measured by EMG of the descended muscles of the left and right upper trapezius.

The workstations offering indirect observation of operations and a negative tilt may be widely used in manual materials handling and office environments. The workstations are particularly helpful in facilitating return to work of those injured at traditional workstations, or in maintaining employment of persons with vision disabilities, and those with neck and spinal diseases, particularly if they are confined to a wheelchair.

4.1   Introduction

From 1991–2006 a large series of parallel studies of ergonomic factors of work health and safety at electronic assembly operations were conducted at several electronic assembly plants in Canada and Ukraine. They were started by V. Venda at Nortel plants in Winnipeg, Calgary, and Montreal, Canada, and then conducted in several Ukrainian cities.

Here are some of the findings:

1. Electronic assembly workers are the first in a long line of different professionals working with computers and telecommunication systems at different stages of production, testing, troubleshooting, repairing, maintenance, programming, and use.

2. For inexperienced people and new employees, electronic assembly may be perceived as light, safe physical work with moderate intellectual requirements. That is why it attracts many people with disabilities, particularly those with hearing and vision disabilities, elderly people, and persons using wheelchairs.

3. Contemporary organization of electronic assembly work requires workers to bend their neck and back, lift hands and shoulders, and maintain a static awkward posture accompanied by high upper trapezius muscle stress and vision strain.

4. Electronic assembly work has very heavy MSD epidemiology. Many assembly workers become disabled by various combinations of neck, back, wrist, and shoulder injuries; some workers have to stay on disability for more than one year.

5. A survey and examination of the current ergonomic design solutions at Nortel and many other electronic assembly companies were conducted. All the solutions were based on the same methodological approach. Designers tried to identify a position of the work surface and printed circuit board (PCB) that might be optimal for eyes, neck,

wrists, arms, back, and shoulders at the same time. None of the solutions were found to provide work safety and comfort.

6. As a result of awkward work postures, many electronic assembly workers suffer from disorders, pain, or MSDs. The existing design solutions do not prevent MSDs, and generally do not allow injured workers to return to the assembly line.

7. A new approach to ergonomic design of assembly workstations is needed.

8. This project is an attempt to use the laws and theory of ergonomics to find a new ergonomic solution that lowers MSD risk and improves work productivity, as well as allowing injured workers, persons with vision disabilities, and those bound to wheelchairs to return to work and maintain appropriate employment in electronic assembly.

4.2 Disabilities caused by MSD

There are two leading workplace factors of MSD in electronic assembly: (1) sustained postures causing static contractions of the neck and shoulder muscles, and (2) combinations of highly repetitive work involving the arm, wrist, and hand, which also affects the musculature of the shoulder and neck region (Aaras, 1994). Existing electronic assembly and material-handling workstations are based on direct observation of manual operations by the worker. This causes bending of neck and body, lifting of arms and shoulders, and flexing of wrists. Even more advanced semiautomatic assembly workstations are based on traditional direct observation of manual operations. Thus, they do not free a worker from lifting arms and shoulders and bending the neck and body. A general view of an electronic assembly workstation is presented in Figure 4.1.

Carpal tunnel syndrome (CTS) is one of the most widely spread MSDs among electronic assemblers because this work in its current form requires constant wrist flexion, similar to the wrist position of computer mouse users. Matias, Salvendy, and Kuczek (1998) studied CTS among computer mouse users. Their study indicated the following: (1) the percentage of the workday spent working with a computer mouse was the most significant factor, accounting for 60% of the variance explaining the causation of musculoskeletal discomfort associated with CTS; (2) a discriminant function with six variables (i.e., work duration, trunk incline, wrist extension, wrist deviation, overall anthropometric measure, and weighted anthropometric measure) correctly classified 73% of the CTS group and 72% of the non-CTS group; (3) using a logistic regression model, the probabilities associated with changes in the predictive variables affecting CTS risk are presented such that increasing the daily work duration from 1 h to 4 h increases the probability of CTS

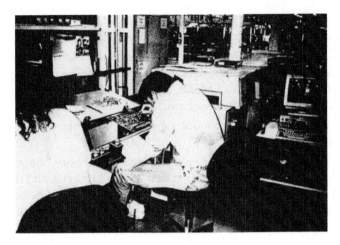

FIGURE 4.1
Existing electronic assembly and material-handling workstations are based on direct observation of manual operations by the worker. This causes bending of neck and body, lifting of arms and shoulders, and flexing of wrists.

risk from 0.45 to 0.92. The results of the study suggest that the main cause of CTS is job design, the secondary (and lesser cause) is posture associated with the workplace design, and the least contributing factor to CTS causation is the individual's anthropometric makeup.

Matias, Salvendy, and Kuczek (1998) focused on general work factors, micromotions, and muscle strain during mouse use. This is the first time all three major aspects were studied simultaneously.

Snook, Vaillancourt, Ciriello, and Webster (1995) used psychophysical methods in studies of repetitive wrist flexion and extension. They showed that along with pinch or power grip force (torque), repetition rates and number of work days a week are the major factors contributing to wrist MSD.

Even more advanced semiautomatic assembly workstations do not free a worker from lifting arms and shoulders and bending neck and body (see Figure 4.2).

A schematic view of a traditional electronic assembly workstation based on direct observation of manual operations is shown in Figure 4.3.

4.3 Laws of Ergonomics

The first law of ergonomics, the law of mutual adaptation (Venda, V.F., 1995; Venda, V.F., and Venda, Yuri V., 1995; V. Venda and N. Venda, 1995; Karwowski and V. Venda, 2000), states that "work efficiency (safety, comfort, productivity) is a bell-shaped function of the factor of mutual adaptation between human work functional structure (strategy) and work environment." Fig-

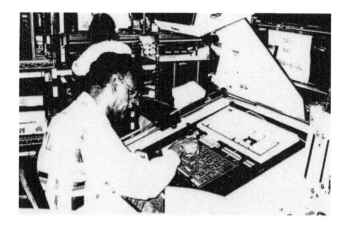

FIGURE 4.2
Even more advanced semiautomatic assembly workstations are based on traditional direct observation of manual operations. Thus, they do not free a worker from lifting arms and shoulders and bending neck and body.

ure 4.4 shows that if the factor of human–machine mutual adaptation, F, is changed (e.g., PCB angle), then work comfort, Q, changes according to a bell-shaped function with $Q = Q_{max}$ if $F = F_{opt}$.

The second law of ergonomics, the law of work structures plurality (Venda, V.F., 1995; Venda, V.F., and Venda, Yuri V., 1995; V. Venda and N. Venda, 1995; Karwowski and V. Venda, 2000), states that "the same work can involve different functional structures (strategies) represented with different bell-shaped functions (curves)." Figure 4.5 shows two bell-shaped curves that may belong to different human organs working together or separately during assembly operations. The first organ (S_1) may have maximal comfort $Q_1 = Q_{1max}$, if $F = F_{1opt}$, but at the same time the second organ (S_2) may suffer extreme discomfort. In contrast S_2 has a maximal comfort Q_{2max} if $F = F_{2opt}$, and this time comfort of the first organ (S_1) Q_1 will be very low.

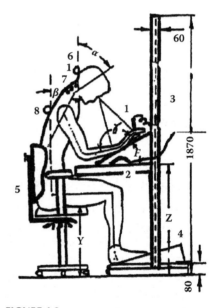

FIGURE 4.3
Schematic view of a traditional electronic assembly workstation based on direct observation of manual operations. 1. PCB to be assembled; 2. table; 3. workstation stand; 4. foot support; 5. chair; 6. head position measurement; 7. angle of bending neck; and 8. back position measurement.

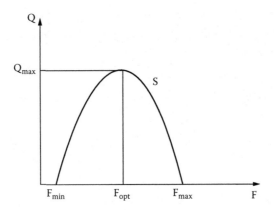

FIGURE 4.4
Graphic representation of the first law of ergonomics.

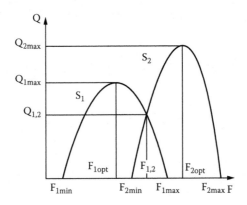

FIGURE 4.5
Graphic representation of the second law of ergonomics.

The third law of ergonomics, the law of interactions and transformations (Venda, V.F., 1995; Venda, V. F., and Venda, Yuri V., 1995; V. Venda and N. Venda, 1995; Karwowski and V. Venda, 2000), states that "maximal work comfort for two interacting or mutually transforming work functional structures is presented as the intersect point of the bell-shaped curves for the two structures" (Figure 4.6).

If two human organs, say eyes and arms, are working together in assembly operations, then Figure 4.6 shows that maximal comfort for both, eyes and arms, is $Q_{1,2}$ when $F = F_{1,2}$ ($Q_{1,2} < Q_{1max}$ and $Q_{1,2} < Q_{2max}$).

The third law states that in certain conditions a separate, independent functioning of two human organs may provide better work comfort for both organs than their joint, interactive functioning.

Based on the laws of ergonomics, it was hypothesized that electronic assemblies may be more comfortable and safe if work conditions for two main groups of organs, for eyes–neck–back and for arms–wrists–shoulders,

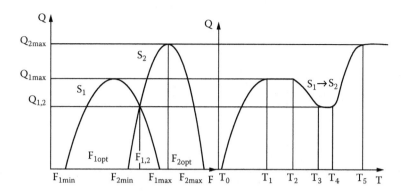

FIGURE 4.6
Graphic representation of the third law of ergonomics.

are studied and optimized in both possible modes, joint and separate functioning. The three laws of ergonomics were successfully used in various international ergonomics design projects (V.F. Venda, N.I. Venda, R.J. Trybus, and E. Averboukh, 2000; V.F. Venda, Nadejda I. Venda, R.J. Trybus, and V.N. Nosulenko, 2000).

4.4 Experimental Study of Optimal Angle of Printed Circuit Board (PCB)

4.4.1 Method

4.4.1.1 Objective

1. To study influence of the angle of work surface on work comfort of eyes–neck–back and of arms–wrists–shoulders if these two groups of organs are working together in traditional style. In this case, the subject (worker) is directly watching the manual assembly operations.

2. To study the influence of the angle of work surface on work comfort of eyes–neck–back and of arms–wrists–shoulders if these two groups of organs are working separately. In this case the subject (worker) is indirectly watching manual assembly operations, and there are two different work surfaces, one for the manual operations involving arms–wrists–shoulders, and another involving eyes–neck–back.

3. Compare work comfort for the both groups of organs for joint and for separate work.

4.4.2 Equipment

Experimental facility included the following:

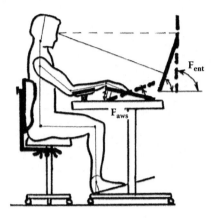

FIGURE 4.7
Schematics of experimental installation to study the influence of visual PCB angle F_{enb} on relative comfort of eyes–neck–back ($Q_{rel-enb}$) and influence of manually manipulated PCB angle F_{aws} on relative comfort of arms–wrists–shoulders.

1. Traditional electronic assembly workstations used at various real plants presented as a schematic in Figure 4.3.
2. Basic experimental assembly workstation (Figure 4.7) with independently changed angles of PCB to be perceived, which determine angles and relative comfort for eyes–neck–back and different angles for PCB to be manipulated by hands. This second flexible rotating panel determined angles and relative comfort for arms–wrists–shoulders.
3. A portable Physiometer PHY-400 by Premed AS, Norway, was used in laboratory experiments and at Nortel plants. PHY allows one to measure muscle strain as EMG, as well as motions and postures of head, arms, and back.
4. An Eye Gaze Development System by LC Technologies was used to analyze visual perception of shop aids (assembly schematics) and PCBs, and to register eye movements and fixations.
5. Work postures were videotaped using two side video cameras focused on subject's profile from left and right.
6. Work operations were videotaped using two more video cameras focused on PCB and subject's motions.

4.4.3 Participants

Participants were 120 engineering students (19–24 years old, 60 females and 60 males).

4.4.4 Procedure

For joint work of the organs the angle of PCB was changed between 0° (horizontal position) and 55° (maximal angle where electronic parts started to fall off the PCB).

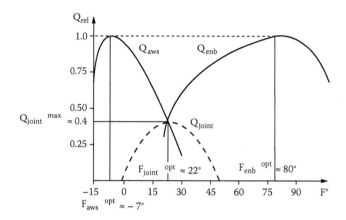

FIGURE 4.8

Laboratory experimental data on influence of angle F_{aws} on relative comfort of arms, wrists, and shoulders Q_{aws} and of angle F_{enb} on relative comfort for eyes, neck, and back Q_{enb}. $Q_{joint} = Q_{enb} + Q_{aws}$.

Each subject was given a random sequence of 35 PCBs with graphic schematics where different numbers of components, from 9 to 32, were to be populated. All experiments (about 3 h each) were videotaped, and then time spent on each PCB was measured from the video tape. Optimal numbers of components found were in the range 13–17. If the number was less than 13, then more time was spent to change PCBs. If the number was larger than 17, the subjects could not recall all operations, referred more often to the shop aids (schematics), and made more assembly errors.

All following experiments to find the optimal angle of PCB and desk height were done with 15 electronic parts inserted into the PCB in each task. Each experiment lasted about 3 h.

Data on work comfort at different PCB angles for the groups of students were very close and presented as a combined result in Figure 4.8. Figure 4.8 represents relative work comfort, Q_{rel}, as a function of the angle of PCB, F.

We intended to find a PCB angle that gives maximum relative comfort for eyes, neck, and back (Q_{enb}) and for arms, wrists, and shoulders (Q_{aws}). Then, a joint (combined) relative comfort Q_{joint} was calculated as $Q_{joint} = Q_{enb} + Q_{aws}$, with the boundary condition that neither Q_{enb} nor Q_{aws} must not equal zero, otherwise $Q_{joint} = 0$.

The maximum relative joint work comfort was found at 22°. At smaller angles, bending of neck and back and difficulties in visually finding the holes on PCBs caused serious discomfort; at larger angles, lifting of arms and shoulders along with wrist flexion were the major factors of discomfort.

At 22° neither eyes and neck, from one side, nor wrists, arms, and shoulders, were in an optimal position, even though work productivity and comfort were maximal for joint work of eyes and arms.

EMG level remained stable at around 20–30% maximum volunteer contraction (MVC) between PCB angles 15° and 30°. Between 0° and 15° it was higher because of very large angles of neck and back bending (measured using goniometers of PHY-400). At angles larger than 22°, upper trapezius muscle strain increased very quickly because of the action of lifting arms and shoulders.

Although the laboratory subjects assessed work comfort at the angle of 22° as a maximal one in comparison with other angles, they complained that there still was a significant discomfort for eyes, neck, arms, shoulders, back, and wrists. EMG measurements of the upper trapezius showed high muscle tension at the level of about 20%MVC.

To check the accuracy of subjects' visual perception of the assembly schematics (shop aid) and corresponding places and holes on the PCB, we used the Eye Gaze Development System.

After completion of the experiments with 120 students, we organized the same experiments with 80 workers at five different assembly plants. The workers were doing their regular job, but with a changing PCB angle every 30 min in a random sequence. Results with workers were similar to those we obtained in the laboratory experiments with students.

Our conclusion is that *optimization of the PCB angle based on joint work of eyes and arms does not lead to an optimal solution for the assembly workstations.*

Because the traditional ergonomic approach did not lead to an optimal solution, we decided to apply the laws of ergonomics and try to find a theoretically based new solution.

This approach required us to study work efficiency Q as a function of various work factors F_j. The first series of experiments was conducted using a flexible workstation (Figure 4.7). We were able to change the angle of the PCB, which was measured as a factor Fe/n influencing relative work comfort for eyes and neck. We also were able to change independently an angle of the PCB manipulated by workers' arms during assembly operations. The experiments yielded two bell-shaped curves. One curve was for work comfort for eyes+neck, $Q_{e/n}$, presented in Figure 4.6 as a function of the angle of the observed PCB. Its optimum is at about 70°–80°. Another bell-shaped curve, $Q_{a/s}$, was obtained for relative comfort of arms and shoulders. Its optimum appeared to be −7°.

Considering the optimum for $Q_{e/n}$, it is obvious that worker's arms, wrists, and shoulders cannot work so high. Besides, the PCB on which electronic parts are being inserted cannot be installed at such large angle, because electronic parts fall down.

Because all existing assembly workstations are designed to prevent falling of the parts and accommodate arms and shoulders, it is not possible to provide the worker with the most comfortable position of eyes and neck (head) (with PCB at about 80°). Therefore, a new principle must be found to provide arm support, prevent wrist flexion and shoulder lifting, and to allow worker's eyes and neck to be in a straight comfortable position.

In the next series of experiments our subjects were asked to keep their heads in a straight neutral position while watching the PCB located at the optimal position found in the previous series. Then we changed the height and angle of the workstation surface supporting the worker's hands (Figure 4.7). The subjects had to assess work comfort for arms, shoulders, and wrists at different heights and angles. Maximal comfort for arms, shoulders, and wrists, Q_a, was found at the angle $-7°$.

It is clear that existing assembly workstations do not allow a negative tilt of work surface because workers cannot see PCB in that position even if they bend their necks to the maximum angle.

This second series of experiments confirmed that some new principle must be found to allow worker's arms and shoulders to be in the most comfortable position.

Therefore, our theoretical approach showed that existing workstations requiring joint work of arms+shoulders+wrists, from one side, and eye+neck, from the other side, seriously compromise work comfort for all these body parts and organs. EMG of descending muscles of the left and right trapezius showed a high level of muscle strain when these body parts and organs work together, in the interactive mode.

The intersect point of the curves $Q_{enb}(F)$ and $Q_{aws}(F)$ models (Figure 4.8) maximal efficiency (work comfort) for the eyes+neck and arms+shoulders working together. This means that using a common work surface for the arms and the eyes leads to a relative maximum work efficiency only at $Q_{ae} = 0.43$ when $F_{ae} = 22°$. All other angles decrease the work comfort of eyes+neck to the left from F_{ae}, and of the arms+shoulders on the right from F_{ae} (Figure 4.8).

Thus, the traditional approach of ergonomic design to assembly workstations, based on coordinating the eyes and arms in assembly operations, should be changed: a comfortable, neutral position of the arms leads to neck flexion, whereas a straight, neutral position of the neck leads to increased lifting of the arms and shoulders.

4.5 Invention of Assembly Workstations with Indirect Observation of Operations and Negative Tilt of Work Surface

The approach based on the laws of ergonomics led us to the following requirement: the design solution for assembly workstations must allow independent operation of the arms and of the eyes to maximize work comfort for eyes, neck, back, shoulders, arms, and wrists. On the basis of ergodynamic principles, we invented and designed a new type of workstation. Our workstation with indirect observation of manual operations and negative tilt of work surface was recognized by International Patent Offices in London (GB) and Geneva (Switzerland) as invention PCT/CA95/00367. Here is a description of the workstation we invented (Figure 4. 9).

The PCB image angle and height are optimized based on our experiments (the angle of the video image on TV screen 1 is about 80°). Therefore, neck and body are in a straight, neutral position.

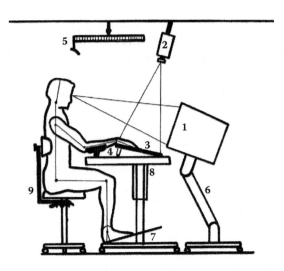

FIGURE 4.9

Schematic view of a new electronic assembly workstation based on indirect observation of manual operations. 1. TV screen, its angle F_{enb} determines relative comfort for eyes–neck–back ($Q_{rel-enb}$); 2. TV camera; 3. PCB to be assembled with possible negative tilt. Its angle, F_{aws}, determines relative comfort for arms–wrists–shoulders; 4. arm support; 5. adjustable light; 6. TV monitor support; 7. adjustable foot support; 8. adjustable table height; 9. adjustable chair.

Work surface (3 in Figure 4.9) height and angle are optimized to support wrists and arms and allow shoulders also to be in a comfortable, relaxed position.

The arm and wrist positions are optimized using negative tilt of work surface 3 and arm support 4. Tilt, height, and other parameters are adjustable to the needs of individual workers.

4.6   Laboratory Testing of the New Workstations

We planned and conducted a large series of comparative tests of our newly invented V-workstations and traditional workstations at ergonomics laboratories and then in an industrial environment at four different assembly plants.

Getting ready for industrial tests, we conducted preliminary laboratory experiments to assess the new design; we registered, measured, and analyzed the positions, motions, and associated muscle strain of upper trapezius using Physiometer-400 connected to a palmtop computer (LX-100 with flash-card, Hewlett-Packard) carried by the subject on a belt to allow mobility. Assembly operations were simulated at the laboratory using existing industrial equipment: workstations, Kan-Ban boxes, lights, foot rests, etc., to match the industrial environment as closely as possible to perform the same experiments later at real assembly plants.

Magnification of the parts assembled was varied to a large degree, from 3 to 12. These workstations allow workers to use direct observation (traditional way) and indirect, or TV, observation; thus, work posture can be changed, further decreasing static muscle strain. Fifty students participated in the laboratory experiments.

Laboratory experiments showed that student subjects were able to learn quickly how to populate electronic parts into PCB, using both direct and indirect vision. However, we could not check if all assembly operations, including soldering, cutting lids, and inspection could be performed using indirect observation. Only full-scale industrial testing with the participation of many experienced workers can answer all the questions.

4.7 Industrial Comparative Testing of Traditional and New Assembly Workstations

The experimental installation used in our industrial tests included two types of traditional assembly workstations, one with a straight table and another with a cut-out table, and two types of new workstations based on the two traditional workstations but with the addition of TV screens and TV cameras.

Eighty assembly workers at four assembly plants participated in the experiments. Each of them worked on the first and second days at the usual workstation with direct observation, but with PHY-400 registration and videotaping. Two types of traditional workstations were used. One of them, coded "T," is one of the widest slide-line workstations with a narrow front. Another traditional workstation, a "cut-out" one coded C, is a stand-alone workstation based on a wide adjustable table with a cut-out that allows the worker to be closer to the PCB, tools, and boxes with parts.

After two 1-day control experiments, each worker worked 10 days at the new workstation. One was based on T with the addition of our TV module, and coded VT-workstation. Another was based on the C-workstation, and coded VC. Our TV module included a TV camera and monitor on the portable stand with adjustable height (Figure 4.9). Therefore, we had two pairs of workstations to be compared: (1) T and VT, and (2) C and VC.

The experiments at the plant lasted 2 months. They were conducted in three shifts: (1) 7 am to 3:30 pm; (2) 3:30 pm to midnight; and (3) midnight to 7 am. The workers were involved in the normal technological assembly process.

We measured *work productivity* as number of PCBs assembled in a shift. Table 4.1 compares T and VT, Table 4.2 compares C and VC, and Table 4.3 compares average for T and C with average for VT and VC.

Table 4.1 shows that productivity at VT increases quickly and becomes equal to or higher than productivity at T. Workers managed to transform all their skills, including soldering, by the fifth to eighth day. Table 4.2 shows the results on the comparison of productivity of work at C and VC. Comparison with Table 4.1 shows that the advanced traditional workstation (C) allows higher productivity than ordinary traditional workstations (T): 53.17 versus

TABLE 4.1

Comparison of Productivity of Assembly Work at Workstations
T and VT

		Day # at VT						
Worker #	Average at T	1	2	3	4	5	6	7
2	31.00	24.00	23.00	35.00	46.00	40.00	37.00	50.00
4	29.00	13.00	35.00	40.00	40.00	42.00	43.00	42.00
6	35.00	36.00	38.00	33.00	46.00	43.00	42.00	44.00
10	44.00	35.00	53.00	52.00	39.00	51.00	47.00	50.00
12	39.00	32.00	40.00	36.00	32.00	44.00	43.00	51.00
Mean	35.60	28.00	37.80	39.20	40.60	44.00	42.40	47.40
Relative	1.00	0.79	1.06	1.10	1.14	1.23	1.19	1.33

TABLE 4.2

Comparison of Productivity of Assembly Work at Workstations C and VC

		Number of PCBs Assembled in a Shift at VC						
					Day #			
Worker #	Average at C	1	2	3	4	5	6	7
1	39.00	17.00	29.00	35.00	35.00	42.00	49.00	52.00
3	59.00	21.00	40.00	50.00	48.00	54.00	62.00	61.00
5	60.00	21.00	55.00	63.00	67.00	60.00	77.00	69.00
7	45.00	26.00	32.00	33.00	42.00	47.00	43.00	52.00
9	57.00	41.00	55.00	54.00	58.00	52.00	59.00	63.00
11	59.00	29.00	46.00	50.00	57.00	50.00	62.00	60.00
Mean	53.17	25.83	42.83	48.50	51.16	52.16	58.66	59.50
Relative	1.00	0.48	0.80	0.91	0.96	0.98	1.10	1.12
T + C versus VT + VC	1.00	0.63	0.93	1.00	1.05	1.10	1.14	1.22

33.80. Our workstation led to a further increase of productivity: 53.17 was
overtaken using VC by the sixth day. The productivity 33.80 was improved
using VT by the second training day. However, productivity 53.17 shown at
C was not reached at VT even by the seventh training day. This means that
the new principle of organizing assembly workstations has more advantages
when compared to advanced traditional workstations.

The EMG of descending muscles of left and right upper trapezius was
measured as %MVC with calibration before each registration. EMG and
positions and angles of motions of neck and back from a straight position
were registered and processed using PHY-400.

Table 4.3 presents a comparison between T and VT, C and VC, and between
averages for T + C and VT + VC, based on processing of the following data:
EMG right and left trapezius, %MVC; head flexion, head side movements,
back flexion, and back side movements (angles, °).

TABLE 4.3

Comparison of EMG and Posture Angle Data for T, VT, C, and VC

Parameter W/S	Right Trapezius	Left Trapezius	Neck Flex	Neck Side	Back Flex	Back Side
T	35	27	28	15	21	7
VT	10	8	5	12	4	9
C	29	25	33	21	17	14
VC	8	6	7	13	6	11
T + C/2	32	26	30.5	18	19	10.5
VT + VC/2	9	7	6	12.5	5	10
T + C(1.0) versus VT + VC	0.28	0.27	0.20	0.69	0.26	0.95

TABLE 4.4

Comparison of Relative Data for Various Workstations

Parameter	Workstation					
	T	VT	C	VC	T + C	VT and VC
Productivity	1.00	1.33	1.00	1.12	1.00	1.225
EMG right tr.	1.00	0.29	1.00	0.27	1.00	0.28
EMG left tr.	1.00	0.30	1.00	0.24	1.00	0.27
Neck flexion	1.00	0.18	1.00	0.21	1.00	0.195
Neck side motions	1.00	0.8	1.00	0.62	1.00	0.59
Back flexion	1.00	0.19	1.00	0.35	1.00	0.27
Back side motions	1.00	1.29	1.00	0.79	1.00	1.04

Note: Data on T, C, and T + C equaled 1.00. Data on VT and VC are for the seventh experimental day.

Comparison of the traditional workstations (T and C) with our new workstations (VT and VC) showed a general advantage for VT and VC in most criteria: EMG of right and left trapezius, head (neck) flexion, and back flexion (see tables 4.3 and 4.4).

Eighty percent of workers who participated in the comparative tests of the two types of the workstations positively assessed improvement in work posture and positions of neck, back, shoulders, arms, and wrists, particularly when negative tilt was offered them, with the newly designed workstation based on indirect observation of the manual operations.

Four assembly workers who had been staying on disability with neck injury more than 1 year agreed to try our workstations under Nortel plant physicians' supervision. They expressed the opinion that this type of workstations probably would not injure them and may help them to return to the assembly work after treatment and rehabilitation are completed. They firmly said that they would not return to the traditional workstations requiring bending the neck and back and lifting arms and shoulders.

4.8 Conclusion

1. Using a theoretical approach based on the laws of ergonomics, a new type of assembly workstation providing for indirect observation of operations and separate optimization of work of eyes, neck, back, shoulders, arms, and wrists was designed and tested. This invention helps improve work posture and lower muscle strain and risk of MSDs at electronic assembly plants. The workstations allow assembly workers to be seated in a straight, neutral position with relaxed shoulders, arms supported, and wrists on the surface with negative tilt.

2. Laboratory and industrial testing of the workstations have proved their advantages in comparison with existing traditional workstations.

3. The workstations allow large magnifications of the product assembled or handled, thus decreasing visual strain and increasing work productivity.

4. The workstations allow workers suffering from repetitive strain injuries of the neck, wrist, and back and those who cannot work at the traditional workstations to return to work.

References

Aaras, A. (1994). The impact of ergonomic intervention on individual health and corporate prosperity in a telecommunications environment. *Ergonomics*, 37, No. 10, 1679–1696.

Caffier, G., Heinecke, D., and Hinterthan, R. (1993). Surface EMG and load level during long-lasting static contractions of low intensity. *International Journal of Industrial Ergonomics*, 12, 77–83.

Carter, B. J. and Banister, W. F. (1994). Musculoskeletal problems in VDT work: a review. *Ergonomics*, 37, 1623–1648.

Faucett, J. and Rempel, D. (1994). VDT-related musculoskeletal symptoms: Interactions between work posture and psychosocial work factors, *American Journal of Industrial Medicine*, 26, pp. 597–612.

Kahn, J. F. and Monod, H. (1989). Fatigue induced by static work. *Ergonomics*, 32, 839–846.

Kilbom, A. (1987). Short and long term effects of extreme physical inactivity-a review. In Knave, B. and Wideback, P.-G. (Eds.). *Work with Display Units*. Amsterdam: Elsevier Science.

Leamon, T. B. (1995). The evolution of ergonomics. *Risk Management*, Vol. 42, No. 9, 47–52.

Matias, A. C., Salvendy, G., and Kuczek, T. (1998) Predictive models of carpal tunnel syndrome causation among computer users, *Ergonomics*, 1998, Vol. 41, No. 2, 213–226.

McAtamney, L. and Corlett, E. N. (1993). RULA: a survey method for the investigation of work-related upper limb disorders. *Applied Ergonomics*, 24(2), 91–99.

Schuldt, K. and Ekholm, J. (1986). Effects of changes in sitting work posture on static neck and shoulder muscle activity. *Ergonomics*, 29, No. 12, 1525–1537.

Schuldt, K., Ekholm, J., Harms-Ringdahl, K., Nemeth, G., and Arborelius U. P. (1986) Effects of changes in sitting work posture on static neck and shoulder muscle activity. *Ergonomics*, 29(12), 1525–1537.

Sauter, S. L., Schleifer, L. M., and Knutson, S. J. (1991). Work posture, workstation design, and musculosceletal discomfort in a VDT data entry task. *Human Factors*, 33(2), 151–167.

Schmitz, W., Newman, L., Carayon, P., and Smith, M. (1997). Emotional demands and musculoskeletal discomfort in telecommunications workers, In *HCI International '97—Proceedings of the 7th International Conference on Human–Computer Interaction*, Salvendy, G, (Ed.) San Francisco, CA August 24–30, 1997, pp. 505–508.

Snook, S. H., Vaillancourt, D. R., Ciriello, V. M., and Webster, B. S. (1995) Psychophysical studies of repetitive wrist flexion and extension. *Ergonomics*, Vol. 38, #7, 1488–1507.

Venda, V. F. and Venda, Y. V. (1991). Transformation dynamics in complex systems, *Journal of Washington Academy of Science*, #4, December.

Venda, V. F. and Hendrick, H. W. (1994) Ergonomics and macroergonomics in analysis of decision making efficiency and complexity, *International Journal of Human–Computer Interaction*, 6(3), 253–274, 1994.

Venda, V. F. (1995) Ergonomics: theory and applications, keynote address for the World Ergonomics Congress IEA'94, *Ergonomics*, 1995, Vol. 38, No. 8, 1600–1616.

Venda, V. F. and Venda, Yuri V. (1995). Dynamics in ergonomics, psychology, and decisions: *Introduction to Ergonomics*. Norwood: Ablex.

Venda, V. and Venda, N. (1995). Industrial and experimental applications of transformation theory and ergonomics. Preprints of the 6th IFAC/IFIP/IFORS/IEA Symposium on Analysis, *Design and Evaluation of Man-Machine Systems*, MIT, June 27–29, pp. 791–797.

Venda, V. F., Venda, N. I., Trybus, R. J., and Averboukh, A., Dynamic usability principles in rehabilitation, work and communication. In: *Proceedings XII Congress IEA/44th Annual Meeting of Human factors and Ergonomics Society*, San Diego, CA 2000.

Venda, V. F., Venda, N. I., Trybus, R. J., and Nosulenko, V. N., Usability of graphic information and navigation maps to senior and disabled drivers. In: *Proceedings XII Congress IEA/44th Annual Meeting of Human factors and Ergonomics Society*, San Diego, CA. 2000.

5

Day-to-Day Monitoring of an Operator's Functional State and Fitness-for-Work: A Psychophysiological and Engineering Approach

Oleksandr Burov

CONTENTS

5.1 Background...89
5.2 Target Setting 91
5.3 Methodology of the Approach..93
    5.3.1 Selection of Informative Psychophysiological Parameters........94
    5.3.2 System Engineering...97
5.4 Methods..98
5.5 Results and Discussion 100
    5.5.1 Correlation between Task Performance and Physiological
        Indices..101
    5.5.2 Fluctuations of Task Performance Time 103
5.6 Conclusions..104
References 105

5.1 Background

Losses of power resources and, hence, financial losses because of operator errors are leading not only to failures, but also to suboptimal actions of operators that result in unnecessary expenses for energy carriers, accelerated deterioration of equipment, additional "load" on ecology, and disruption of the manufacturing process. Reduction of such losses even by 1% is equivalent to savings of billions of dollars.

An analysis of the structure of operator errors was carried out at various enterprises in Ukraine (Gerasimov, 1996) and has shown that:

- Insufficient level of knowledge and lack of skills of operators explain up to 25% of failures and disruption of production.
- Nonergonomic design of the equipment and industrial rooms of power plants accounts for almost 30% of disruptions.
- Discrepancy between operator's psychophysiological states and occupational requirements causes about 35% of performance errors.
- Mistakes in organizational designing and management, mistakes in task planning, etc., explain about 10% of errors.

Much attention was given to issues of training and retraining of power plant personnel. Special training centers were created, curricula were developed, and training sessions were carried out by skilled experts at all kinds of power plants—nuclear, thermal, water-power engineering, and dispatching services. Professional selection of operators has been conducted on the basis of the depth of knowledge and skills. The greatest attention is paid to this component of the operators' expertise, according to the literature and practice.

However, it appears that this has not been enough to adequately reverse losses. More than one-third (35%) of errors were made by operators who had sufficient knowledge and skills, passed the regular annual check of professional knowledge, but whose abilities did not completely meet the requirements of their chosen occupation. As a consequence, such operators could not fully apply their knowledge and skills, even if their motivation to work was high.

As the research carried out at the Institute for Occupational Medicine (Kiev) has shown, the job of thermal power plant operator demands a high level of responsibility, and significant attention and operative memory (Kundiev et al., 1982). The operator must process considerable data and information coming from different sources: devices, chief of the shift, chief operator, other operators, and the subordinate personnel.

In contrast to the operator charged with continuous tracking, the work of a power unit operator includes two separate parts: intense supervision over technological processes (60 to 70% of working hours) and active actions (30 to 40% of working hours). If it is necessary to maintain stable unit performance, the operator's work becomes monotonous. Thus, in the course of the work shift, nervous activity characterized by fluctuations in simple visual-motor reaction, stability and switching attention, and decrease of short-term memory is seen.

In the same research it has been shown that in 25% of failures and disruptions of technology, the operators completely met the requirements of the occupation, had good qualifications and motivation, and were also sufficiently experienced. In such cases it was rather difficult to give any rational explanation for their errors as well as actions of their colleagues. If one takes into account fluctuations of the operators' psychophysiological condition, an explanation of these failures becomes obvious. Unfortunately, not enough attention is given to studying this phenomenon, and managers of industries, as a rule, ignore this fact despite its being established in studies.

It is well known from the annual medical inspection of power plant personnel that the specificity of operator's work results in increased stress and distress, thus increasing risk to their health and causing chronic diseases in some cases. Examination of 125 operators of the fossil power plant revealed the following: stomach and duodenum ulcer (32% of examinees), disorders of the cardiovascular system (43%), and weak eyesight (nearly 50%) (Zaychuk et al., 1996). Thus, as a result of the discrepancy in psychophysiological parameters of the operator's state and occupational requirements, not only do reliability and safety of work suffer but also health of the person, which in turn influences work outcome.

We strongly believe that this is a result of an operator's need to work effectively even when his functional state does not meet the requirements of the working situation. Nonoptimal effort leads to greater stress and overregulation of the psychophysiological reactions. Work effectiveness cannot be secured without health, individual safety, and optimal motivation to work, i.e., job satisfaction. Activity theory has given many useful solutions to problems of human behavior, especially over the last decade, for general (Bedny and Karwowski, 2004a, b) and specific activity (Zarakovsky and Kazakova, 2004).

Operators' work in the power industry needs to use a conceptual model of the technological process, i.e., cognitive activity using mental models. Because human fitness-for-work cannot always be at a high level and varies irrespective of the level of knowledge and motivation of an operator, it is necessary, first of all, to allocate adaptively functions of control among operators and the automated control system and, second, to predict adverse changes in human fitness-for-work, i.e., their violation of the allowed bounds. The solutions to these two problems serve to achieve the overall objective of providing an adaptive and reliable control system that in the long term guarantees profitability of a complex technical system such as a power plant.

5.2 Target Setting

The most complicated problem is an assessment of human fitness-for-work (the operator's functional state) and prediction of its change. Successful resolution of this problem determines the reliability of the information that may be used in systems of adaptive automation. Traditionally, fitness-for-work is determined by means of measurement of physiological parameters and by the analysis of parameters of physiological conditions of human activity. Although these methods result in high accuracy and objectivity, there are fundamental limitations:

- Exact measurement of physiological parameters demands additional equipment and standardized conditions of measurement, including certain restrictions on mobility of the person and on the condition of biological gauges; it is difficult to achieve this under real-world conditions, at least at the plants in Ukraine.

- Prediction opportunities of electrophysiological parameters are limited to 10–15 min (Karpenko et al., 1984), so, to measure an operator's state during a whole shift, continuous monitoring is necessary; i.e., electrodes will have to be used for the whole working period, which is impractical in real-world industrial conditions.

As is known from published data, even in the most successful such experiments, even with the participation of trained experts, the maximum accuracy of the prediction does not exceed 70%. Automatic systems for the prediction of operator's functional states are not known to the author.

As direct monitoring of physiological parameters of mental work is rather difficult under real-world production conditions, there are three possible methods for the assessment and prediction of professional fitness-for-work of an operator:

1. Measurement and estimation of professional work parameters.
2. Recording of operator behavior by external means, for example, analysis of speech, video control, etc.
3. Selection of those psychophysiological parameters that would be informative in relation to operator activity and would allow one to construct a sufficiently accurate prognosis for a working day (activity), on the basis of measurements prior to the start of work.

The first method is based on knowledge of the technology associated with the operator's performance and a quantitative estimation of efficiency of his actions. However, such an estimation in the case of power plant operators is difficult, in principle, because the operator's basic duties are to supervise functioning of the equipment, mental estimation of the technological process according to conceptual model, decision making in case of process deviation from the normal mode or control manipulations by facilities, if necessary.

The second method is even less realistic because it is difficult to formalize and automate the evaluation of the operator's performance efficiency.

The third method is insufficiently developed; however, it looks the most plausible. The basis for this assumption is the availability of a fruitful scientific hypothesis—the theory of functional systems (Anokhin, 1973) and its theoretical advancement (Burov and Gerasimov, 2006)—as well as experimentally confirmed facts of a strong relationship between fluctuating (oscillatory) components of the perceptual and cognitive tasks performance parameters and physiological conditions during this activity (Burov, 1986).

The purpose of this chapter is to develop the methodology, principles of construction, practical implementation, and industrial validation of the information technology (computer system) for not only day-to-day assessment, but prediction of operator fitness-for-duty in the example of dispatch-

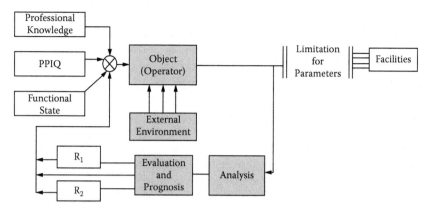

FIGURE 5.1
Functional structure of operator reliability control.

ers of various types of power plants—nuclear, thermal, and water-power plants, and power grids.

5.3 Methodology of the Approach

A functional structure of a system that controls effectiveness of another human–machine system has been developed by a macroergonomic approach. The human–machine system control is achieved by use of the system for day-to-day control (monitoring) of an operator's state, which is synthesized as a set of major factors, determining readiness of the operator to work (Figure 5.1):

- Professional knowledge and skills
- $Z = \{z_1, z_2, ..., z_l\}$, conditions of equipment controlled by the operator
- $S = \{p_1, p_2, ..., p_m\}$, operator functional state (psychophysiological status)
- $P = \{p_1, p_2, ..., p_m\}$, and parameters of environment
- $V = \{v_1, v_2, ..., v_n\}$,

which affect both equipment and the operator's functional state.

Control of the human operator's reliability can be achieved by various methods:

- Revealing significant health deviations
- Estimation of functional reserves of the operator
- Periodic control of the functional state
- Continuous control of the functional state with the purpose of correction when parameters of functioning cross the allowable limits

From the point of view of management theory, the working system is a complex system, many processes in which cannot be adequately described. Influences on object of control (operator) represent superposition of factors Z, S, P, and V, which affect target parameters $Y = \{y_1, y_2, ..., y_r\}$; these parameters can stay within allowable bounds only during certain intervals of time τ after which they might adversely affect the functioning of the system, or even cause its destruction.

Any organism/biosystem may be characterized by a set of parameters (target coordinates) that determine a certain functional level of the organism.

The complexity of the problem of human reliability control consists in determination of the stability of target coordinates when the system is in state y_i, influencing the reliability of the entire system; such a state is a function of all variables, $y_i = \Phi_i(Z,S,P,V)$.

As parameters of the functional state and the environment can vary over the period of several days and within a working day, then $P = P\ (t)$, $V = V\ (t)$ and $y_i = \Phi_i(Z,S,P(t),V(t))$. Thus, to maintain professional fitness-for-work of the operator at a required level, it is necessary to estimate and predict changes of his or her functional state daily and, if necessary, within a working day.

With this purpose, the system that controls operator reliability includes the analyzer of target parameters, allowing one to assess the parameters of the human state, their conformity to required restrictions, and to evaluate whether or not to correct the current functional state.

The methodology developed is a synthesis of two systems. One system determines the choice of informative parameters of operator fitness-for-duty by the macroergonomic decision on their use, and the other system provides for automatic measurement of parameters and associated algorithms.

Thus, in the methodological attitude, the approach assumes solving of two interconnected tasks: informative psychophysiological parameters selection and system engineering.

5.3.1   Selection of Informative Psychophysiological Parameters

Selection of informative psychophysiological parameters of a human fitness-for-duty is carried out when an operator is performing his or her work. Such parameters should describe an operator state prior to or concurrent with the start of activity, but describe his or her fitness-for-work within the whole working day (mission).

From the point of view of Anokhin's theory of functional systems, it is not only parameters of a person's functional state that influence parameters of mental work, but productivity of mental work also affects, in turn, physiological maintenance, which is the key factor of the functional system that determines performance level of the work. Results of the work will affect its physiological maintenance at the next stage and will reflect existing changes. Thus, it is possible to expect that dynamics of mental work form a functional state of an organism, and reflect this state in its parameters.

Previous studies have confirmed this hypothesis (Burov, 1986). They included studies of physiological parameters, recorded as examinees performed operations at the experimental board. The mental work consisted of recognition of combinations of displayed stimuli (complex visual-motor response). Task performance time was 180 min; the total number of trials performed depended on subjects' test rate and varied from 5000 to 5600. Thus, the performance time and accuracy for each trial were fixed. In parallel, were recorded interbeat intervals R-R, time of pulse wave dissemination RPs, and electrodermal activity (skin-galvanic reaction) EDA. All parameters (both test performance and physiological) were averaged for 20-min intervals shifted by 5 min in comparison with the previous one. In addition, for test performance time, spectral (periodical) components have been calculated using the Fourier transformation.

The hypothesis about the relationship of some physiological parameters (R-R, RPs, EDA) with the oscillatory structure of the information processing was checked, and two bands of oscillatory components were investigated: from 40 s to 20 min and from 4 s to 2 min.

The repeatability of the relationship between fluctuating time-performance (spectral) components and physiological parameters was checked when modeling various functional states at three levels of activation:

1. Without pharmacological effect
2. Using 0.05 g aminazine (15 min prior to the start of the experiment) to simulate an inhibition of activation
3. Using 0.2 g caffeine (15 min prior to the start of the experiment) to simulate an activation

The regularity revealed—an increase of the relationship force and stability between a level of physiological parameter and parameters of task performance time when using its oscillatory structure ($R = 0.85–0.93$, $p < 0.01$)—was confirmed not only for pulse rate, but also for tone of vessels (by mean value of pulse wave spread) and EDA.

Thus, it is possible to assume that in a fluctuating structure of information processing is reflected a number of parameters of passive mental work. This conclusion can be coordinated to the notion that the organization of physiological maintenance of mental work is a functional system of work activity (Navakatikian, 1980) in which the result of mental work exerts a modulating influence on the dynamics of its physiological maintenance and plays the role of a backbone factor. Because different physiological systems have an unequal lag effect in responding to the modulating influence of the backbone factor, as far as changing their activity is concerned, they, in turn, will create fluctuations of productivity of mental work with their inherent response times. We believe this explains the oscillatory structure of information processing and its strong relationship to parameters of passive maintenance of mental work.

TABLE 5.1

Common Oscillations for Different Subjects and Functional States

	Periods(s)	Frequencies Studied (Hz)
RR-periods known:	5–9 s (waves of 1st order)	0.11–0.2
	15–20 s (waves of 2nd order)	0.05–0.07
	30–40 s (waves of 3rd order)	0.025–0.033
	210–240 s	0.004–0.005
	360–540 s (catecholamine-related "waves")	0.0019–0.0028
Performance	40–60 s	0.017–0.025
	60–80 s	0.0125–0.017
	100–120 s	0.008–0.01
	120–150 s	0.007–0.008
	210–240 s	0.004–0.005
	360–540 s	0.0019–0.0028

It has been shown that the correlation factor of physiological parameters' values with oscillatory components in 20-min bands is higher than in 2-min bands. From known cyclic processes of physiological parameters in a 20-min interval in comparison with 2-min, there are added only cyclic manifestations of hormonal systems activity, in particular a hormonal link of sympathoadrenal system and system hypothalamus-hypophysis-adrenal glands (Karpenko, 1976). It is possible to assume that the oscillatory structure of task performance time in the 20-min band reflects activity of some hormonal systems (catecholamine, glycocorticoids) that have a high relationship with efficiency of mental work, as well as the activity of a brain appropriate to superslow fluctuations of brain potentials (Aladzhalova et al., 1975).

Essential interindividual distinctions were found out: the oscillatory components' pattern differs with various people in identical experiments. However, in some time bands they are common for all subjects (Table 5.1).

There is no doubt that the efficiency of mental work depends on the state of the physiological systems maintaining it. Our data allow us to establish which characteristics of mental work reflect its physiological maintenance. Such characteristics are parameters of the oscillatory structure of the task performance time.

The results obtained have permitted the following conclusions:

- Results of the perceptual test performance, which are reflected in oscillatory characteristics of the test processing, contain the information both on success of mental work and on its vegetative maintenance.

- For practical purposes, it is meaningful to use a psychophysiological test containing oscillatory components in a range of the periods from 1 to 10 min, which probably reflects activity of regulatory hierarchies down to some hormonal systems.

- For a better understanding of mechanisms of the relationship between mental work and its physiological maintenance in various functional states, it is expedient to study the effect of deficiency of time for quality of activity by modeling restrictions of task performance time.

5.3.2 System Engineering

System engineering is concerned with measurement, accumulation, analysis, and use of informative psychophysiological parameters with the purpose of predicting the operator fitness-for-duty.

The choice of a regulator and influence on the object of control are determined by the fact that the preshift control of a functional state represents a control system of operator reliability on the basis of information models of dynamics of psychophysiological parameters, and regulating influence on object of control (operator) may be realized in two ways (Figure 5.1). One method is correction of the functional state by the operator himself by means of psychophysiological (autotraining, special complex of physical exercises) or pharmacological (tea, tonics, valerian, etc.) methods. The other way—organizational—may include replacement of a worker by another operator if necessary.

Possible ways to use the information concerning operator functional state are considered in methodical recommendations (Karpenko et al., 1985).

The choice of type of regulating influence depends on accuracy and reliability of the functional state assessment. Therefore, when a system for operator reliability control is designed, one should give special attention to functional state description. Taking into account the fact that the human organism can be described as a complex hierarchical control system using for the regulation both continuous and discrete controlling signals, the exact assessment should characterize a human as a set of hierarchical management systems of various levels. Collection of information needs to use an automated information system collecting the information both about "prior" human states, and an information model of his or her functional states changes.

The functional states' assessment at the preshift control is based on the measurement of certain parameters describing regulatory realignments in the human organism. At the same time, use of the computer allows one to carry out psychophysiological performance tests to estimate the operator's state and get a prognosis of his fitness-for-work for the coming shift. The opportunity of the preshift control is based on an assumption that, in regular modes of operation, the operator's functional state within a working day varies less than circannual rhythm and is determined by realignments in the organism at the moment of the start of work. In that case there is an opportunity to build the fitness-for-work prognosis on the basis of the analysis of functional realignments in conditions of mental work modeling results of the psychophysiological performance test using an automated information system.

To solve this problem, principles of system design (Burov, 1991; Burov and Chetvernya, 1996) were developed:

- Test techniques used to evaluate psychophysiological parameters of an operator's fitness-for-work should be aimed at evaluating not the absolute levels, but the time structure of task performance.
- Information concerning a human condition should characterize him or her as a complex system.
- Parameters of psychophysiological maintenance and its sets should be described by an adequate adaptive model.
- Psychophysiological test control should be exercised with the data collected in the system database with an adaptive algorithm.

The preshift control system has been designed on the basis of these principles and its serviceability was validated at the fossil power plant. Such an approach—to use information technology for operator safety at the workplace—was new, at least, in the power industry. Therefore, they developed the methodology and organizational principles of the predicting system operation as well (Burov, 2005).

The methodological characteristic property of our approach to use the preshift control is that it is carried out not *prior* to the start of operator work, but *right after* he or she takes over a post. It stands to reason that general professional training of the operator and his or her adaptation to operating conditions at the enterprise are necessary to form the functional system of activity (FSA), i.e., stable psychophysiological mechanisms for maintenance of stable, reliable, and effective professional work, providing an optimum effort and the maximal fitness-for-work. The operator is supposed to control the complex units only when he can work independently and reliably. And it is achieved after gaining the necessary knowledge and skills, which are impossible without formation of the FSA.

As any professional work is the result of its FSA, it is necessary to estimate human psychophysiological parameter values under conditions of performing the particular duty, i.e., parameters appropriate to the FSA. Hence, test performance that is adequate for professional work should be carried out after the operator has taken over a post, but does not carry out control of the equipment.

In other words, the preshift control is not so much *preshift*, as *starting intrashift*. In practice, it means that the test performance session should be executed 30–45 min after the formal start to performance of official duties by the operator. The correctness of this position was validated during industrial operation of the system by accuracy of the forecast of operator fitness-for-work.

5.4 Methods

In laboratory research we have studied the following:

1. Effect on perceptual test performance pace of the effort of physiological systems for mental work maintenance and reproducibility of high self-descriptiveness of oscillatory structure of test performance time concerning physiological parameters change when modeling functional states by various levels of time limits for the task performance.
2. Parameters related to the test performance when subjects performed a cognitive test each working day for one month.

Eight volunteers aged 20–30 years participated in the first research, which included four experiments:

- Training (adaptation to the test and to conditions of activity)
- Research, 3 h length, with the constant fixed time of 4 s between trials
- Research, 3 h length, with the constant fixed time of 3 s between trials
- Research, 3 h length, with the constant fixed time of 2 s between trials

In all 180-min experiments subjects were asked to carry out the same test, to look for the missing figure. On the VDT screen of the computer the sequence of numbers of natural lines (from 0 up to 7) in the causal order was shown to subjects. They were asked to find the missing digit in a line and to press the corresponding button. Task performance time (TPT) and accuracy (both "wrong answer" and "no answer") were recorded.

In parallel the following were measured:

- Interbeat intervals RR (Neb's method);
- Time of pulse wave spread RPs (from the heart up to a middle finger of the left hand);
- Integral of skin-galvanic reaction EDA (Tarkhanov's method) from palm-back surfaces of the left wrist.

In the second research four computer users (operators and programmers) carried out daily computer test performance at the start and at the end of a working day for 10 min. The rest of the time they performed work according to their professional duties. Research proceeded for 1 calendar month.

Computer test performance consisted in performance of two test tasks by duration of 5 min each. The first test was the same as in the previous case. The second, cognitive, test was to rearrange figures in increasing order. The choice to use this test was dictated by the fact that according to the available data, the more difficult the mental work, the more sensitive are its results for changes in the functional state (Nikandrova, 1980).

The test task was displayed to the subject on the VDT screen and consisted of four nonrepeating figures from 0 up to 9, in the causal order. The test presented the subject with a series of numbers with the task of identifying the number of operations necessary to arrange the numerals in increasing order.

The basic operation was to interchange two neighboring digits step by step. So, the question was to determine the total number of interchanges (steps) to get the digits in increasing order.

For example, the line displayed was

7 4 8 1

Operations to be performed:

4 7 8 1
4 7 1 8
4 1 7 8
1 4 7 8

Total number of operations equals four. The subject should type "4."

Subjects were asked to work as fast and accurately as possible. Once a response was made, the next trial was presented ("autopace"). Time of the task performance and accuracy (both "wrong answer" and "no answer") were recorded.

Before the start of test performance, heart rate (HR) and blood pressure, systolic BPs, and diastolic BPd were recorded.

Processing of the received data included two stages:

- Selection and filtration of the recorded data:
  - The artifact exceeding possible values was substituted for that value
  - "Jumping out" values were replaced with the average value (with the exception of all similar points—artifacts)
  - Filtration with smoothing was done when necessary
- Calculation of the basic statistics (mean value M, standard deviation σ, asymmetry factor As, excess factor Ex)
- Time series analysis—spectral analysis of tasks performance time
- Statistical analysis—correlation, multiregression, stepwise regression analysis

5.5 Results and Discussion

Repeatability of the high self-descriptiveness of oscillatory structure of test performance time was studied with regard to change in physiological parameter values when modeling functional states. The modeling was carried out by "time pressure" at three levels of fixed time for the task performance.

TABLE 5.2

Correlation of RR-Intervals with Parameters of Task
Performance Time

Subject	Work Load Type	Task Performance Time Indices			
		M	Σ	M + σ + As + Ex	Oscillations**
A	1	0.16	0.15	0.45*	0.82*
	2	0.07	−0.12	0.51	0.88*
	3	0.05	−0.33	0.61	0.86*
B	1	−0.84	−0.43	0.88*	0.85*
	2	−0.48	−0.42	0.66*	0.86*
	3	0.31	0.35	0.51*	0.64*
C	1	−0.80*	−0.81	0.82*	0.91*
	2	0.48	0.49	0.72	0.82*
	3	−0.68	−0.23	0.73	0.78*
D	1	0.72*	0.77*	0.87*	0.89*
	2	0.39*	0.18	0.48	0.85*
	3	0.15	0.02	0.43	0.85*
E	1	0.14	0.17	0.56*	0.83*
	2	−0.24	0.33	0.36	0.72*
	3	0.25	−0.28	0.67*	0.77*

Note: Type of the workload 1,2,3—constant tasks' pace: 4 s; 3 s; 2 s; * Significance ($p < 0.01$) is marked; **four informative spectral components selected by stepwise analysis.

5.5.1 Correlation between Task Performance and Physiological Indices

Correlation analysis was performed as follows.

There are values of paired correlation factors between mean value (M) of the task performance time TPT and mean value of RR-intervals (Table 5.2, column 3). From the table, the relationship value of average RR-intervals with mean TPT essentially depends on a human functional state, time "pressure," and character of activity and varies for different examinees. Criteria for estimation of mental fitness-for-work based only on average levels of physiological parameters are not reliable and not very useful for practical purposes.

As in our prior research with pharmacological impact, characteristics of the relationship between stability of information processing (a standard deviation) and average levels of physiological parameters are also unstable, not always authentic, and essentially individual.

The complex characteristic of TPT frequency distribution has a greater relationship with parameters of physiological maintenance of mental work (at the simultaneous account of the first four central moments of the TPT statistical distribution—average mean M, standard deviation σ, factors of asymmetry As, and excess Ex). It is clear from the data in the third column

TABLE 5.3

Correlation of Values of Hemodynamics Parameters with the
TPT Parameters in the Cognitive Test Conducted over 1 Month

Measured Parameter	Subjects	Task Performance Time Indices			
		M	σ	M + σ + As + Ex	Oscillation Components **
HR	6	−0.20	−0.16	0.29	0.69*
	7	−0.03	−0.07	0.07	0.66*
	8	−0.07	−0.21	0.29	0.42
	9	−0.12	−0.21	0.28	0.48*
BPs	6	0.36	0.18	0.38	0.85*
	7	−0.20	−0.24	0.41	0.70*
	8	−0.18	−0.08	0.20	0.57*
	9	0.09	0.09	0.35	0.44
BPd	6	0.36	0.27	0.39	0.58*
	7	−0.02	−0.18	0.44	0.62*
	8	−0.24	−0.31	0.39	0.61*
	9	0.11	0.16	0.27	0.51*

Note: *Significance ($p < 0.01$) is marked.

that changes of TPT pace totally characterized by parameters of statistical distribution are, as a rule, highly related with parameters of physiological maintenance of mental work.

At the same time, as has already been noted, the significant part of dispersion falls under a small number of oscillatory components. They were chosen from all TPT spectral components by stepwise multiregression analysis that had the greatest relationship with mean value of RR-intervals, R = 0.82 on average (Table 5.2, column 6). Factors of multicorrelation in all functional states at all examinees were high enough and similar.

As before, similar results were obtained both for tone of vessels (the factor of multicorrelation R = 0.73 … 0.93), and for integral of skin-galvanic reaction (R = 0.75 … 0.94), and for hemodynamic parameters (Table 5.3).

Thus, earlier results on the high self-descriptiveness of oscillatory structure parameters of information processing in relation to parameters of physiological maintenance of mental work were confirmed by performance of monotonous tests during 3-h research; however, functional states were modeled by means of various levels of time pressure instead of pharmacological means. This result allows the assumption that a good and stable relationship between time structure of TPT and parameters of the physiological "cost" of this activity is inherent in all people and in any functional state. In other words, the stable relationship looks like a general mechanism of psychophysiological maintenance of operator work.

5.5.2   Fluctuations of Task Performance Time

In the second laboratory research the hypothesis about high self-descriptiveness of oscillatory structure of test performance time concerning physiological parameters change was checked using the more difficult cognitive test and test performance over 1 month; i.e., examinees appear in various functional states, and repeatedly. It was possible to expect a decrease of the relationship level owing to averaging results on sets of functional states. However, results showed that the same law was at work:

- The low and doubtful level of correlation relationship on use of average values of parameters
- A little bit higher relationship when using joint parameters of TPT distribution density
- High and authentic relationship between TPT oscillatory components and physiological parameters measured

The fact of high and reliable relationship allows one to assume that parameters of the cognitive test organized in a similar way may be highly informative and reliable when used in the system for the daily check of operators' fitness-for-work.

It is necessary to note that the model of relationship did not always include the same oscillation components in each case, though in some frequency bands they were repeated practically for all examinees and under various conditions of the experiment (Burov, 1991b).

It is possible to note essential interindividual distinctions—for different people in identical experiments the model of the relationship includes various periodic components. However, in some time spans they are repeated for all examinees and correspond to the earlier data (Table 5.1). So, in the first experiment for all examinees 2 to 3 components fall into intervals of 2.5 to 4 min and 60 to 80 s.

In experiments with medium and high paces, the common spans of the periods also come to light. More detailed analysis of the spans found specific concurrence of these periods to well-investigated "waves" for some physiological parameters. For example, for the heart rhythm (Baevsky, 1979) there have been revealed respiratory waves of 1st order (5 to 10 s), 2nd order (15 to 30 s), and 3rd order (1 to 2 min); Traube–Gering "waves" are revealed for a tone of vessels (10 to 30), and they present in blood pressure and plethysmogram as well.

In parameters of tissue oxygenation the "waves" of various origins and the period are recorded. In relation to a mental activity there are waves with the periods 25 to 40s and 40 to 60 s. In neurophysiologic researches (Aladzhalova, 1975) there have been revealed fluctuations of brain potentials in spans 1.5, 2, 2.5, 3.5, and 6+ min.

Thus, the known data from the literature on the relationship of physiological maintenance with the neurophysiologic parameters of mental work are supplemented by the fact of reflection of physiological maintenance as a result of mental work.

Oscillation components of TPT have the same natural character as physiological parameters, but exhibit more complex patterns. The factorial analysis of the periodic structure of TPT reveals from 8 to 12 factors; for heart rate, 4 to 6 factors; for tone of vessels, 5 to 7; and for EDA, 2–3 (Karpenko et al., 1984). It confirms the assumption that human mental activity is complex and is a system characteristic of the dynamic state of the operator, because it reflects an apparently higher level of the control organization in a living system (a human being) in comparison with "subsystems" that are control systems and regulations of a lower level (physiological systems).

5.6 Conclusions

A preliminary study in the power industry has shown the considerable influence of working conditions on operators' health. Most of the observed operators had chronic diseases. The analysis of those results suggested that the conformity of isolated hygienic parameters to required norms is not sufficient for maintenance of operators' health. It is ample evidence of the combined influence of industrial environment factors, including high neuroemotional impact. The changes of health level and operators' professional capacity for work (fitness-for-work) depend on the human ability to adapt to influences of industrial environment and job tasks, namely, to support an optimal level of mental strain.

The starting point for this research was a hypothesis that variations in operator performance were associated with changes in mental load and that aspects of mental activity were reflected in physiological measures. When the psychophysiological state of the operator is optimally adjusted to the task requirements, there will be a strong association between parameters reflecting the dynamics of task performance and psychophysiological state measures. The dynamics of mental activity (and particularly of its biorhythm structure) has a central role in the analysis of fitness-for-work and maintenance of functional status. To assess fluctuations in the operators' working efficiency, an integrated approach is required that focuses on the covariation of measures reflecting the quality of task performance as well as the psychophysiological state of the individual. Regarding task performance, such measures focus on rate and accuracy of information processing with respect to a physiological metric involving various cardiovascular parameters and other measures.

We were able to determine a predictive relationship between parameters of mental activity and physiological maintenance. This relationship is most strongly seen in rhythmic characteristics of these processes rather than in stable baselines or tonic levels of response. The most marked oscillation components of task performance time are 0.0125 to 0.017 Hz, 0.008 to 0.01 Hz, 0.007 to 0.008 Hz, 0.004 to 0.005 Hz, and 0.0019 to 0.0028 Hz. We may assume that a breakdown in these rhythms reflects the strain of adjustments to demanding activity and stress, and hence, may be used for evaluation and prediction of psychophysiological regulation of activity and of its effects on operational effectiveness.

According to data from laboratory and field research, these oscillations reflect both the physiological "cost" of mental workload and the professional success of operator work (Burov, 1989a, b).

The basic methodological features of the computer-based system developed to assess and to predict operator fitness-for-duty are the following:

- An individual-based approach to evaluation of operational efficiency and psychophysiological state, which allows one to establish individual norms.

- Use of a temporal biorhythmical structure of task performance as a correlate of a human psychophysiological condition, on the one hand, and his or her professional serviceability on the other.

- The system is self-adjusting and uses the adaptive model of an operator's psychophysiological state.

The basic principles of the system for operator fitness-for-duty individual monitoring were developed and validated, and allowed one to create a new class of systems for preshift (preflight, pretrip, etc.) control on the basis of a human psychophysiological state assessment and prediction (Burov, 2006) with a level of accuracy prediction up to 98% (Burov and Gerasimov, 2006).

The systems implement organizational (performance, prognostication, recommendations for managers), informational (state assessment and prognosis of fitness-to-work), and technical (software) support. The development of psychophysiological indicators used in the systems is based on the recognized importance of control and organizational processes in influencing human performance.

Positive results of systems implementation at power plants (thermal and water-power plants operators, dispatchers of power grid) by both objective indices (technical and economical parameters) and expert marks allow one to use the approach developed to meet other industrial needs.

References

Aladzhalova, N. A., Leonova, N. A., and Rusalov, V. M. (1975). About parity of features of aim formation at accomplishment of automated mental operations and super slow brain potentials. *Fiziol Cheloveka*, 5, 739–745. Russian.

Anokhin, P. K. (1973). Principle questions of the general theory of functional system. Principles of the system organization of function. Moscow: *Science*, 5–61. Russian.

Bedny, G. Z. and Karwowski, W. (2004a). A functional model of the human orienting activity. *Theoretical Issues in Ergonomic Science*. July–August 2004, 5, 4, 255–274.

Bedny, G. Z. and Karwowski, W. (2004b). Meaning and sense in activity theory and their role in study of human performance. *Ergonomia*, 26, 2, 121–140.

Burov, A. Yu. (1986). Evaluation of the functional status of operators according to indices of their mental work capacity. *Fiziol Cheloveka*, 2, 281–288. Russian.

Burov, A. Yu. (1989a). Biorhythmic aspects of the duty efficiency of generating units operators. *Energetics and Electrification*, 1, 27–56. Russian.

Burov, A. Yu. (1989b). Psychophysiological correlates of operators fitness-for-work at power plants. *Energetics and Electrification*, 2, 24–29. Russian.

Burov, A. Yu. (1991a). The principles of creation and functioning of automated information systems for pre-shift valuations of operators fitness for work. *Cybernetics and Computer Engineering*, 90, 29–33. Russian.

Burov, A. Yu. (1991b). Model selection to predict the fitness-for-work of power plants operators. *Cybernetics and Computer Engineering*, 90, 94–97. Russian.

Burov, A. Yu. and Chetvernya, Yu. (1996). Methodological Principles of Psychophysiological Forecasting of Operators' Reliability. In: *Advances in Occupational Ergonomics and Safety I. Proceedings of the XIth Annual International Occupational Ergonomics and Safety Conference*. Vol. 2, (44–47). Ohio: U.S.A. International Society for Occupational Ergonomics and Safety.

Burov, O. (2005). Information environment: an opportunity for social labour rehabilitation vs. new ergonomic problems. In: *Ergonomia niepełnosprawnym w zmieniajacym sie otoczeniu I w rehabilitacji*. Pod red. J. Lewandowsiego, J. Lacewics-Bartoszewskiej. Wydawnictwo Politechniki Łodzkiej, 2005. 98–105.

Burov, O. Yu. and Gerasimov, B. M. (2006). The systems for a human capacity control and prediction. *Scientific and Technical Information*. 2006, 2, 27–30. Ukrainian.

Burov, O. (2006). Development and industrial use of computer systems for operators fitness-for-duty check. *Ergonomia*, 28, 1, 33–45.

Gerasimov, A. V. (1996). Psychophysiological factors of predisposition erogenous decisions and acts. Problems of personnel training in power industry. (106–109). *Book of Abstracts*. Kiev. Russian.

Karpenko, A. V. (1976). Excretion catecholamine, corticosteroids with urine and heart rhythm structure at mental work of various intensity. Unpublished doctoral dissertation. Kiev: KRIHOH. Russian.

Karpenko, A. V., Burov A. Yu., Kalnish, V. V., Kapshuk, A. P., Grigorus, A. G., Grigorjantz, T. N., and Bobko, N. A. (1984). Control of current reliability of work of operators on the basis of psychophysiological criteria. *Nuclear Power Plants*, 7, 156–162. Russian.

Karpenko, A. V., Burov, A. Yu., Gureev, V. A., and Rishkevich, A. I. (1985). Methodical recommendations on use of systems for pre-shift psychophysiological check of operational personnel fitness-for-duty for power plants. Moscow: Ministry of power industry of the USSR, 18. Russian.

Kundiev, Y., Navakatikian A., and Buzunov, B. (1982). *Work Hygiene and Physiology at Thermal Power Plants*. Moscow: Medicina, 226. Russian.

Navakatikian, A. O. (1980). Mechanisms and criteria of nervously emotional effort at mental work. *Occupational Hygiene of Work Diseases*, 6, 5–12. Russian.

Nikandrova, L. R. (1980). A comparative estimation of psychophysiological tests self-descriptiveness at different levels of CNS activation. *Occupational Hygiene and Professional Diseases,* 7, 29–36. Russian.

Zarakovsky, G. M. and Kazakova, Y. K. (2004). Systemic principals of research of functional states of the gas-man operator while working in desert. *Theoretical Issues in Ergonomic Science,* July–August 2004, Vol. 5, No. 4, 338–357.

Section II

Modular Processes in Mind and Brain

6

Identification of Mental Modules

Saul Sternberg

CONTENTS

Abstract.. 112
6.1   Modules and Modularity... 112
6.2   The Process Decomposition Method 114
    6.2.1   Separate Modifiability, Process-Specific Factors, Selective
           Influence, and Functional Distinctness 114
    6.2.2   Processes and Their Measures, Pure and Composite, and
           Combination Rules 114
    6.2.3   Overview of Examples and Issues 117
6.3   Isolation of a Timing Module in the Rat.. 118
6.4   Mental Processing Stages Inferred from Reaction Times.................... 119
    6.4.1   The Method of Additive Factors... 119
    6.4.2   Selectivity of the Effect of Sleep Deprivation in a Process
           with Stages.. 120
    6.4.3   Support for Stages from Analysis of a Neural Process 123
6.5   Signal Detection Theory: Sensory and Decision Modules 123
    6.5.1   The Finding of Only Partial Modularity in Traditional
           Sensory Experiments.. 123
    6.5.2   Modularity of Sensation and Decision When
           Reinforcement Ratio Is Controlled... 125
6.6   Process Decomposition versus Task Comparison................................. 127
    6.6.1   Two Methods Compared 127
    6.6.2   Finding Which Process Is Influenced by a Manipulation 128
    6.6.3   Donders' Subtraction Method: Task Comparison with a
           Composite Measure... 128
6.7   Brief Summaries of Three Other Examples 129
    6.7.1   Evidence for Modular Spatial-Frequency Analyzers from
           Selective Adaptation.. 129
    6.7.2   Evidence for Modular Spatial-Frequency Analyzers from
           the Detectability of Compound Gratings................................... 129
    6.7.3   Modular Processes for Learning and Motivation..................... 131
6.8   Discussion... 131
Acknowledgments 132
References 132

Abstract

One approach to understanding a complex process or system begins with
an attempt to divide it into *modules*: parts that are independent in some
sense and have different functions. In this chapter, a method for the modular decomposition of mental and neural processes is discussed. Several
applications of process decomposition to mental processes, using behavioral measures, are presented. (Applications to neural processes, using
brain measures and brain manipulations, which lead to the identification
of neural modules, are presented in Chapter 7.) Of the applications in this
chapter, two are well established (signal detection theory applied to discrimination data and the method of additive factors applied to reaction-
time data) and lead to the identification of *mental modules*. I argue that
the process decomposition method discussed here, in which the criterion
for modularity is *separate modifiability*, is superior for modular decomposition to the more frequently used task comparison procedure. Finally, three
additional applications are briefly described, to indicate the wide range of
organisms, measures, and modules to which process decomposition can
be applied.

6.1   Modules and Modularity

The first step in one approach to understanding a complex process or system is to attempt to divide it into *modules*: parts that are independent in
some sense and have different functions.* Early in the last century, scientific psychology, dominated by behaviorism, emphasized the directly
observable relations between stimuli and responses, and devoted little
effort to describing the perception, memory, and thought processes that
intervene. During the second half of the century, there was a change in the
kinds of questions that psychologists asked and in the acceptable answers.
This change was influenced by the growth of computer science, which
persuaded psychologists that programming concepts might be acceptable as precise descriptions of information processing by people as well
as machines. Also, the software-hardware distinction supported the legiti-
macy of theories couched in terms of abstract information-processing operations in the mind rather than only neurophysiological processes in the
brain. In the "human information processing" approach, complex activities
of perception, decision, and thought, whether conscious or unconscious,
came to be conceptualized in terms of functionally distinct and relatively

---

* A module may itself be composed of modules.

independent ("modular") subprocesses responsible for separate operations such as input, transformation, storage, retrieval, and comparison of internal representations—modules whose arrangement was expressed in systematic flowcharts.* A useful distinction that is encouraged by comparing the mind to a digital computer is that between processors and the processes they may implement. The existence of functionally specialized processors is a sufficient condition, but not a necessary one, for functionally distinct processes, the subject of this chapter.

The rise in the 1980s of parallel distributed processing might seem to conflict with the idea of modular organization of processes, but it need not: such models "do not deny that there is a macrostructure," and are intended to "describe the internal structure of the larger (processing) units" (Rumelhart et al., 1986). Furthermore, even starting with a relatively unstructured neural network, there is reason to believe that over time and with experience it will develop functionally specialized processing modules and, hence, functionally specialized processes (Jacobs, 1999).

A method for the modular decomposition of two kinds of complex process, mental and neural, along with several examples of its application from the psychology and cognitive neuroscience literature, is described in this chapter and in Chapter 7.† A mental-process module is a part of a process, functionally distinct from other parts, and investigated with behavioral measures. Such modules will be denoted **A**, **B**, etc. A neural-process module is a part of a neural process, functionally distinct from other parts, and investigated with brain measures. Such modules will be denoted α, β, etc. (When either kind of module is meant, **A**, **B**, etc., will be used.) The distinction between processes and processors is sometimes overlooked. Processes occur over time; their arrangement is described by a flowchart. In contrast, processors are parts of a physical or biological device (such as the brain); their arrangement can often be described by a circuit diagram.

---

* Heuristic arguments for the modular organization of complex biological computations have been advanced by Simon (1962, 2005) and, in his "principle of modular design," by Marr (1976). Marr (p. 485) argued that "any large computation should be split up and implemented as a collection of small subparts that are as nearly independent of one another as the overall task allows. If a process is not designed in this way, a small change in one place will have consequences in many other places. This means that the process as a whole becomes extremely difficult to debug or to improve, whether by a human designer or in the course of natural evolution, because a small change to improve one part has to be accompanied by many simultaneous compensating changes elsewhere."

† I discuss and defend the method, describe its antecedents, illustrate it with a dozen applications to mental and neural processes, and further discuss its inferential logic in Sternberg (2001).

6.2 The Process Decomposition Method

6.2.1 Separate Modifiability, Process-Specific Factors, Selective Influence, and Functional Distinctness

Much thinking by psychologists and brain scientists about the decomposition of complex processes appeals either implicitly or explicitly to *separate modifiability* as a criterion for modularity: Two (sub) processes **A** and **B** of a complex process (mental or neural) are modules if and only if each can be changed independently of the other.[*] One purpose of this chapter is to explicate, by example, the notion of separate modifiability and the conditions under which one can assert it. To demonstrate separate modifiability of **A** and **B**, we must find an instance of *selective influence*; that is, we must find experimental manipulations (factors) *F* and *G* that influence **A** and **B** selectively; i.e., such that **A** is influenced by *F* but is invariant with respect to *G*, whereas **B** is influenced by *G* but is invariant with respect to *F*. Usually, one starts with hypotheses about what the component processes are, and about corresponding *process-specific factors* that are likely to influence them selectively. Separate modifiability of **A** and **B** is also evidence for their *functional distinctness*; information about what a process does is provided by the sets of factors that do and do not influence it; if two processes had the same function, they would be influenced by the same factors.[†]

6.2.2 Processes and Their Measures, Pure and Composite, and Combination Rules

How do we demonstrate that a process is influenced by a factor or invariant with respect to it? We know only about one or more hypothesized *measures* M_A of process **A**, not about the process as such. Depending on the available measures, there are two ways to assess separate modifiability of **A** and **B**. Suppose we have *pure measures* M_A and M_B of the hypothesized modules: A pure measure of a process is one that reflects changes in that process only. Examples include the duration of a process and the discriminability parameter of signal-detection theory. To show that *F* and *G* influence **A** and **B** selectively, we must demonstrate their selective influence on M_A and M_B; that is, we must show that M_A is influenced by *F* and invariant with respect to *G*, and vice versa for M_B. If F_j has two levels, $j = 1, 2$, the *effect* of *F* on M_A is a difference:

$$effect\ (F) = M_A\ (F_2) - M_A\ (F_1) \tag{6.1}$$

---

[*] This criterion for modularity seems to be far weaker than the set of module properties suggested by Fodor (1983), according to whom modules are typically innate, informationally encapsulated, domain specific, "hard-wired," autonomous, and fast. Fodor's domain specificity appears to imply separate modifiability.

[†] Such double dissociation of subprocesses should be distinguished from the more familiar double dissociation of tasks, discussed in Section 6.6. For comments on the distinction see Sternberg (2003).

TABLE 6.1

Inferential Logic for Pure Measures

Joint Hypothesis

H1: Processes **A** and **B** are modules (separately modifiable).

H2: M_A, M_B are pure measures of **A**, **B**.

Prediction

We may be able to find factors F and G that influence M_A and M_B selectively:

$$p_1: M_A \leftarrow F, \ p_2: M_B \nleftarrow F, \ p_3: M_B \leftarrow G, \ p_4: M_A \nleftarrow G$$

Alternative Results

We find factors F, G that influence M_A and M_B selectively.	We fail to find such factors.

Inferences

Support for joint hypothesis $H1 + H2$.	Refutes one/both of *H1*, *H2*, *or* we did not look enough for F, G.

(For factors with multiple levels the effect can be regarded as a vector of differences associated with successive levels.) The logic for inferring separate modules when we hypothesize that we have pure measures is shown in Table 6.1.* The p_k's mentioned in the table are four properties of the data; $M_A \leftarrow F$ should be read as "M_A is influenced by F"; $M_A \nleftarrow F$ should be read as "M_A is not influenced by F."

The influence and invariance requirements are both critical. Unfortunately, it is seldom appreciated that persuasive evidence for invariance cannot depend solely on failure of a significance test of an effect; such a failure could merely reflect variability and low statistical power.†

---

* When the hypotheses about **A** and **B** are sufficiently detailed to specify particular process-specific factors that should influence them selectively, this leads to an alternative formulation of the inferential logic, in which the specification of F and G is included in the joint hypothesis, with the remainder of the reasoning adjusted accordingly. For a discussion of such alternatives, see Sternberg (2001, Section A.2.3).

† Perhaps the most common error of interpretation in psychology is to assert the nonexistence of an effect or interaction merely because it fails to reach statistical significance. In evaluating a claim that an effect is null, it is crucial to have at least an index of precision (such as a confidence interval) for the size of the effect. An alternative is to apply an *equivalence test* that reverses the asymmetry of the standard significance test (Berger and Hsu, 1996; Rogers et al., 1993). In either case we need to specify a critical effect size (depending on what we know and the particular circumstances) such that it is reasonable to treat the observed effect as null if, with high probability, it is less than that critical size.

TABLE 6.2

Inferential Logic for a Composite Measure with Summation as the Combination Rule

---

Joint Hypothesis

H1: Processes **A** and **B** are modules (separately modifiable).

H3: Contributions u_A, v_B of **A**, **B** to M_{AB} combine by *summation*.

Prediction

We may be able to find factors F and G that influence **A** and **B** selectively:

$$p_1': u_A \leftarrow F, \; p_2': v_B \not\leftarrow F, \; p_3': v_B \leftarrow G, \; p_4': u_A \not\leftarrow G$$

and jointly influence no other process.

If so, their *effects* on M_{AB} will be *additive*.

Alternative Results

We find factors F and G with *additive effects* on M_{AB}.	We fail to find such factors.

Inferences

Support for joint hypothesis $H1 + H3$.	Refutes one/both of $H1$, $H3$, *or* we did not look enough for F, G.

Instead of pure measures, suppose we have a *composite measure* M_{AB} of the hypothesized modules—a measure to which they both contribute. Examples of composite measures are the event-related potential (ERP) at a particular point on the scalp (which may reflect several ERP sources in the brain), and mean reaction time, $\overline{RT}$ (which may depend on the durations of several processes). To support a hypothesis of selective influence, in this case, we must also know or have evidence for a *combination rule*—a specification of how the contributions of the modules to the measure combine. With pure measures, factorial experiments (rather than separate experiments for different factors) are desirable, because they provide tests of generality. With a composite measure, however, factorial experiments are essential, to assess how the effects of the factors combine; unfortunately, they are rare.

The logic for inferring separate modules using a composite measure when we either know or hypothesize that the combination rule is *summation* is shown in Table 6.2.*

To understand Table 6.2, it is important to keep in mind what the effect of a factor is, and what it means for effects of different factors to be *additive*. To

---

* Although properties $\{p_k\}$ (Table 6.1) apply to observable quantities, the analogous properties $\{p_k'\}$ (Table 6.2) apply to contributions to a composite measure that are not directly observable.

simplify the discussion, let us assume that there are two factors, each with just two levels. Let u and v be the contributions of processes **A** and **B** to M_{AB}. If summation is the combination rule, we have $M_{AB} = u + v$. If **A** and **B** are selectively influenced by factors F and G, we have

$$M_{AB}(F_j, G_k) = u(F_j) + v(G_k), \tag{6.2}$$

where, for example, $u(F_j)$ is a function that describes the relation between the level of F and the contribution of **A** to M_{AB}. In general, we regard u, v, and M_{AB} as random rather than deterministic variables, and may sometimes wish to average sets of such variables.

Using $\bar{M}_{AB}$, $\bar{u}$, and $\bar{v}$ to indicate the means of these random variables, it is convenient that with no further assumptions, Equation 6.2 implies*

$$\bar{M}_{AB}(F_j, G_k) = \bar{u}(F_j) + \bar{v}(G_k). \tag{6.3}$$

In what follows, I treat the levels of factors as ordered, which permits describing changes in level as "increases" or "decreases." From Equation 6.2 it is easy to show that F and G are *additive factors*: the combined effect on M_{AB} of increasing the levels of both F and G is the sum of the effect of increasing only F and the effect of increasing only G:

$$\begin{aligned} effect\ (F, G) &\equiv M_{AB}(F_2, G_2) - M_{AB}(F_1, G_1) = [u(F_2) + v(G_2)] - [u(F_1) + v(G_1)] \\ &= [u(F_2) - u(F_1)] + [v(G_2) - v(G_1)] \equiv effect(F) + effect(G). \end{aligned} \tag{6.4}$$

Equation 6.2 also implies that the effect of each factor will be invariant over levels of the other. Thus,

$$\begin{aligned} effect(F \mid G = G_k) &\equiv M_{AB}(F_2, G_k) - M_{AB}(F_1, G_k) \\ &= [u(F_2) + v(G_k)] - [u(F_1) + v(G_k)] = [u(F_2) - u(F_1)], \end{aligned} \tag{6.5}$$

regardless of G_k. A given measure may be pure or composite, depending on the hypothesized modules of interest. This attribute of a measure is part of the joint hypothesis that is tested as part of the process decomposition method.

6.2.3   Overview of Examples and Issues

Much of what follows consists of descriptions of three examples of the process decomposition method. There are also brief summaries of three additional examples in Section 6.7 that convey the breadth of the method. Examples will be referred to by the section numbers in which they are first discussed. Thus, the example discussed in Section 6.3 will be called "Ex. 6.3."

Three of the examples involve factorial experiments with two factors. The exceptions are Ex. 6.3 and Ex. 6.7.1, in which the effects of the two factors

---

* Matters are not so simple for other combination rules, such as multiplication; see Section 6.7.2.

are studied in separate experiments, and Ex. 6.4.2, in which we consider the effects of three factors. In all cases the factors have been selected because it is hoped that they will be "process specific"; i.e., that they will selectively influence only one of the two or more processes that are hypothesized to underlie performance of the task.

The method of additive factors (AFM) is discussed in Section 6.4.1, and applied in Section 6.4.2 to the problem of locating the effect of a manipulation of interest (sleep deprivation) within a pair of already established mental modules, one for encoding the stimulus, the other for selecting the response. In Section 6.6.1 I comment on the conflict between the conclusion from the AFM that the sleep deprivation effect is selective and a claim, based on the more popular task comparison method, that its effect is global.

As discussed in Section 6.5, signal detection theory has provided a widely applied method for isolating and measuring sensory processes in tasks that also involve decision processes, but has in general failed to isolate those decision processes. Ex. 6.5.2 illustrates a variant of the method, in which the decision factor that is controlled, and that has been used to successfully demonstrate the modularity of decision as well as sensory processes in an experiment with pigeons, is not the traditional payoff matrix.

In Section 6.6, process decomposition is contrasted with the more familiar *task comparison* method, in connection with the sleep deprivation example. Unlike task comparison, which is often used in a way that requires various assumptions (including modularity) to be made without test (Shallice, 1988; Rumelhart et al., 1986; Sternberg, 2001, Appendix A.1), the process decomposition method incorporates such a test.

In Section 6.7 I provide brief summaries of three additional applications of the process decomposition method to the analysis of mental processes, examples that illustrate its wide range of organisms, measures, combination rules, and modules. Included with Ex. 6.7.2 is an outline of the inferential logic associated with a multiplicative combination rule.

6.3   Isolation of a Timing Module in the Rat

In the *peak procedure* as used by Roberts (1981), a rat that has learned to press a lever for food is presented randomly with two kinds of trial, each starting with the onset of a signal such as a light. On *food trials*, food is "primed" after a fixed interval (such as 20 s) called *time of food*, or *TF*, in the sense that the first lever press after *TF* causes food delivery and signal offset. On *empty trials* no food is delivered, and after 2 or 3 min the signal turns off, independent of the rat's behavior. After a rat is trained, its response rate during an empty trial rises and then falls, reaching a maximum at about the time when food would become available on food trials. Results from this procedure and others suggest that animals have a clock process, **C**, that measures such time intervals. Roberts' goal was to *isolate* this process, that is, to show that **C** can be modified and measured separately from the remainder of the stimulus-response path, which I shall call the *response process* **R**. Among other functions, **R** uses information from **C** in its control of response rate.

For each condition, Roberts started with the function that relates mean response rate to time from signal onset, the *response-rate function*. This function first rises to a peak and then falls. Many of the properties of such a function might be pure measures; examples are the maximum rate (*peak rate*), the time at which the peak rate is achieved (*peak time*), the width of the function (*spread*), and the rate toward the end of empty trials (*tail rate*). Roberts was interested in the possibility that processes **C** and **R** are modules associated with pure measures M_C (peak time) and M_R (peak rate), respectively.

In one experiment (1981, Exp. 1) Roberts randomly mixed trials on which the signals were a light and a sound. He varied two factors, *TF* on food trials, defined earlier, and the proportion of trials on which food was delivered (*probability of food, PF*). During Phase 1 of the experiment *PF* was fixed at 0.8 while *TF* varied: *TF* = 20 s with one of the two signals, *TF* = 40 s with the other. During Phase 2, *TF* was fixed at 20 s while *PF* varied: *PF* = 0.2 with one of the two signals, *PF* = 0.8 with the other. In both phases the two signals occurred randomly and equally often. In a group of ten rats, the mean effects on peak time of *TF* (Phase 1) and *PF* (Phase 2) were 20 ± 1 s and 1.1 ± 1.0 s, respectively, whereas the mean effects of those factors on peak rate were 1.3 ± 2.5 responses/min and 42 ± 5 responses/min, respectively. Thus, peak time (M_C) was influenced by *TF* but invariant with respect to *PF*, whereas peak rate (M_R) was influenced by *PF* but invariant with respect to *TF*. These findings of selective influence (and others; Roberts, 1993, 1998) support the hypothesis that the process controlling response rate can be partitioned into two different modules (**C** and **R**) that control peak time (M_C) and peak rate (M_R), respectively, and that M_C and M_R are pure measures of these modules.

6.4 Mental Processing Stages Inferred from Reaction Times

6.4.1 The Method of Additive Factors

If a process can be partitioned into subprocesses arranged in stages, then the RT becomes an example of a composite measure with summation as the combination rule; thus, if two factors *F* and *G* change *RT* but influence no stages in common (*selective influence*), their effects on mean reaction time should be additive, as described in Table 6.2. Conversely, if factors *F* and *G* interact, so that *G* modulates the effect of *F* rather than leaving it invariant, then *F* and *G* must influence at least one stage in common. Suppose, then, that we have a process in which RT measurements have revealed two or more factors with additive effects. This supports* the hypothesis that the process contains subprocesses arranged sequentially, in stages, with each of the factors influencing a different subprocess selectively.† Thus, one approach to searching for the modular decomposition of a complex process is the method of additive factors (AFM), which involves determining whether two or more factors have additive or interacting effects on mean RT.‡ The application discussed in the section that follows exemplifies how, by combining inferences from a pattern of additive and interacting factor effects, it is possible to reveal more complex processing structures.

6.4.2 Selectivity of the Effect of Sleep Deprivation in a Process with Stages

One of the most provocative applications of the additive factor method is described by Sanders et al. (1982, Exp. 1), and leads to the controversial conclusion that the effect of sleep deprivation is selective (process-specific) rather than global. What follows is a simplified description of their experiment and findings.

What permits asking the question of process specificity of the effect of a factor is a feature of this example that distinguishes it from the others: the inclusion of more than two factors. The stimuli were the single digits "2," "3," "4," and "5"; the responses were their spoken names, "two," "three," "four," and "five." Four factors were manipulated, each at two levels: The first was *stimulus quality* (SQ); the digits, presented as dot patterns, could be *intact*, or *degraded* by adding other dots. The second was the *mapping familiarity* (MF) from digits to names; it could either be *high* (respond to each digit with its name) or *low* (respond to "2," "3," "4," and "5" with "three," "four," "five," and "two," respectively). The third was *sleep state* (SLP), which was either *normal* (data taken during the day after a normal night's sleep) or *deprived* (data taken during the day after a night awake in the lab). Test sessions occurred in both the morning and afternoon, creating a fourth two-level factor, *time of day* (TD). The $2^4 = 16$ conditions were run in separate blocks of trials. The measure was the mean RT for trials with correct responses. For simplicity the data shown in Figure 6.1 have been averaged over levels of TD.

Other studies (see Section 7.3.3 in Chapter 7) had already suggested that SQ and MF were likely to influence two different processing stages selectively,

---

* As in testing any set of hypotheses, the degree to which a hypothesis is supported by confirmation of a prediction is diminished to the extent that alternative hypotheses that generate the same prediction are plausible (Howson and Urbach, 2006). In this case the alternative contenders (some discussed in Roberts and Sternberg, 1993) do not involve processors arranged in stages, but they do involve different processes that are selectively influenced by the factors, i.e., they involve modules. To discriminate among such alternatives, aspects of the data other than mean RT may have to be considered.

† Given an assumption about the durations of different stages that is stronger than zero correlation but weaker than stochastic independence, stages plus selective influence implies numerous properties of other features of the RT distributions in addition to their means (Sternberg, 1969; Roberts and Sternberg, 1993).

‡ See Sternberg (1969, 1998a), and Section 6.4.3.

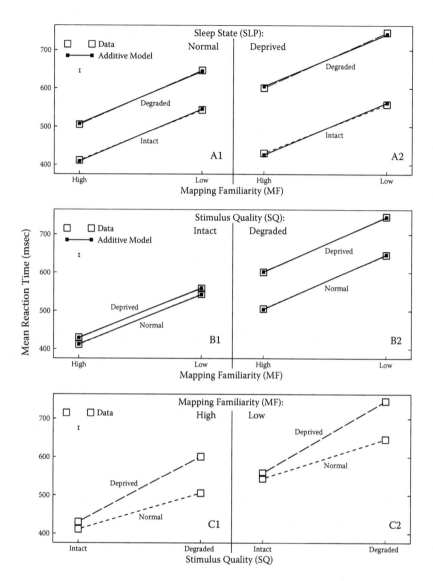

FIGURE 6.1

Means over the two levels of *time of day, TD*. Each pair of panels shows the same $2 \times 2 \times 2 = 8$ data points, plotted in different ways. Each point is the mean of about 300 *RT*s from each of 16 subjects. A fitted additive model is also shown in each of the top four panels. Mean absolute deviations of data from model are 3.3 ms (Panels A1, A2) and 1.0 ms (Panels B1, B2). Because basic data are no longer available, values were obtained from Figure 1 of Sanders et al. (1982). For the same reason, neither within-cell nor between-subject measures of variability are available. The ±*SE* bars were therefore determined by separating the data by *TD*, fitting a model that assumes the additivity of *MF* with *SQ*, *SLP*, and *TD*, and using the deviations (7 df) to estimate *SE*.

processes that might be described as stimulus encoding (**S**) and response selection (**R**).*

The results in Figure 6.1 consist of the mean *RT*s from the $2 \times 2 \times 2 = 8$ conditions. Panels A1 and A2 show that at each level of *SLP* there are additive effects on $\overline{RT}$ of *SQ* and *MF*. This evidence supports the following:

(1) Performing the task involves at least two modules, arranged as stages.

(2) Factors *SQ* and *MF* influence no stages in common.

Panels B1 and B2 show that at each level of *SQ* there are additive effects on *RT* of *MF* and *SLP*; that is, the extra time a subject takes to execute an unfamiliar S-R association rather than a well-learned one is invariant over sleep states, rather than being increased by sleep deprivation. This evidence lends further support to (1) and also supports the following:

(3) Factors *SLP* and *MF* influence no stages in common.

Panels C1 and C2 show that at each level of *MF* there are interactive effects of *SQ* and *SLP*: Increasing the level of *SLP* has a far greater effect on *RT* when the stimulus is degraded (98 ms) than when it is intact (17 ms); that is, sleep deprivation modulates the effect of the difficulty of stimulus encoding. This evidence supports the following:

(4) Factors *SLP* and *SQ* influence at least one stage in common.

Taken together, the three pieces of evidence support a theory according to which the process used to perform the task contains at least two modules, **S** and **R**, these modules are arranged as stages, and among the factors *SQ*, *MF*, and *SLP*, *SQ* and *SLP* influence **S**, whereas *MF* alone influences **R**. It is reasonable to suppose that the stimulus is identified during **S**, and the response selected during **R**. (This is suggested by the nature of the factors *SQ* and *MF* that influence them.) The AFM has thus led us to the surprising conclusion that whereas *SLP* influences stimulus encoding, it does not influence response selection.†

Unlike some other applications of composite measures, the findings from this experiment not only demonstrate separate modifiability and thereby permit us to partition the S-R path into two modules (here, stages) **S** and **R** (the former selectively influenced by *SQ*, the latter by *MF*), but also extend that analysis, providing an example of localizing the influence of a third fac-

---

* This conclusion (separately modifiable sequential processes, or stages) is further strengthened by analyses of complete RT-distributions rather than just *RT* means, from similar experiments (Sternberg, 1969, Sec. V; Roberts and Sternberg, 1993, Exp. 2).

† Electrophysiological evidence that confirms the selectivity of the effect of *SLP* can be found in Humphrey et al. (1994).

tor *SLP* in one of the identified modules, **S**, thereby further characterizing **S** and **R**.

6.4.3   Support for Stages from Analysis of a Neural Process

Stages are functionally distinct operations that occur during nonoverlapping epochs such that the response occurs when they have been completed. In a process with two stages, the stream of operations between stimulus and response can be *cut* at some point, separating Stage **A** (the processes before the cut) from Stage **B** (the processes after the cut). If **A** and **B** are influenced selectively by factors *F* and *G*, respectively, this means that *F* (*G*) can have an effect only before (after) the cut. In most analyses based on behavioral data, the cut is merely hypothetical, inferred from the additivity of factor effects. However, in the neural process analysis discussed in Section 7.2.1 of Chapter 7, the cut corresponds to a particular observable neural event and divides the neural process into two subprocesses whose durations are selectively influenced by the two factors. This more direct observation of processing stages supports the inferences from the RT data.

6.5   Signal Detection Theory: Sensory and Decision Modules

6.5.1   The Finding of Only Partial Modularity in Traditional Sensory Experiments

The most influential approach to deriving pure measures of two processes underlying the performance of a task is the one associated with signal detection theory, SDT (Swets et al., 1961; Macmillan and Creelman, 2005). At the heart of this approach is the recognition that even simple psychophysical tasks involve decision processes as well as sensory mechanisms. Consider a psychophysical experiment in which two types of trials are randomly intermixed, each with a slightly different light intensity. On one type of trial the brighter light, S_T (the target stimulus) is presented; on the other type the dimmer light, S_{NT} (the nontarget stimulus) is presented. The observer's task is to respond with either R_T ("it was the target") or R_{NT} ("it was the nontarget"). On each trial, according to SDT, the observer forms a unidimensional internal representation of the stimulus (X_T and X_{NT}, for S_T and S_{NT}). Because S_T is brighter than S_{NT}, X_T will tend to be larger than X_{NT}. It is also assumed, however, that because of external and internal noise, X_T and X_{NT} are random variables with distributions rather than being fixed constants, and that because S_T and S_{NT} are similar, these distributions overlap. It is the overlap that creates the discrimination problem for the observer. In what follows, I shall assume that the distributions of X_T and X_{NT} are Gaussian with equal variances (supported in the experiment described below, but often false), and also that S_T and S_{NT} trials are equally frequent.

According to SDT the value of X on a trial results from the operation of a sensory process **S**; this value is then used by a decision process **D** to select one of the two responses, selecting R_T if X exceeds a criterion and selecting R_{NT} otherwise. The subject's choice of criterion determines the direction and magnitude of response bias.

The data from such an experiment can be described by four numbers arranged in a 2×2 matrix, where the rows correspond to the two trial types S_T and S_{NT} and the columns correspond to the two responses R_T and R_{NT}. In the top row are the proportions of the target (S_T) trials that elicited each response, which estimate $Pr\{R_T \,|S_T\}$ (the true positive or "hit" probability) and $Pr\{R_{NT} \,|S_T\}$ (the false negative or "miss" probability). In the bottom row are the proportions of the nontarget (S_{NT}) trials that elicited each response, which estimate $Pr\{R_T \,|S_{NT}\}$ (the false positive or "false alarm" probability) and $Pr\{R_{NT} \,|S_{NT}\}$ (the true negative or "correct rejection" probability). From such a matrix, two measures can be derived: One is d', presumed to be a pure measure of the sensory process **S**, and proportional to $\bar{X}_T - \bar{X}_{NT}$, which increases with discriminability. The other is an estimate of the criterion, presumed to be a pure measure of the decision process **D**.

Many factors have been used in attempts to influence **S** and **D**, some expected to influence just a sensory process (sensory-specific or *s-factors*), and others expected to influence just a decision process (decision-specific or *d-factors*). Stimulus features such as the luminance difference between S_T and S_{NT} are examples of s-factors used to influence the measure $M_S = d'$. In studies with human observers, 2×2 payoff matrices, containing positive or negative values associated with the four possible outcomes on a trial, have been used as a factor (*PM*) to influence the response bias associated with **D**. In analogous studies with animals, partial reinforcement is often used: a food reward is provided for only a fraction of the correct responses of each kind; let these fractions be π_T and π_{NT} for R_T and R_{NT}.

In terms of the inferential logic of Table 6.1, the SDT approach has met with only partial success: d' has been found to be sensitive to s-factors (such as the luminance difference mentioned earlier), and invariant with respect to d-factors (such as the payoff matrix), which argues for $M_S = d'$ being a pure measure of **S**. From the viewpoint of a psychophysicist investigating **S**, the SDT approach can thus be extraordinarily helpful. However, despite attempts to find a pure measure M_D of the decision process—in particular, a measure that reflects response bias—none has been found that is invariant with respect to s-factors (as well as being sensitive to d-factors) in experiments with either humans (Dusoir, 1975, 1983) or animals (Alsop, 1998). The SDT approach thus seems to provide only *partial modularity*.

Why has SDT failed in this respect? The possibility considered by Alsop (1998) and by others whose work he reviews is that the problem may be a failure of the assumption that the payoff matrix *PM* influences **D** selectively and controls response bias. A few animal studies suggest that response bias is controlled, not by the conditional probabilities described by the payoff matrix, but by the distribution of rewards over the two alternative responses (sometimes described as the *reinforcement ratio, RR*).

One way to characterize *RR* is as $Pr\{R_{NT} \mid Reward\}$, the proportion of the total number of rewards (for both kinds of correct responses) that are given for R_{NT}. Assuming equal frequencies of S_T and S_{NT}, $Pr\{R_{NT}|Reward\} = r/(1 + r)$, where r, the ratio of the expected numbers of rewarded R_{NT} responses to rewarded R_T responses, is given by $[\pi_{NT} Pr\{R_{NT} \mid S_{NT}\}]/[\pi_T Pr\{R_T|S_T\}]$. As well as depending on the values of π_T and π_{NT} specified by the payoff matrix, the ratio r, and hence *RR*, also depends on discriminability and the current criterion. Controlling or manipulating the payoff matrix thus provides only partial control of a factor (*RR*) that may influence **D** selectively, but whose level is also affected by s-factors. Fortunately, it is possible to control *RR* itself instead of the payoff matrix.

6.5.2 Modularity of Sensation and Decision When Reinforcement Ratio Is Controlled

In a luminance-discrimination experiment with six pigeons, McCarthy and Davison (1984) used a linked concurrent pair of variable-interval (VI) schedules to control *RR*. On each trial in a series, one of two light intensities appeared on the center key of three keys; these two trial types were equally frequent. The correct response was to peck the left key (R_T) if the center key was "bright," and to peck the right key (R_{NT}) if it was "dim." Correct responses were reinforced with food, with a mean probability of about 0.37, controlled by the VI schedules. Two factors were varied orthogonally: The *luminance ratio* (*LR*) of the two lights was varied by letting the dimmer luminance be one of five values, including, for the most difficult *LR* level, a value equal to the brighter luminance. The *reinforcement ratio* (*RR*), described by $Pr\{R_{NT}|Reward\}$, could be one of three values, 0.2, 0.5, or 0.8.* There were thus $5 \times 3 = 15$ conditions. For each bird, each condition was tested for a series of consecutive daily sessions until a stability requirement was satisfied; the data analyzed came from the last seven sessions in each condition (about 1060 trials per condition per bird).

For each condition and each bird, the data can be summarized by two proportions, $Pr\{R_T \mid S_T\}$ and $Pr\{R_T \mid S_{NT}\}$. If the distributions of X_T and X_{NT} are Gaussian with equal variances, and $z(.)$ is the z-transform of a proportion (the inverse Gaussian distribution function), then the ("ROC") curve traced out when $z(Pr\{R_T \mid S_T\})$ is plotted against $z(Pr\{R_T \mid S_{NT}\})$ as *RR* is changed from 0.2 to 0.5 to 0.8, is expected to be linear with slope 1.0. Examination of the set of 30 such curves (6 birds $\times$ 5 levels of *LR*) supports this expectation, and hence the equal-variance Gaussian model.

Given such support for the model, suitable estimators for the discriminability and criterion measures for each condition are

---

* If $Pr\{R_{NT} \mid Reward\} = 0. 2$, for example, the ratio r defined earlier is 1/4; for each rewarded R_{NT} response there are four rewarded R_T responses, encouraging a liberal (low) criterion for R_T.

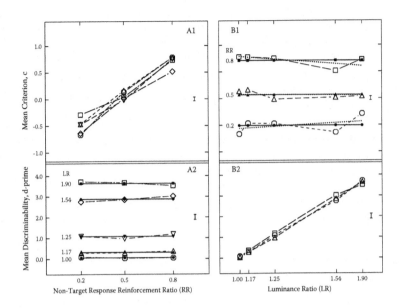

FIGURE 6.2

Mean effects of *Reinforcement Ratio, RR =Pr{R_{NT} |Reward}* (Panels A1, A2) and *Luminance Ratio, LR* (Panels B1, B2) on criterion $\hat{c}$ (Panels A1, B1) and on discriminability $\hat{d}'$ (Panels A2, B2) are shown by unfilled points and light lines. *RR* and *LR* levels have been scaled so as to linearize their mean effects on $\hat{c}$-value and $\hat{d}'$-value (Panels A1 and B2). Filled points and heavy solid lines in Panels A2 and B1 represent fitted models in which $\hat{c}$ and $\hat{d}'$ are invariant with respect to factors *RR* and *LR*, respectively. The dotted lines in Panel B1 represent a fitted model with a multiplicative interaction of the two factors (which is not statistically significant). The ±*SE* error bars reflect estimates of the variability of each plotted point after removing mean differences between birds. Plotting symbols correspond from top panels to bottom, but not from left to right; the plotted *y*-values are the same from left to right. (Data from McCarthy, D. and Davison, M. 1984. Isobias and alloiobias functions in animal psychophysics. *J. Exp. Psychol. Anim. Behav. Process.* 10: 390–409.; basic data kindly provided by B. Alsop.)

$$\hat{d}' = z(Pr\{R_T |S_T\}) - z(Pr\{R_T |S_{NT}\}) \text{ and } \hat{c} = [z(Pr\{R_T |S_T\}) + z(Pr\{R_T |S_{NT}\})]/2$$

The origin for the criterion measure is the midpoint between X_{NT} and X_T; the sign of the criterion thus expresses the direction of the bias. Means over birds of these two measures are shown in Figure 6.2.

The left side of the figure shows that although the criterion responds strongly to factor *RR* (Panel A1), the often-demonstrated invariance of *d'* with respect to d-factors is also persuasive here (Panel A2): there is neither a main effect of *RR* on *d'*, nor is there any modulation by *RR* of the effect of *LR*. The invariance model fits well. Thus, we have evidence for the hypothesis that although *RR* is potent, as shown by its influence on the criterion *c* (hence on **D**), it leaves invariant our measure *d'* of discriminability (and, hence, of **S**).

Effects of *LR* are shown on the right side of the figure. Panel B2 shows the orderly effect of *LR* on *d'*; discriminability ranges widely, from *d'* ≈ 0 to *d'* ≈ 3. 6. Panel B1 shows that to a good approximation the criterion is uninfluenced by *LR*.

Overall, this study, along with others (Alsop, 1998), seems promising in its suggestion that by using something other than the payoff matrix as a d-factor in signal detection experiments, the separate modifiability of **S** and **D** can be demonstrated. Although these data are from pigeons, Johnstone and Alsop (2000) have shown with humans that when *RR* is controlled, the criterion *c* is approximately invariant over changes in discriminability.

This example illustrates how important is the choice of factors. When the measures are $M_S = d'$ and $M_D = c$, and the factors are the s-factor *LR* (luminance ratio) and the d-factor *PM* (payoff matrix), the modularity of **S** and **D** is hidden: M_D depends on *LR* as well as *PM*. In contrast, if the factors are *LR* and *RR*, the pigeon data suggest that to a good approximation they influence M_S and M_D selectively, thus demonstrating the separate modifiability of **S** and **D**. One way to think of this is that M_D depends on *LR* as well as *PM* because they both affect *RR* if it is permitted to vary; if we control *RR* rather than *PM*, we find M_D to be invariant with respect to *LR*, just as M_S is invariant with respect to d-factors.

6.6   Process Decomposition versus Task Comparison

6.6.1   Two Methods Compared

The above applications exemplify a process decomposition method whose goal is to divide the complex process by which a particular task is accomplished into modular subprocesses. The factor manipulations are not intended to produce "qualitative" changes in the complex process (such as adding new operations or replacing one operation by another), which may be associated with a change in the task, just "quantitative" ones that leave it invariant.* The task comparison method is a more popular approach to understanding the structure of complex processes. Here, one determines the influence of factors on performance in different tasks, rather than on different parts of the complex process used to carry out one task. The data pattern of interest is the selective influence of factors on tasks, i.e., the single and double dissociation of tasks. Although it may achieve other goals, task comparison is inferior to process decomposition for discovering the modular subprocesses of a complex process or investigating their properties: The interpretation of task comparison often requires assuming a theory of the complex process in each task; the method includes no test of such assumptions. In contrast, process decomposition requires a theory of only one task, and, as illustrated by the previous examples, incorporates a test of that theory. An illustration of the task comparison method with a discussion of its limitations is provided in Section 7.4.1 of Chapter 7.

---

* See Section 7.4 in Chapter 7, and Sternberg (2001, sections A.2.1 and A.9.2).

6.6.2   Finding Which Process Is Influenced by a Manipulation

As discussed in Section 6.4.2, it is in the context of the evidence for stages
selectively influenced by SQ and MF that the interaction of SLP with SQ
and its additivity with MF can be interpreted. The implication—that sleep
deprivation has effects that are process specific—contradicts the conclusion of Dinges and Kribbs (1991), according to which there is "a generalized effect of sleepiness on all cognitive functioning." They based their
conclusion not on the process decomposition approach, but on results of
the task comparison method, and the finding that performance is impaired
by sleep deprivation in a wide range of tasks. If Dinges and Kribbs (1991)
were correct, SLP should influence both **S** and **R**. Increasing the level of
SLP should therefore exacerbate both kinds of difficulty: SLP should interact with both SQ and MF by amplifying the effects of both, contrary to
what Sanders et al. found. The problem for the task comparison method
in this application is perhaps the high likelihood that all tasks in which
performance can be measured involve some perceptual operations, i.e.,
processes akin to **S**.

Process decomposition might also be more fruitful than task comparison
in investigating the mechanism of action of different drugs, for similar reasons: Even if a drug influences processes in class **A** and not in class **B**, it may
be difficult to find any task that does not involve processes in both classes.
This may be why, for example, in a study that used task comparison to determine which processes are affected by clonidine versus temazepam (known
to have different pharmacological mechanisms) in which 11 tasks were used,
each drug produced statistically significant effects on all except one of the
tasks (Tiplady et al., 2005). Using the process decomposition approach, if factor F (*G*) is known to selectively influence process **A** (**B**) in a task, and process
A (**B**) is selectively influenced by Drug 1 (Drug 2), then both drugs would
affect task performance, but the effect of F (*G*) on that performance would be
modulated only by Drug 1 (Drug 2).

6.6.3   Donders' Subtraction Method: Task Comparison
with a Composite Measure

Perhaps the most venerable version of the task comparison method is
Donders' *subtraction method* (Donders, 1969; Sternberg, 1998b, Appendix 2)
for two tasks with measures $M_1 = \overline{RT_1}$ and $M_2 = \overline{RT_2}$. The joint hypothesis
consists, first, of a pair of task theories that specify the constituent processes
of each task, and second, a combination rule:

H1 (Task Theory 1): Task 1 is accomplished by process **A**.

H2 (Task Theory 2): Task 2 is accomplished by **A** and **B**, where **A** is
identical, at least in duration, to the corresponding process in Task 1.
(That is, addition of **B** satisfies a "pure insertion" assumption.)

H3 (Combination Rule): Contributions $u = T_A$ of **A** and $v = T_B$ of **B** to $M_2 = RT_2$ combine by *summation* (Table 6.2), as implied by Donders' assumption that **A** and **B** occur sequentially, in stages.

Given these hypotheses, it follows that $M_1 = \overline{RT_1}$ is an estimate $\hat{T}_A$ of the mean duration of **A**, $\overline{RT_2}$ is an estimate $\hat{T}_A + \hat{T}_B$ of the sum of the mean durations of **A** and **B**, and therefore, by subtraction, $\hat{T}_B = \overline{RT_2} - \overline{RT_1}$ provides an estimate of the mean duration of **B**. It is a serious limitation of the subtraction method that it usually embodies a test neither of the combination rule nor of pure insertion; that is, H2 and H3 are assumed, but not tested.

Despite this problem, an analog of Donders' method has also often been used with brain activation measures, as discussed in Section 7.4.2 of Chapter 7.

6.7 Brief Summaries of Three Other Examples

In this section I provide brief summaries of three additional examples that show the diversity of applications of the process decomposition approach. The first example, concerned with visual detection of different spatial frequencies, makes use of pure measures. The second supports the same conclusion, using a composite measure with multiplication as the combination rule; I include a discussion of the inferential logic for this case. The third, concerned with learning and motivation in rats, also uses a composite measure with multiplication as the combination rule.

6.7.1 Evidence for Modular Spatial-Frequency Analyzers from Selective Adaptation

The thresholds for sinusoidal gratings of sufficiently different high and low spatial frequencies respond selectively to adaptation by high- and low-frequency gratings. This supports the hypothesis that there exist modular analyzers that detect the high and low frequencies, of which the thresholds for high and low frequencies are pure measures (Graham, 1970; Sternberg, 2001, Section 9).

6.7.2 Evidence for Modular Spatial-Frequency Analyzers from the Detectability of Compound Gratings

Imagine a task in which a subject says "yes" when either or both of two detection processes respond, and says "no" if neither detector responds. For the present discussion we ignore the complication introduced by "guessed" yes responses that may occur when neither detector responds: The subject says "no" if and only if both detectors fail to respond. If the behavior of the two detectors is uncorrelated, then the probability of this joint event is the product of the individual nonresponse probabilities for the two detectors. $Pr\{\text{"no"}\}$ is thus a composite measure of the two detectors, with a multiplicative combination rule.

To describe the consequences of a multiplicative combination rule for a composite measure, I need to introduce the idea of a proportional effect, or *p.effect*. We saw in Section 6.2.2 that the *effect* of a factor on a measure is defined as a *difference* (for a factor with two levels), as in Equation 6.1, or can be defined as a vector of differences (for a factor with multiple levels). Similarly, the *p.effect* of a factor on a measure M_A is defined as a *ratio* (for a factor with two levels):

$$p.effect(F) = M_A(F_2) / M_A(F_1),\tag{6.6}$$

or a vector of ratios (for a factor with multiple levels). Suppose we have a composite measure with a multiplicative combination rule:

$$M_{AB}(F_j, G_k) = u(F_j) \times v(G_k).\tag{6.7}$$

To derive the equivalent of Equation 6.3 from Equation 6.7, however, requires us to assume that the contributions u and v from processes **A** and **B** to M_{AB} are uncorrelated. In that case, it follows from Equation 6.7 that

$$\bar{M}_{A,B}(F_j, G_k) = \bar{u}(F_j) \times \bar{v}(G_k).\tag{6.8}$$

By analogy to Equation 6.4, it can then be shown that

$$p.\,effect(F, G) = p.\,effect(F) \times p.\,effect(G).\tag{6.9}$$

If the p.effects of the factors are multiplicative, as in Equation 6.9, this supports the joint hypothesis that processes **A** and **B** are separately modifiable, that their contributions to M_{AB} combine by multiplication, and that their contributions are uncorrelated. The inferential logic in this case is outlined in Table 6.3.

This observation was exploited in a famous experiment in which the detectors were hypothesized spatial-frequency analyzers sensitive to different frequency bands. Sachs et al. (1971) independently varied the contrasts (F, G) of widely separated frequencies in a compound grating whose presence subjects had to detect. They found that increasing the contrast of one of the frequencies in the compound reduced $Pr\{\text{nondetect}\}$ by a constant factor, consistent with Equation 6.9, thus supporting the joint hypothesis. Their findings provided important early evidence for modular spatial-frequency analyzers and for the multiplicative combination rule.

TABLE 6.3

Inferential Logic for a Composite Measure with Multiplication as the Combination Rule

---

Joint Hypothesis

H1: Processes **A** and **B** are modules (separately modifiable).

H4: Contributions u_A, v_B of **A**, **B** to M_{AB} combine by *multiplication*.

H5: Contributions of **A** and **B** to M_{AB} are uncorrelated.

Prediction

We may be able to find factors F and G that influence **A** and **B** selectively:

$$p_1': u_A \leftarrow F, \; p_2': v_B \not\leftarrow F, \; p_3': v_B \leftarrow G, \; p_4': u_A \not\leftarrow G$$

and jointly influence no other process.

If so, their *proportional effects* on M_{AB} will be *multiplicative*.

Alternative Results	
We find factors F and G with *multiplicative p.effects* on M_{AB}.	We fail to find such factors.

Inferences	
Support for joint hypothesis $H1 + H4 + H5$.	Refutes one/both of *H1, H4, H5, or* we did not look enough for F, G.

6.7.3   Modular Processes for Learning and Motivation

Roberts (1987) has drawn attention to the orderliness of response rate in factorial experiments across a remarkable range of factors and animal species. Several of the data sets he examined showed multiplicative effects of two factors on response rate. One of these cases is provided by Clark's (1958) experiment on the effects of *hours of food deprivation* (a motivational factor with seven levels) and *frequency of feeding* (a learning factor with three levels) on the rate of bar pressing by rats under variable-interval reinforcement schedules. The p.effect of each factor was found to be invariant over levels of the other, supporting the existence of uncorrelated modular processes influenced selectively by the two factors, and a multiplicative combination rule.

6.8   Discussion

In this chapter the process decomposition method is introduced, along with six examples of its application to the analysis of mental processes, using behavioral data. The applications range widely, with humans, pigeons, and rats as subjects, and with reaction time, discriminability, detection thresholds, and response rate among the measures. The applications include composite measures with combination rules of summation and multiplication, as well as pure measures, and they include a large variety of factors in experiments, most of which use factorial designs. It was argued that for purposes of discovering processing modules, process decomposition is superior to the more popular task comparison method. In Chapter 7, the process decomposition method is applied to experiments using brain measures, with the goal of identifying neural modules.

Comments and issues related to both chapters 6 and 7, as well as suggestions for further reading, are considered at the end of Chapter 7.

Acknowledgments

For providing unpublished details of their data I thank Brent Alsop, Jacob Nachmias, and Seth Roberts. For helpful comments on the manuscript I thank Allen Osman. For numerous enlightening discussions I thank Seth Roberts.

References

Alsop, B. 1998. Receiver operating characteristics from nonhuman animals: Some implications and directions for research with humans. *Psychonomic Bull. and Rev., 5*, 239–252.

Berger, R. L. and Hsu, J. C. 1996. Bioequivalence trials, intersection-union tests and equivalence confidence sets. *Stat. Sci.,* 11, 283–319.

Clark, F. C. 1958. The effect of deprivation and frequency of reinforcement on variable-interval responding. *J. Exp. Anal. Behav.,* 1: 221–228.

Dinges, D. F. and Kribbs, N. B. 1991. Performing while sleepy: Effects of experimentally-induced sleepiness. In *Sleep, Sleepiness and Performance,* Monk, T. H. (Ed.), 97–128. London: Wiley.

Donders, F. C. 1969. Over de snelheid van psychische processen [On the speed of mental processes]. *Onderzoekingen gedaan in het PhysiologischLaboratorium der Utrechtsche Hoogeschool, 1868–1869, Tweede reeks, II,* 92–120, 1868. Transl. by Koster, W. G. In *Attention and Performance II,* Koster, W. G. (Ed.), *Acta Psychol.,* 30: 412–431.

Dusoir, A. E. 1975. Treatments of bias in detection and recognition models: A review. *Percept. Psychophys.* 17: 167–178.

Dusoir, A. E. 1983. Isobias curves in some detection tasks. *Percept. Psychophys.* 33: 403–412.

Fodor, J. 1983. *The Modularity of Mind: An Essay on Faculty Psychology.* Cambridge, MA: MIT Press.

Fodor, J. 2000. *The Mind Doesn't Work That Way: The Scope and Limits of Computational Psychology*. Cambridge, MA: MIT Press.

Graham, N. 1970. Spatial frequency channels in the human visual system: Effects of Luminance and pattern drift rate. *Vis. Res.* 12: 53–68.

Howson, C. and Urbach, P. 2006. *Scientific Reasoning: The Bayesian Approach*. 3rd ed., Chicago: Open Court.

Humphrey, D. G., Kramer, A. F., and Stanny, R. R. 1994. Influence of extended wakefulness on automatic and nonautomatic processing. *Hum. Factors*, 36: 652–669.

Jacobs, R. A. 1999. Computational studies of the development of functionally specialized neural modules. *Trends Cognit. Sci.* 3: 31–38.

Johnstone, V. and Alsop, B. 2000. Reinforcer control and human signal-detection performance. *J. Exp. Anal. Behav.* 73: 275–290.

Macmillan, N. A. and Creelman, C. D. 2005. *Detection Theory: A User's Guide*, 2nd ed. Mahway, NJ: Erlbaum.

Marr, D. 1976. Early processing of visual information. *Phil. Trans. Roy. Soc. Lond. B.* 275: 483–524.

McCarthy, D. and Davison, M. 1984. Isobias and alloiobias functions in animal psychophysics. *J. Exp. Psychol. Anim. Behav. Process.* 10: 390–409.

Roberts, S. 1981. Isolation of an internal clock. *J. Exp. Psychol. Anim. Behav. Process.* 7: 242–268.

Roberts, S. 1987. Evidence for distinct serial processes in animals: The multiplicative-factors method. *Anim. Learn. Behav.* 15: 135–173.

Roberts, S. 1993. Use of independent and correlated measures to divide a time discrimination mechanism into parts. In *Attention and Performance XIV: Synergies in Experimental Psychology, Artificial Intelligence, and Cognitive Neuroscience – A Silver Jubilee*, Meyer, D. E. and Kornblum, S (Eds.), 589–610. Cambridge MA: MIT Press.

Roberts, S. 1998. The mental representation of time: Uncovering a biological clock. In Scarborough, D. and Sternberg, S. (Eds.) *An Invitation to Cognitive Sci.*, Vol. 4: *Methods, Models, and Conceptual Issues*. (pp. 53–105) Cambridge MA: MIT Press.

Roberts, S. and Sternberg, S. 1993. The meaning of additive reaction-time effects: Tests of three alternatives. In *Attention and Performance XIV: Synergies in Experimental Psychology, Artificial Intelligence, and Cognitive Neuroscience – A Silver Jubilee*, Meyer, D. E. and Kornblum, S. (Eds.), 611–653. Cambridge, MA: MIT Press.

Rogers. J. L., Howard, K. I., and Vessey, J. T. 1993. Using significance tests to evaluate equivalence between two experimental groups. *Psychol. Bull.* 113: 553–565.

Rumelhart, D. E., McClelland, J. L., and the PDP Research Group. 1986. *Parallel Distributed Processing: Explorations in the Microstructure of Cognition*, Vol. 1: Foundations. Cambridge, MA: MIT Press.

Sachs, M. B., Nachmias, J., and Robson, J. G. 1971. Spatial-frequency channels in human vision. *J. Opt. Soc. Am.* 61: 1176–1186.

Sanders, A. F., Wijnen, J. L. C., and Van Arkel, A. E. 1982. An additive factor analysis of the effects of sleep loss on reaction processes. *Acta Psychol.* 51: 41–59.

Shallice, T., *From Neuropsychology to Mental Structure*, Cambridge University Press, Cambridge, 1988.

Simon, H. A. 1962. The architecture of complexity. *Proc. Am. Phil. Soc.* 106: 467–482.

Simon, H. A. 2005. The structure of complexity in an evolving world: The role of near decomposability. In *Modularity: Understanding the Development and Evolution of Natural Complex Systems*, Callebaut, W. and Rasskin-Gutman, D. (Eds.), *ix–xiii*. Cambridge, MA: MIT Press.

Sternberg, S. 1969. The discovery of processing stages: Extensions of Donders' method. In *Attention and Performance II*, Koster, W. G. (Ed.), *Acta Psychol.*, 30: 276–315.

Sternberg, S. 1998a. Discovering mental processing stages: The method of additive factors. In *An Invitation to Cognitive Science*, Vol. 4: *Methods, Models, and Conceptual Issues*, Scarborough, D. and Sternberg, S. (Eds.), 703–863. Cambridge, MA: MIT Press.

Sternberg, S. 1998b. Inferring mental operations from reaction-time data: How we compare objects. In *An Invitation to Cognitive Science*, Vol. 4: *Methods, Models, and Conceptual Issues*, Scarborough, D. and Sternberg, S. (Eds.), 365–454. Cambridge, MA: MIT Press.

Sternberg, S. 2001. Separate modifiability, mental modules, and the use of pure and composite measures to reveal them. *Acta Psychol.* 106: 147–246.

Sternberg, S. 2003. Process decomposition from double dissociation of subprocesses, *Cortex*, 39: 180–182.

Swets, J. A., Tanner, W. P., Jr., and Birdsall, T. G. 1961, Decision processes in perception. *Psychol. Rev.* 68: 301–340.

Tiplady, B., Bowness, E., Stien, L., and Drummond, G. 2005. Selective effects of clonidine and temazepam on attention and memory. *J. Psychopharmacol.* 19: 259–265.

7

Identification of Neural Modules

Saul Sternberg

CONTENTS

Abstract...136
7.1 Overview of Examples and Issues 136
7.2 Decomposing Neural Processes with the Lateralized Readiness
Potential...138
    7.2.1 Serial Modules for Preparing Two Response Features 138
    7.2.2 Analysis of the Reaction-Time Data in Ex. 7.2.1......................139
    7.2.3 Parallel Modules for Discriminating Two Stimulus
    Features 141
7.3 Neural Processing Modules Inferred from Brain Activation Maps...144
    7.3.1 Introduction..144
    7.3.2 Modular Processes in Number Comparison 145
    7.3.3 Modular Processes for Stimulus Encoding and Response
    Selection 146
7.4 Process Decomposition versus Task Comparison................................152
    7.4.1 Two Tactile Perception Tasks 152
    7.4.2 An Analog of Donders' Subtraction Method Applied to
    fMRI Data...154
7.5 The Use of Transcranial Magnetic Stimulation (TMS) to
Associate Mental Processes with Brain Regions.................................155
    7.5.1 Introduction..155
    7.5.2 Visual Search and TMS...156
    7.5.3 Number Comparison and rTMS 157
7.6 Evidence from the Event-Related Potential (ERP) for Modular
Processes in Semantic Classification: Brief Summary........................158
7.7 Comments and Issues 159
    7.7.1 Introduction..159
    7.7.2 Task-General Processing Modules 159
    7.7.3 Quantitative versus Qualitative Task Changes 159
    7.7.4 Specialized Processors and Modular Processes.......................160
    7.7.5 Relation between Mental and Neural Modules 160
    7.7.6 Separate Modifiability as a Criterion for Modularity..............160
    7.7.7 Implications of Brain Metabolism Constraints........................161
7.8 Further Reading 161

Acknowledgments 161
References 162

Abstract

In Chapter 6, I introduced the process decomposition approach to the understanding of complex processes, with separate modifiability as the criterion for modularity, and discussed examples of its application to mental processes, using behavioral data. In the present chapter we consider applications of this method of modular decomposition to neural processes, using brain measures and brain manipulations. One application involves electrophysiological brain measurements and leads to the identification of *neural modules*. Two applications use both neural (event-related potential [ERP] or functional magnetic resonance imaging [fMRI]) and behavioral (reaction time [RT]) measurements and lead to the identification of neural and mental modules that correspond. Three of the examples raise questions without answering them: One involves both neural (fMRI) and behavioral (RT) measurements, but leads to the identification of only neural modules. Two promising but incomplete applications are attempts to relate brain regions to behaviorally defined processing modules, using a brain manipulation (transcranial magnetic stimulation, TMS), but suffer from flaws in design or analysis. As a contrast to process decomposition, an example of task comparison involving TMS is presented, and its limitations discussed. Finally, an additional application using ERPs is briefly described.

7.1   Overview of Examples and Issues

As discussed in Chapter 6, factorial experiments are necessary for the process decomposition method when the measures are composite, and are desirable when the measures are pure. Because such experiments have been rare among studies involving brain measurements or brain manipulations, there are relatively few data sets involving such measurements in which separate modifiability can be tested and process decomposition applied. For this reason, I have had to include some flawed or limited examples among those that will be discussed here. Indeed, the final two examples (7.5.2 and 7.5.3) are intriguing and tantalizing cases where the method could have been used, but was not, because of incompleteness of either the design or the analysis. As in the previous chapter, examples will be referred to by the section numbers in which they are first discussed.

In Section 7.2, two applications based on electrophysiological measurements at the scalp, treated as pure measures and shown to be such, are discussed. In both cases, two neural modules are identified: In Ex. 7.2.1, in which the modules are associated with preparation of two different aspects of the response, they are found to operate successively, as *stages*, thus providing support for the inference of stages based on additivity of effects on mean RT in the method of additive factors, discussed in Section 6.4 of Chapter 6. In the discussion of the RT data from Ex. 7.2.1, the RT is treated as a composite measure and shown to lead to the identification of two mental modules that correspond to the neural modules inferred from the electrophysiological data. (*Correspond to* means that their durations are influenced selectively by the same factors and to the same extent.)

In Ex. 7.2.3, in which the inferred neural processing modules are associated with encoding two different aspects of the stimulus, they are found to operate in parallel.

To the extent that there is localization of function in the brain, so that each of several modular processes are implemented in disjoint regions, and to the extent that the level of activation in a region varies with changes in the process it implements, the level of such activation can function as a pure measure of the process. In Section 7.3, I discuss two examples in which fMRI signals in different brain regions were measured for this purpose; in both cases, RTs were measured as well. In Ex. 7.3.2 (number comparison), in which RT measurements in a similar experiment had already indicated separate stages for encoding the test number and comparing it to the target, both the new RT data and the fMRI data support this analysis, suggesting mental and neural modules that correspond. However, the fact that the direction of the effect of the encoding factor on the fMRI response differs in different brain regions raises interesting questions of interpretation. In Ex. 7.3.3 (which employed a choice-reaction paradigm with four stimulus-response pairs), the fMRI data support the hypothesis of modular neural processes for stimulus encoding and response selection, but unlike earlier observations of effects on $\overline{RT}_j$ from several similar paradigms, the effects of the two factors on $\overline{RT}_j$ interact rather than being additive, raising questions about interpretation. One unexpected and potentially important finding in this example is the additivity of effects of the encoding and response selection factors on the fMRI measure in the two brain regions where both factors were found to have effects.

In Section 7.4, I contrast process decomposition with the more familiar *task comparison* method, introduced in Chapter 6, by describing an example of task comparison based on the effects of repetitive TMS. Unlike task comparison, which is often used in a way that requires various assumptions (including modularity) to be made without test (Shallice, 1988; Rumelhart et al., 1986; Sternberg, 2001, Appendix A.1), the process decomposition method incorporates such a test.

The final two primary examples involve the effects of transcranial magnetic stimulation (TMS) on RT measurements. The goal is to relate brain regions to mental modules. In these experiments, TMS in certain brain regions is found to increase the $\overline{RT}_j$ without otherwise disrupting performance. The presence and absence of TMS in such a region can be regarded as two levels of a factor. If this factor influences $\overline{RT}_j$, the brain region to which TMS is applied probably plays some role in performance of the task. However, the potential of this method to associate brain regions with component processes is realized only when the TMS factor is used together with other, process-specific factors to determine which of their effects, if any, are modulated by TMS. For visual search (Ex. 7.5.2), the absence of variation of a within-task search-specific factor precluded doing so. For number comparison (Ex. 7.5.3), a comparison-specific factor was varied, but no focused test of the interaction of this factor with TMS was conducted.

In Section 7.6, I provide a brief summary of an additional example of application of the process decomposition method in a factorial experiment involving ERP measurements. One feature of this example is that, rather than being a part of the hypothesis, the nature of the measure (composite) and the combination rule (summation) are given by the physics of volume conduction in the brain.

7.2   Decomposing Neural Processes with the Lateralized Readiness Potential

7.2.1   Serial Modules for Preparing Two Response Features

Consider a trial in a choice-reaction experiment in which two alternative responses are made by the two hands. Recall that the part of the motor cortex that controls a hand is contralateral to that hand. When enough information has been extracted from the stimulus to permit selection of the hand, but before any sign of muscle activity, the part of the motor cortex that controls that hand becomes more active than the part that controls the nonselected hand. This asymmetric activity can be detected as an increase in the average over trials of the difference between electrical potentials (ERPs) at the two corresponding scalp locations.* Let $A_{mc}(t)$ (an index of *motor-cortex asymmetry*) express this difference as a function of time after stimulus onset (if "stimulus locked"), or as a function of time before the overt response (if "response locked"). $A_{mc}(t)$ is normally zero, but becomes positive when the response hand is selected; the increase of such asymmetry is called the *lateralized-readiness potential (LRP)*. The onset time of the *LRP* is thus an estimate of the time at which the side of the response (left or right) has been selected. Having an estimate of this time makes it possible to ask whether the neural

---

* As in some other brain measurements (e.g., PET, fMRI), averaging over trials is often required for the measures to be interpretable, because of the poor S/N ratio. Here, the "noise" is due partly to neural events unrelated to the task being performed, whose contributions are eliminated by combining subtraction of the prestimulus baseline level with an averaging process that reveals only those events that are consistently time-locked to the stimulus or the response.

process responsible for selecting the side of the response is separate from the neural process responsible for preparing other aspects of the response.

This possibility was exploited in an experiment by Smulders et al. (1995). It was a two-choice RT experiment with single-numeral stimuli mapped on left-hand and right-hand responses. Two factors were varied orthogonally across blocks of trials: stimulus quality (SQ_j: digit intact versus degraded) and response complexity (RC_k: one keystroke—simple—versus a sequence of three keystrokes—complex—made by fingers of the responding hand).* In addition to the composite measure $\overline{RT}_{jk}$, Smulders et al. (1995) measured the onset time of the LRP, based on both stimulus-locked (LRP_s) and response-locked (LRP_r) averaging of the $A_{mc}(t)$ functions. Let T_{sjk} (measured with respect to the stimulus) and T^*_{rjk} (a negative quantity, measured with respect to the response) be the corresponding LRP onset times, and let $T_{rjk} = RT_{jk} + T^*_{rjk}$. T_{sjk} and T_{rjk} are then different estimates of the same time point between stimulus and response; let $T_{\cdot jk}$ be their mean. Averaging over the four conditions, $RT.. = 416$ ms and $T... = 264$ ms. If α is the neural process from stimulus to LRP onset, and β is the neural process from LRP onset to response, these values give us measures of their durations (pure measures of α and β) averaged over the four conditions: $D_{\alpha..} = T... = 264$ ms and $D_{\beta..} = RT.. - T... = 152$ ms.

Shown in the A and B panels of Figure 7.1 are the estimated durations of processes α and β separated by condition: $D_{\alpha jk} = T_{\cdot jk}$, and $D_{\beta jk} = RT_{jk} - T_{\cdot jk}$. Because T_{sjk} and T_{rjk} give similar estimates for effects of the two factors on D_α and D_β, the estimates are based on $T_{\cdot jk}$. The results show that the two factors SQ and RC have selective effects on D_α and D_β. This supports the hypothesis that in this situation the LRP onset indeed defines a boundary or *cut* between two neural modules that function sequentially, as stages, consistent with the reasoning in Table 6.2, Chapter 6 (see Section 6.4.3 of Chapter 6). In contrast to D_α and D_β, which are hypothesized (and confirmed) to be pure measures, RT_{jk} shown in Panel C is a composite measure. With two stages between stimulus and response, the RT reflects contributions from both. In the next section I consider the RT data from this experiment and how they relate to the LRP data.

7.2.2 Analysis of the Reaction-Time Data in Ex. 7.2.1

The conclusion in Section 7.2.1 is that the time between the stimulus digit and the response (the RT) can be partitioned into two intervals, from stimulus to LRP (duration D_α, influenced by SQ but not RC), and from LRP to response (duration D_β, influenced by RC but not SQ). Consider just the mean reaction time, measured under the four factor-level combinations:

$$\overline{RT}_{jk} = \bar{D}_{\alpha j} + \bar{D}_{\beta k} = \bar{D}_\alpha(SQ_j) + \bar{D}_\beta(RC_k). \tag{7.1}$$

---

* Because there were two levels of each of two factors, and all combinations of levels of the two factors were used (a "complete factorial" experiment), the experiment contained four conditions.

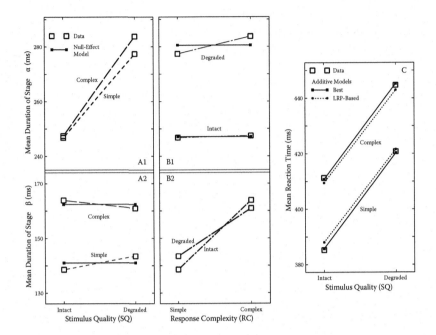

FIGURE 7.1

Means over 14 subjects of data from Smulders et al. (1995). Estimated durations $D_{\alpha jk}$ of Process/Stage α, from stimulus presentation to LRP onset (panels A1, B1) and $D_{\beta jk}$ of Process/Stage β, from LRP onset to response (panels A2, B2). These are shown as functions of *stimulus quality* (SQ_j; panels A1, A2) and of *response complexity* (RC_k; panels B1, B2). Data in panels A1 and A2 are separated by the level of RC; those in panels B1 and B2 are separated by the level of SQ. Also shown in panels A2 and B1 are null-effect models. Main effects of SQ on D_α and D_β are 34 ± 6 ms (Panel A1) and 1 ± 8 ms (Panel A2); the corresponding main effects of RC are 4 ± 8 ms (Panel B1) and 21 ± 7 ms (Panel B2). The RT data (discussed in Section 1.4.3) are shown in Panel C, together with two fitted models. One (unbroken lines) is the best-fitting additive model (mean absolute deviation 0.5 ms); the other (dotted lines) is an additive model based on estimates of process durations from the LRP data (mean absolute deviation 1.8 ms).

Suppose we had only those data and not the LRP data. It follows from partitioning of the $\overline{RT}$ shown in Equation 7.1 that the combination rule for the contributions of α and β is summation. As discussed in Section 6.2.2 and 6.4.1 of Chapter 6, that and the selective influence of factors SQ and RC on neural processes α and β imply that the effects of SQ and RC on RT are additive. In general, factors that selectively influence the durations of different sequential components of the $\overline{RT}$ must have additive effects on the composite measure. Thus, if a hypothesis asserts that the RT in a particular task is the duration of two modular mental processes **A** and **B** arranged sequentially (as stages) with durations T_A and T_B, and selectively influenced by factors F and G, we should expect:

$$\overline{RT}(F_j, G_k) = \overline{T}_A(F_j) + \overline{T}_B(G_k). \tag{7.2}$$

The goodness of fit of the parallel unbroken lines in Figure 7.1C confirms the expectation of additivity for the RTs in the experiment by Smulders et al. and supports the joint hypothesis of Table 6.2 of Chapter 6. (The interaction contrast of SQ and RC was a negligible 2 ± 5 ms.) However, because in this case we also have measures of the durations of neural modules (from the analysis of the LRP data), we can go further: we can ask whether the mental modules derived from the purely behavioral analysis of the composite measure correspond to the neural modules inferred from the LRP-based pure measures.

Thus, suppose that the mental modules **A** and **B** responsible for the additive RT effects are implemented by the neural modules α and β, demarcated by the LRP. Then, not only should the same factors influence them selectively, but also the sizes of their effects should be the same. Agreement among the effect sizes can be examined by assuming that the two factors indeed have perfectly selective effects on α and β, and by using the appropriate subset of the LRP data to "predict" the pattern of the $\{\overline{RT}_{jk}\}$. Thus, we should be able to use just the data in Figure 7.1A1 (averaging over RC levels) to obtain the estimates $\hat{D}_\alpha$ (SQ_1) and $\hat{D}_\alpha$ (SQ_2). Similarly, we should be able to use just the data in Figure 7.1B2 (averaging over SQ levels) to obtain the estimates $\hat{D}_\beta$ (RC_1) and $\hat{D}_\beta$ (RC_2). If $RT = D_\alpha + D_\beta$, we have the "predictions" $RT^*_{jk} = \hat{D}_\alpha$ $(SQ_j) + \hat{D}_\beta (RC_k)$ for the four conditions.* The dotted lines in Figure 7.1C show that the agreement is good: $\overline{RT}_{jk} \approx RT^*_{jk}$. Numerically, the effects of SQ and RC on the composite measure $\overline{RT}$ are 35 ± 3 ms and 25 ± 7 ms, close to their mean estimated effects (34 and 21 ms) on the sum of the pure measures $\hat{D}_\alpha + \hat{D}_\beta$.

Another illustration of the independent use of two methods is provided by Ex. 7.3.2: pure measures to ask about the structure of a neural process, and a composite measure to ask about the structure of a corresponding mental process.

7.2.3 Parallel Modules for Discriminating Two Stimulus Features

Consider a situation in which two different features of the same stimulus must be discriminated to determine how to respond. Are modular neural processes involved in doing so? And, if so, how are they organized temporally? Osman et al. (1992) devised a clever way to ask these questions, using the LRP. On each trial, the visual stimulus had two features. Its *position* (left versus right, which was rapidly discriminated) indicated which response to make should a response be required. Its *category* (letter versus digit, which was discriminated more slowly) indicated whether this was a "Go" trial (on which the selected response should be activated) or a "NoGo" trial (on which no response should be made). Under these conditions, the LRP occurs even on trials with no overt response, and with an onset that is indistinguishable from the LRP on "Go" trials.

---

* This way of deriving the $\{RT^*_{jk}\}$ forces their means into agreement: $RT^*_{..} = \overline{RT}_{..}$; the question of interest is whether the differences among the two sets of four values agree.

Event 1 will refer to the onset of the LRP; the latency T_1 of Event 1 can thus be used to indicate when the stimulus location has been discriminated and the response selected. Let us denote this *location discrimination* process by α. Normally (on "Go" trials), A_{mc} continues to rise until the overt response is initiated. If a "NoGo" signal tells the subject not to respond, however, A_{mc} starts falling. The time at which $A_{mc}(t; NoGo)$ diverges from $A_{mc}(t; Go)$—the latency T_2 of *Event 2*—can thus be used to indicate when the category (the NoGo signal) is discriminated and response preparation ceases. Let us denote this *category discrimination* process by β. T_1 and T_2 are hypothesized to be pure measures of their respective processes.

Events 1 and 2 indicate the completion of processes α and β. Can response preparation start when the location but not the category of the stimulus has been discriminated? And, if so, can category discrimination proceed in parallel with response preparation? To answer such questions, Osman et al. (1992) examined the effects of two factors: One (in Exp. 1) is *Go-NoGo Discriminability, GND*, which should influence β; it could be *high* (letter and digit with dissimilar shapes, GND_1) or *low* (similar shapes, GND_2). The other factor (in Exp. 2) is the spatial compatibility of the stimulus-response mapping, *MC*, which should influence α; it could be *compatible* (respond with the hand on the same side as the stimulus, MC_1) or *incompatible* (respond with the hand on the opposite side, MC_2).

Idealizations of the resulting $A_{mc}(t)$ functions are shown in Figure 7.2, and the observed values of T_1 and T_2 in Figure 7.3. Each of the four panels of Figure 7.2 shows the pair $A_{mc}(t; Go)$ and $A_{mc}(t; NoGo)$ for one condition. The two latency measures for a condition were derived in different ways from this pair of $A_{mc}(t)$ functions: The latency T_1 of Event 1 (onset of the *LRP*) is the time at which the *sum* of the two $A_{mc}(t)$ functions reliably exceeds baseline. The latency T_2 of Event 2 (divergence of the Go and NoGo *LRPs*) is the time at which their *difference* reliably exceeds zero. In Exp. 1, *GND* influenced T_2 (by 43 ms) but not T_1 (compare Figures 7.2A1 and 7.2A2, and see Figure 7.3A). The absence of any effect of *GND* on T_1 ($T_1 \approx 170$ ms on both *Go* and *NoGo* trials) shows that the location of the stimulus, but not its category, controlled the start of response preparation. Because the stimulus influenced response preparation before both of its features were discriminated, these findings from Exp. 1 demonstrate the transmission of "partial information" from the perceptual process to the response process.

In one of the conditions of Exp. 2, the stimulus-response (S-R) mapping was incompatible, which was expected to delay selection of the response. To ensure that stimulus location had an opportunity to influence response preparation on *NoGo* trials in both conditions, it was important to delay the Go-NoGo discrimination.* Osman et al. (1992) therefore reduced letter-digit shape discriminability so as to increase T_2 to about 350 ms. In this experiment, *MC* influenced T_1 (by 121 ms) but not T_2 (compare Figures 7.2B1 and

---

* It is this requirement that would have made it difficult to implement a suitable factorial experiment.

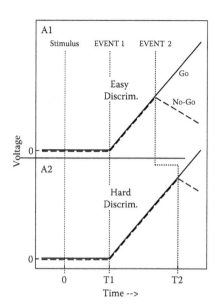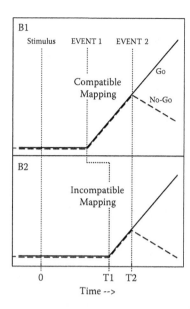

FIGURE 7.2

Schematic idealized asymmetry functions $A_{mc}(t)$ from Osman et al. (1992). Event 1 is the *LRP* onset; Event 2 is the onset of the divergence of $A_{mc}(t; Go)$ from $A_{mc}(t; NoGo)$. Panels A1 and A2: Asymmetry functions from Exp. 1, in which *Go-NoGo Discriminability* could be high $(GND = GND_1)$ or low $(GND = GND_2)$. Panels B1 and B2: Asymmetry functions from Exp. 2, in which the *stimulus-response mapping* could be compatible $(SRM = SRM_1)$ or incompatible $(SRM = SRM_2)$.

7.2B2, and see Figure 7.3B). Increasing the level of mapping difficulty from MC_1 to MC_2 therefore *reduced* the interval between Event 1 and Event 2.

Taken together, the two experiments show that *MC* and *GND* influenced the two measures T_1 and T_2 selectively, supporting the hypothesis that they are pure measures of two different modular processes (see Table 6.1 in Chapter 6). The results also show how processes α and β are arranged. Suppose they were arranged sequentially, as stages. Prolonging the first of two stages by Δt ms should delay completion of the second by the same amount: the prolongation Δt should be *propagated* to the completion time of the next stage. If we assume equal delays between completion of each process and its effect on $A_{mc}(t)$, then the order of process completions would be the same as the order of Events 1 and 2. The finding (Exp. 2; Figure 7.3B) that the effect of *MC* on T_1 is not propagated to T_2 would then be sufficient to invalidate a stage model. If we relax the equal delays assumption, permitting us to assume the opposite order of process completions, then the propagation property requires that any effects on T_2 propagate to T_1, contrary to what was found (Exp. 1; Figure 7.3A) for the effect of *GND*. The alternative to a *stages* arrangement is that processes α and β operate in parallel, such that the *RT* (on Go trials) is determined by the completion time of the slower of the two processes. Such an arrangement is consistent with the further finding (from Exp. 2) that the effect of *MC* on $\overline{RT}$ (16 ms) is dramatically smaller than its effect

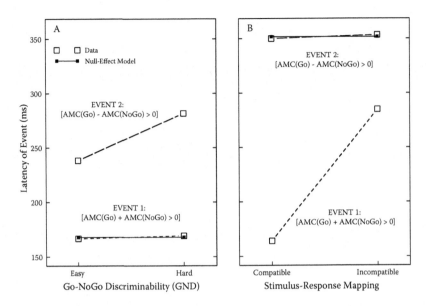

FIGURE 7.3

Mean effects of two factors on the latencies of Events 1 and 2 in Osman et al. (1992). Panel A: Effects of *Go-NoGo Discriminability, GND* (Exp. 1, $N = 6$). Its effect on T1 is 2.5 ± 5.0 ms; its effect on T2 is 43 ± 14 ms; the difference between these effects is 41 ± 11 ms $(p \approx 0.01)$. Panel B: Effects of *stimulus-response mapping, SRM* (Exp. 2, $N = 6$). Its effect on T_1 is 121 ± 17 ms, and T_2 is 3.3 ± 8.8 ms; the difference between these effects is 129 ± 27 ms $(p \approx 0.01)$. Also shown are null-effect models for T_1 in Panel A and T_2 in Panel B.

on T_1 (121 ms). This can happen because regardless of how much the duration of α is shortened by decreasing the level of MC, response initiation on a Go trial must await the appropriate completion of β as well.

These findings about *RT* contrast to those of Ex. 7.2.1, in which pure measures based on the LRP provide evidence for a serial arrangement of two neural processes. In that case, unlike this one (as we saw in Section 7.2.2), a composite behavioral measure *(RT)* leads to a similar analysis of corresponding mental processes. In the present example, however, T_1 is a measure of a process (response selection, separated from response execution) for which there may be no pure behavioral measure, and whose contribution to the composite *RT* measure may be large or small, depending on the level of *GND*.

7.3 Neural Processing Modules Inferred from Brain Activation Maps

7.3.1 Introduction

Modular neural processes can be discovered by applying process decomposition to the kinds of activation measures provided by PET and fMRI. Suppose there is localization of function, such that two such processes α and β are implemented by different processors P_α and P_β in nonoverlapping brain regions R_α and R_β. Then, activation levels in R_α and R_β are pure measures of α and β, and, with sufficiently precise data and factors that influence the processes selectively, separate modifiability is easy to demonstrate. However, if α and β are implemented by different neural processors, P_α and P_β (or by the *same* processor $P_{\alpha\beta}$) in *one* region, $R_{\alpha\beta}$, then the activation level in $R_{\alpha\beta}$ is a composite measure that depends on both α and β, and to test separate modifiability we must know or show how their contributions to the activation measure are combined.*

Both of the following examples involve tasks in which modular mental processes were expected, based on earlier evidence from RT measurements. Consider two such processes, **A** and **B**, influenced selectively by *F* and *G*. In such cases it is tempting to ask whether the operations carried out by these inferred mental processes are implemented by modular neural processes α and β in anatomically separate brain regions R_α and R_β. Because of their effectiveness in assessing the level of activity in localized brain regions, PET and fMRI are good techniques for this purpose. Then, because process α should be influenced selectively by *F*, the activation level of region R_α should vary with *F*, but not with *G*, and conversely for region R_β. Thus, the existence of regions whose fMRI signals are influenced selectively by *F* and *G* provides evidence for modular *neural* processes that correspond to the modular *mental* processes inferred from the behavioral data. If this were found, it would support the modular decomposition inferred from the behavioral data and would also support the conclusion that the processors that implement **A** and **B** are anatomically localized.

For both of the following examples, the modular decomposition into two processing stages inferred from RT data suggested a new experiment which would incorporate fMRI measurements to search for corresponding neural processes. And in both cases, concurrent RT data were taken, along with the fMRI data. Whereas the RT data in the first example confirmed earlier findings, results in the second example did not, probably because of paradigm differences, which raises interesting questions about the fMRI findings.

7.3.2 Modular Processes in Number Comparison

In an experiment by Pinel et al. (2001), subjects had to classify a sequence of visually displayed numbers, *k*, as being greater or less than 65. One factor was *notation* (*N*): the numbers *k* could be presented as Arabic numerals (e.g.,

---

* For example, if the combination rule is *summation* (sometimes assumed without justification) and if factors *F* and *G* influence α and β selectively, then the effects of *F* and *G* will be additive. Finding such additivity in a factorial experiment (as in Ex. 7.3.3) would support the combination rule in that brain region, as well as selective influence. If summation is assumed erroneously, selective influence would be obscured: the effect of each factor would appear to be modulated by the level of the other.

"68") or number names (e.g., "SOIXANTE-HUIT"). The other was *numeri-cal proximity (P)*, defined as the absolute difference $|k - 65|$, and grouped into three levels. The interesting phenomenon here is the "symbolic distance effect": the smaller the value of P (the closer the proximity), the slower the response. A similar experiment (Dehaene, 1996) had shown additive effects of N and P on $\overline{RT}$; this was interpreted to indicate two modular subpro-cesses arranged as stages: encoding (**E**), influenced by N, which determines the identity of the stimulus and is slower for number names than numbers, and comparison (**C**), influenced by P, which uses the stimulus identity in performing the comparison and is slower for closer proximities. In the new experiment, in which fMRI as well as RT measurements were taken, most of the 16 brain regions examined whose activation was influenced by N or by P were influenced significantly by only one of them, consistent with two separately modifiable neural processes ε and γ that are implemented by separately localized processors. When we average absolute effect sizes and SEs over the regions of each type, we find that for the nine N-sensitive regions, the N effect was $0.17 \pm 0.05\%$ (median p-value $= 0.01$), whereas the P effect was $0.06 \pm 0.08\%$; for the seven P-sensitive regions, the P effect was $0.32 \pm 0.10\%$ (median p-value $= 0.01$), whereas the N effect was $0.04 \pm 0.04\%$.

The fMRI data from three well-behaved regions are shown in Figure 7.4B, C, and D. The concurrently collected RT data (Figure 7.4A) replicated the ear-lier study, suggesting that we associate the neural modules ε and γ with the mental modules **E** and **C**; it is important that the mental and neural modules be selectively influenced by the same factors. However, although the direc-tion of the effect of P was the same in all the brain regions it influenced, the direction of the effect of N was not: the change from numeric to ver-bal notation (which increased RT) increased activation in some regions (e.g., Figure 7.4C) and decreased it in others (e.g., Figure 7.4D).*

7.3.3 Modular Processes for Stimulus Encoding and Response Selection

A common finding has emerged from several studies of choice-reaction time (one of them was discussed in Section 6.4.2 of Chapter 6), using various experimental arrangements and various realizations of the factors SQ (stim-ulus quality) and stimulus-response mapping difficulty (either MF, mapping

---

* Without requiring it, this finding invites us to consider that there are two qualitatively dif-ferent encoding processes ε, one for each notation, rather than "one" process whose settings depend on N. This possibility is now supported by the observation that "the notation fac-tor affects the circuit where information is processed, not just the intensity of the activity within a fixed circuit" (Dehaene, 2006, personal communication). If so, we have a case where a change in the level of a factor (here, N) induces a task change (one operation replaced by another; see Section 6.6 in Chapter 6), but evidence for modularity emerges nonetheless. Whereas the (multidimensional) activation data from such a simple (two-factor) experiment can support a claim of operations replacement, based on the idea that the processes imple-mented by different processors are probably different, if any (unidimensional) RT data could support such a claim, it would require a more complicated experiment, such as one that showed a suitable modulation, by N, of the effect of a third factor.

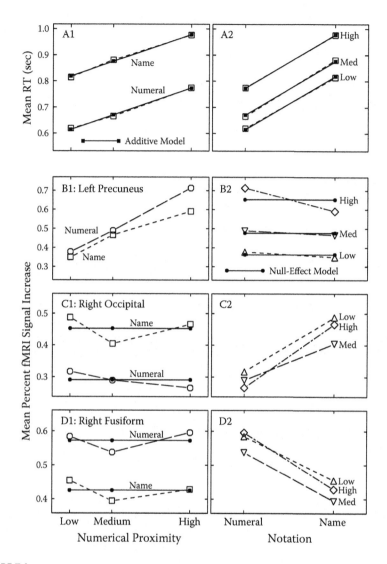

FIGURE 7.4

Reaction time and selected brain activation data from Pinel et al. (2001). The same data are plotted on the left as functions of *P* (proximity), with *N* (notation) the parameter, and on the right as functions of *N*, with *P* the parameter. Means over nine subjects of median *RT*s for correct responses are shown in Panel A, with a fitted additive model. The three levels of *P* have been scaled to linearize the main effect of *P* on *RT*; this effect, from low to high *P*, is 159 ± 24 ms, whereas the main effect of *N* is 204 ± 34 ms. SEs are based on variability over the nine subjects. The difference across levels of *N* between the simple effects of *P* from low to high (a measure of interaction) is a negligible 4 ± 20 ms. (The SE may be inflated by unanalyzed condition-order effects.) Mean activation measures from three sample brain regions, relative to an intertrial baseline, are shown in panels B, C, and D, accompanied by fitted null-effect models in panels B2, C1, and D1. Shown in panels B1, C1, and D1, the main effects of *P* (from low to high, using fitted linear functions) are 0. 29 ± 0. 09% (*p* ≈ 0. 01), −0. 03 ± 0. 03%, and 0. 00 ± 0. 04%. Shown in panels B2, C2, and D2, the main effects of *N* are −0. 06 ± 0. 06%, 0. 16 ± 0. 05% (*p* ≈ 0. 01), and −0.15 ± 0. 05% (*p* ≈ 0. 02).

familiarity, or *MC*, spatial mapping compatibility): these studies have shown that stimulus quality and mapping difficulty have additive effects on $\overline{RT}$, consistent with the idea that there are two processes, arranged in stages, that are selectively influenced by these factors. (These studies include Biederman and Kaplan, 1970, after a session of practice; Frowein and Sanders, 1978; Roberts and Sternberg, 1993, Exp. 2; Sanders, 1977, Exp. II; Sanders, 1980, Exp. 3; Sanders et al., 1982, Exp. I; and Schwartz et al., 1977, Exp. 2). The notion is that in initiating a response to a stimulus, the stimulus must first be identified (one stage, **S**) and then, given the identity, the response must be determined (a second stage, **R**).

Using a new choice-reaction task with their versions of *SQ* and *MC*, Schumacher and D'Esposito (2002) measured fMRI in several brain regions concurrently with RT. In their task, the stimulus was a row of four circular patches, one patch brighter than the others. The response was to press one of four keys, depending on which of the patches was the brighter one. The two factors, each at two levels, were the discriminability of the brighter patch from the others (*SQ*), and the spatial compatibility of the patch-to-key mapping (*MC*). Each subject was tested under all four combinations of factor levels, with measurements of fMRI signals in six different brain regions, and of *RT*. Unlike Ex. 7.3.2, in every region where a factor had an effect on the fMRI signal, the "more difficult" level of that factor—the level that produced the longer *RT*—also produced the larger fMRI signal. Figure 7.5 shows that in one of the regions, only *SQ* had a reliable effect (Panel A), in two of the regions only *MC* had reliable effects (panels B and C), in two regions both factors had reliable effects (panels D and E), and in one region neither factor had a reliable effect (Panel F). The selective effects found in three of the regions (where the fMRI signal was influenced by one of the factors but not the other) are consistent with **S** and **R** being implemented, at least in part, by anatomically distinct populations of neurons.*

If **S** and **R** are the only processes involved in the task, as suggested by the earlier RT experiments, then there should be no process influenced by both factors. But, in two brain regions (panels D and E), both factors were found to have effects. However, it is possible that these regions each contain two specialized populations of neurons, each of which is influenced selectively by a different one of the two factors. If so, the total amount of neural activity in each of these regions would be influenced additively by the two factors. Alternatively, if processes **S** and **R** are sequential (arranged in stages), as suggested by the RT data from the earlier experiments, the same neural processor in the same region could contribute to the implementation of both processes, but, again, it is plausible that the summed neural activity would

---

* It is noteworthy and requires explanation that in each of the five cases where an effect is not statistically significant, it is nonetheless in the same direction as in those cases where the effect is significant. Is this because the neural populations that implement the **S** and **R** processes are incompletely localized, or because the measured regions do not correspond to the populations, or for some other reason?

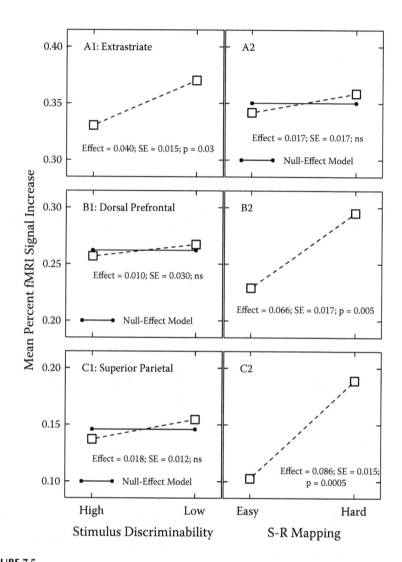

FIGURE 7.5

Brain activation data relative to a fixation baseline for the six regions measured by Schumacher and D'Esposito (2002). Mean main effects over nine subjects (eight for anterior cingulate) of stimulus discriminability and S-R mapping in each region, with null effect models shown for each nonsignificant effect.

Continued.

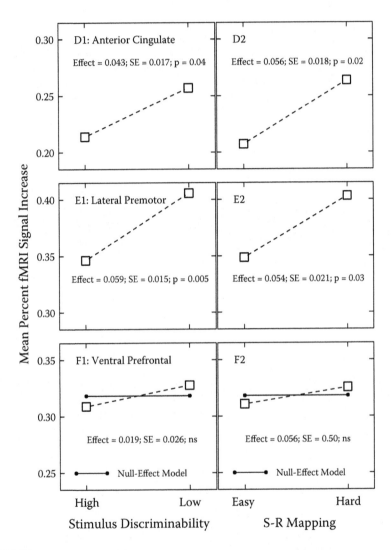

FIGURE 7.5
Continued.

be influenced additively by the two factors. As shown in panels A and B of Figure 7.6, the effects of SQ and MC on the fMRI signal were remarkably close to being perfectly additive. Additive effects on the amount of neural activity by itself does not imply additive effects on the fMRI signal, but the supplementary assumptions required for this implication are relatively weak.* Thus, the additivity of the effects of SQ and MC on the fMRI signal supports additivity of their effects on the amount of neural activity, which in turn supports the idea that separate processes within the regions shown in panels D and E contribute to the implementation of **S** and **R**.

In contrast to the earlier findings described above with various experimental arrangements, the effects of the two factors on $\overline{RT}$ in the Schumacher–D'Esposito task were unfortunately not additive; as shown in Figure 7.6C, there was a reliable interaction: the effect of raising the level of each factor was greater when the level of the other factor was higher (an "overadditive" interaction); such an interaction was found in the data for eight of the nine subjects. This finding seems inconsistent with the fMRI data, all of which support the idea that no neural process is influenced by both factors. One possibility is that there is such a process, but it happens not to be localized in any of the six regions that were examined, which suggests that stronger inferences require sampling of more brain regions.[†,‡]

---

* One condition under which additivity of amount of neural activity would be reflected by additivity of the fMRI (blood-oxygenation level dependent, BOLD) signal would be if the fMRI signal were linear in amount of neural activity. There is debate about whether such linearity obtains, and about whether the fMRI signal displays linear temporal summation (Heeger et al., 2000; Heeger and Ress, 2002). With respect to linearity of the relationship, however, it is important that differences in the fMRI signal strength between experimental and control conditions are very small fractions of that signal strength. For example, the mean value of the largest difference observed in the Schumacher–D'Esposito experiment is 0.43%. Even if the BOLD signal (B) grew nonlinearly with amount of neural activity, N (e.g., as a function of the form $B = 1 - e^{-N}$), the approximation to linearity for such small fractions would be good. Also, because of the possibility that **S** and **R** occur sequentially, the goodness of temporal summation may seem important. However, to a first approximation this does not matter: For example, suppose a failure of temporal summation in which the B/N ratio is smaller for the later process **R** than for the earlier process **S**: Additivity of the effects of SQ and MC on N would still be reflected by the additivity of their effects on B.

† Schumacher and D'Esposito (2002) suggest that such a process might occur only under the stress of a subject's being in the scanner, and not under normal conditions. On the other hand, a whole-brain analysis of the fMRI data did not reveal any additional task-sensitive regions (E. Schumacher, personal communication, November 27, 2006), and RT data from the practice session, outside the scanner, showed a nonsignificant interaction of about the same size and in the same direction.

‡ There is an unresolved puzzle about these data that suggests that it would be valuable to replicate this experiment, using a procedure that is known to produce additive effects on $\overline{RT}$. The large SEs associated with the very small mean interaction contrasts for the data shown in panels D and E reflect the fact that the variability of the interaction contrast over subjects is quite large—so large relative to the mean that the reported F-statistics in both cases were 0.00. Indeed, relative to the variability, the reported interaction contrasts were significantly ($p < 0.05$) too small.

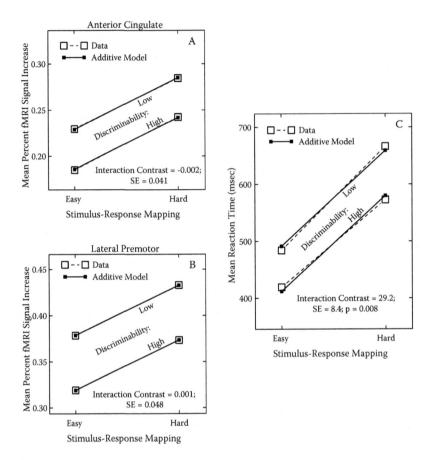

FIGURE 7.6
Panels A and B show mean simple effects of the two factors on the brain activation measure in the two regions where both main effects were significant in Schumacher and D'Esposito (2002). Also shown are fitted additive models and the interaction contrasts that measure the badness of fit of these models. Panel C shows simple effects of the two factors on *RT*, and the corresponding interaction contrast. (The mean effect of discriminability when the mapping was easy versus hard is 64 versus 94 ms, respectively, a difference that is 37% of the smaller main effect.)

7.4 Process Decomposition versus Task Comparison

7.4.1 Two Tactile Perception Tasks

Process decomposition was contrasted with Task Comparison in Section 6.6 of Chapter 6. Here, I present an elegant example of task comparison, but one that is subject to the usual limitations. It is a study by Merabet et al. (2004) of the effects of repetitive transcranial magnetic stimulation (rTMS) of different brain regions on subjective numerical scaling of two tactile perceptual dimensions. Both tasks involved palpation by the fingers of one hand of a set

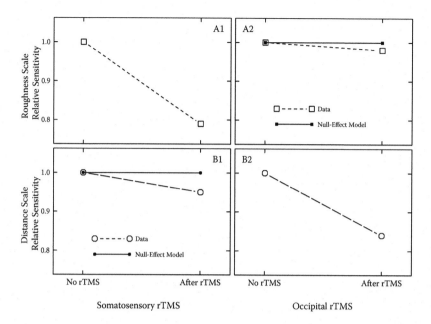

FIGURE 7.7

Selective effects on two subjective scaling tasks of repetitive transcranial magnetic stimulation of two brain regions. Mean sensitivity of scale values from 11 subjects relative to their non-rTMS scales are shown for the scaling of roughness (panels A1, A2) and distance (panels B1, B2), and for rTMS of somatosensory (rTMS$_s$; panels A1, B1) and occipital (rTMS$_o$; panels A2, B2) cortex. Also shown are null-effect models in panels A2 and B1. Effects on roughness scaling, measured by $1 - b$, are $1 - b_{rs} = 0.21 \pm 0.07$ (Panel A1; $p = 0.02$) and $1 - b_{ro} = 0.02 \pm 0.03$ (Panel A2). Effects on distance scaling are $1 - b_{do} = 0.16 \pm 0.07$ (Panel B2; $p = 0.04$) and $1 - b_{ds} = 0.05 \pm 0.04$ (Panel B1).

of tactile dot arrays with varying dot spacings. The judged dimensions were roughness (r) in one task and distance between dots (d) in the other. Where rTMS had an effect, it reduced the sensitivity of the obtained scale values to the differences among dot arrays. One measure of *relative sensitivity* is the slope, b, of the linear regression of post-rTMS scale values on non-rTMS scale values. If there were no effect, we would have $b = 1.0$; the effect of rTMS is measured by $1 - b$. The data (Figure 7.7) indicate that performance in the roughness-judgment task is influenced by rTMS of the contralateral somatosensory cortex rTMS$_s$ (Panel A1), but negligibly by rTMS of the contralateral occipital cortex rTMS$_o$ (Panel A2), whereas performance in the distance-judgment task is influenced by rTMS$_o$ (Panel B2), but negligibly by rTMS$_s$ (Panel B1), a double dissociation of the two tasks.*

---

* Subscripts d and r refer to the two tasks; subscripts s and o refer to the two stimulated brain regions. SEs are based on between-subject variability. Also supporting the claim of double dissociation, the differences, $b_{ro} - b_{rs}$ and $b_{ds} - b_{do}$ are significant, with $p = 0.01$ and $p = 0.04$, respectively. However, because non-rTMS measurements were made only before rTMS rather than being balanced over practice (a fault in the experimental design), straightforward interpretation of the slope values requires us to assume that effects of practice on those values were negligible.

Plausible theories might include, for each task, processes for control of stimulus palpation (α_d, α_r), for generation of a complex percept (β_d, β_r), for extraction of the desired dimensions (γ_d, γ_r), and for conversion of its value into a numerical response (δ_d, δ_r). Any or all of these processes might differ between tasks. The striking findings indicate that the members of one or more of these pairs of processes depend on different regions of the cortex. A weak pair of task theories might also assert that γ_d and γ_r depend on the occipital and somatosensory cortex, respectively, but if nothing is said about the other processes, this would be insufficient to predict the results. A stronger pair of task theories might add the assumptions that α_d and α_r are identical, that β_d and β_r are identical, and that δ_d and δ_r are identical. This pair of theories would predict the results, which would then also suggest that none of processes α, β, or δ is sensitive to either rTMS$_s$ or rTMS$_o$, indicating that they are implemented by processors in neither of the stimulated regions. Unfortunately, the findings do not bear on the validity of such hypothesized task theories, weak or strong, or even on the question whether the operations in either task can be decomposed into modular subprocesses such as α, β, γ, and δ, which exemplifies the limitations of the task-comparison method.

7.4.2   An Analog of Donders' Subtraction Method Applied to fMRI Data

Despite the problems mentioned in Section 6.6.3 of Chapter 6, an analog of Donders' method has often been used with brain activation measures (e.g., Petersen et al., 1988; Lie, et al., 2006). Suppose we are interested in studying a subprocess β of a complex neural process. If β were implemented by a localized processor P_β in brain region R, then the level of activation of R might be a pure measure of the subprocess. However, suppose instead that P_β is not localized (Haxby, 2004), and the activation measure $M_{\alpha\beta}$ is a composite measure that reflects contributions of u from subprocess α as well as v from β. (Evidence for contributions from two processes was found for two of the brain regions measured in Ex. 7.3.3.)

Let $M_{\alpha\beta,1}$ and $M_{\alpha\beta,2}$ be the values of the measure in Task 1 and Task 2, respectively, and let us posit the following joint hypothesis, which consists, first, of a pair of task theories that specify the constituent processes of each task and, second, a combination rule:

H1 (Task Theory 1): Task 1 is accomplished by process α.

H2 (Task Theory 2): Task 2 is accomplished by α and β, where the activation produced by α is identical to that in Task 1 (that is, addition of β satisfies a "pure insertion" assumption).

H3 (Combination Rule): Contributions u from α and v from β to $M_{\alpha\beta,2}$ combine by *summation*.

In this case, possible justifications of summation as the combination rule include an assumption that α and β are implemented by different populations of neurons in R that contribute independently to the brain activation measure, as well as linearity of the function relating the fMRI signal to the total amount of neural activity. (Observations of fMRI additivity, as in Ex. 7.3.3, support summation as the combination rule, but unfortunately, brain-imaging experiments have seldom employed the factorial design required for making such observations.) The preceding hypotheses imply that $M_{\alpha\beta,1}$ and $M_{\alpha\beta,2} - M_{\alpha\beta,1}$ are estimates of u and v, respectively, and can thus play the roles of pure measures of α and β. However, as in the case of RT, having these measures provides no test of the joint hypothesis.*

7.5 The Use of Transcranial Magnetic Stimulation (TMS) to Associate Mental Processes with Brain Regions

7.5.1 Introduction

Why does sleep deprivation increase reaction time (RT)? An answer to this question was provided by Ex. 6.4.2 in Chapter 6, using a task in which the additivity of the effects of two factors (SQ and MF) on $\overline{RT}$ had furnished evidence for two modular processes, arranged in stages. A factorial experiment that incorporated two levels of sleep state as well as variation in SQ and MF to determine which of these factor effects is modulated by sleep state led to the conclusion that sleep deprivation slows responses by interfering with the stimulus encoding process but not with response selection.

In recent years it has been discovered that a single pulse of TMS at certain times and in some brain regions can prolong the RT in some tasks without reducing accuracy very much (Walsh and Pascual-Leone, 2003). This opens the intriguing possibility of employing TMS within the method of additive factors, just as sleep deprivation was used by Sanders et al. (1982).† By using the TMS manipulation of region R in a factorial experiment while varying

---

* If summation proves to be incorrect as the combination rule, other analytic strategies may be available. For example, suppose measured activation were shown to be a decelerating function of the amount of brain activity. If it was a logarithmic function, we would have $M_{\alpha\beta} = log\,(u + v)$, and the subtraction method could be applied to the transformed activation measure $M' = exp\,(M_{\alpha\beta}) = u + v$. Alternatively, the function might be $M_{\alpha\beta} = 1 - exp[-(u + v)]$; here, it can be shown that $M_{\alpha\beta} = log(1 - M_{\alpha\beta}) = f\,(u) + f\,(v)$, and again, the subtraction method could be applied to the transformed activation measure. Note, however, that if one of these relations holds for individual trials, it does not necessarily follow that it holds for the corresponding means.

† An important advantage of TMS over measures of brain activation (Section 1.6) in determining which brain regions are involved in implementing a process is that whereas activation of a region in conjunction with process occurrence does not mean that such activation is *necessary* for that process, interference with a process by stimulation of a region does indicate the necessity of that region for the process to occur normally, just as does interference by a lesion in that region (Chatterjee, 2005).

the levels of other factors that are believed to influence different process-ing stages selectively—and determining which effects of the other factors on $\overline{RT}$ are modulated by TMS—it may be possible to learn whether region R is involved in the implementation of any of those processing stages. For exam-ple, suppose there is a task in which we find that factors F and G have addi-tive effects on $\overline{RT}$, from which we infer separate stages, **A** (influenced by F) and **B** (influenced by G). Now we add *TMS(R)* (TMS of region R) as a third factor, and ask whether it interacts with F or G. In the ideal result of such an experiment, the TMS factor (1) would have an effect on RT, (2) would interact with (modulate the effect of) one of the other factors, say, G, and (3) would not interact with (would have an effect that was additive with) the effect of the other factor, say, F. We would then have evidence that region R is involved in the implementation of **B**, but not of **A**. Such findings would strengthen the inferences made without TMS about the organization of the underlying processes, and would identify brain region R as being associated with one process (such as **B**) and not the others.* Unfortunately, although there are several interesting studies employing TMS in this new approach, there is as yet no fully satisfactory one; both of the following applications have missed the opportunity to make such observations, because of inadequate experi-mental design in Ex. 7.5.2, and incomplete analyses of data in Ex. 7.5.3.

7.5.2 Visual Search and TMS

In visual search for a conjunction of features, as in many search tasks, it is usually observed that the $\overline{RT}$ for both present and absent responses increases approximately linearly with the number of elements to be searched (the "dis-play size"), suggesting a process of serial comparison of the search target with the displayed elements. If a linear function is fitted to such data, the slope of the function is often interpreted to reflect the time per comparison, whereas the intercept reflects the summed durations of residual processes whose durations are not influenced by display size.† In the search for a single feature, unlike the search for a feature conjunction, RT may increase very little or not at all with display size, and the intercepts may also differ from those for conjunction search. Thus, feature and conjunction search appear to differ in residual processes as well as in the search process itself.

---

* Note, however, that if the effect of *TMS(R)* is time specific, as is likely with single-pulse TMS, it may not be straightforward to determine which other factors it interacts with. Thus, in the present example, suppose that it is region R that implements process **B**, and that **B** follows **A**. Because effects on the duration of **A** influence the starting time of **B** (and hence the time of TMS relative to **B**), a change in the level of F might modulate the effect of *TMS(R)* on **B**, and hence the effect on *RT*, leading to the erroneous conclusion that R is involved in the imple-mentation of **A**. This argues for the use of repetitive TMS (rTMS) before the task is performed in such studies, rather than a single TMS pulse during the task.

† Which intercept is appropriate depends on details of the search process, and may differ for target-absent and target-present trials, where, e.g., it might be the zero- and one-intercept, respectively.

Ashbridge, Walsh, and Cowey (1997) examined the effect of TMS of the parietal cortex during feature and conjunction search. In a preliminary experiment in which TMS was not used but display size was varied, along with search type (feature versus conjunction), their observations conformed to the foregoing description. However, in the experiment in which they applied TMS, they studied only one level of the display-size factor. They found TMS to interact with search type, having a substantially greater effect on conjunction than feature search. However, they missed an opportunity: If they had varied display size, thus obtaining a measure of its effect (the slope) with and without TMS, their findings could have told us whether TMS produced its effect by influencing the serial comparison process and/or residual processes. Without the display-size factor, we do not know which subprocess is responsible for the effect of parietal TMS on conjunction search.

There may be several differences between the complex processes that underlie conjunction search and feature search. So, although finding that TMS influences one type of search but not the other is probably telling us something important, it is not clear which difference between processes is responsible. In the language of this chapter, the search-type factor may determine which of two different tasks is being carried out, rather than varying something about the same task. And because theories for the two tasks are not detailed enough to specify exactly how the associated complex processes differ, interpretation of the differential effects of parietal TMS on performance of the two tasks (and the inference from this of the role of the parietal cortex in the serial-comparison process in conjunction search) requires speculation.

7.5.3 Number Comparison and rTMS

In Section 7.3.2 I discussed an experiment in which subjects had to classify visually displayed numbers as being greater or less than 65. RT measurements, supported by fMRI data, indicated that performance in this task depends on (at least) two processing stages, one to determine the identity of the comparison stimulus, and the other to perform the comparison. The processing stage of particular interest—number comparison—is that which produces the effect of proximity: responses are slower for stimulus numbers that are closer to the criterion. Goebel, Walsh, and Rushworth (2001) found that repetitive TMS (rTMS) applied to the left or right angular gyrus influenced the RTs in such an experiment. Do these effects mean that the angular gyrus is involved in implementing the number comparison process? Alternatively, it might be involved in another process that contributes to the RT. They varied two other factors in addition to the presence of rTMS (and the brain region to which it was applied): (a) the magnitude and (b) the sign of the difference between the stimulus number and the criterion. (In the experiment considered in Section 1.6.1 the data were collapsed over the sign of the difference, and the magnitude—the absolute value—of the difference was termed "proximity".) The shape of the function on each side of the criterion is sufficiently close to linear so that it would be reasonable to use the slope of a fitted linear function as the measure of the proximity effect. The measures needed to assess the proximity effect are the slopes of the functions for positive and negative differences ($k > 65$ and $k < 65$) that relate $\overline{RT}$ to the difference, $|k - 65|$. To determine whether TMS influences the comparison process, we need to know whether it modulates the proximity effect, i.e., whether it changes either or both of these slopes. Effects of rTMS on the *heights* of these functions might reflect influences of TMS on encoding or response processes, as well as on the comparison process.

In principle, the data to answer this question were collected in these experiments. (I say "in principle" because the data could be sufficiently imprecise to preclude drawing either of the two interesting conclusions: that the angular gyrus does or does not play a role in implementing the comparison process.) However, unfortunately, a focused test of the effect of TMS on the relevant slopes was not conducted. We thus have a case where the design of the experiment seems ideal for examining whether and how TMS modulates the proximity effect, which would indicate an effect on the number comparison process, but the appropriate analysis has not yet been reported.

7.6 Evidence from the Event-Related Potential (ERP) for Modular Processes in Semantic Classification: Brief Summary

At any particular time, the ERP at any point on the scalp is a composite measure of all the neural processors ("sources") in the brain that are active at that time. Furthermore, the physics of volume conduction tells us that the combination rule is summation. Hence, unlike most other cases, the combination rule is not a part of the hypothesis that must be tested. Consider two modular neural processes α and β implemented by processors P_α and P_β and factors F and G that influence them selectively. It follows that the effects of F and G on the ERP will be additive at all scalp locations. Furthermore, if P_α and P_β are at different locations in the brain, the topographies of the effects of F and G (the way their sizes vary with location on the scalp) will differ.

Kounios (2007) exploited this in a study of the effects of priming on the semantic classification of spoken nouns. The words consisted of *primes* and *probes*. The two factors were the *semantic relatedness* of the probe to the preceding prime, and the number of immediate repetitions of the prime before the probe, which determined the *semantic satiation* of the prime. The measures of interest were the ERPs elicited by the nontarget (nonresponse) probes at several locations on the scalp. The hypothesis was that relatedness and satiation influence two modular neural processes, α and β, selectively. For the epoch from 600 to 800 ms after the probe the predictions were confirmed: at each location the factor effects were additive, supporting the modularity of processes α and β. Furthermore, the topographies of the two effects differed, consistent with P_α and P_β being localized differently within the brain.

7.7 Comments and Issues

7.7.1 Introduction

This chapter is concerned with applications of the process decomposition method introduced in Chapter 6 to the problem of identifying neural modules, using brain measurements or manipulations. Examples that make use of fMRI, ERP, and LRP data are considered, and I describe the use of a brain manipulation (TMS) to relate neural processors to mental processes. Together with the applications discussed in Chapter 6, these examples raise a number of issues, some of which are mentioned in the following subsections.

7.7.2 Task-General Processing Modules

One plausible expectation is that different tasks are accomplished by different subsets of a small set of "basic" modular processes. To test this expectation we need a reasonable number of tasks for which persuasively successful decompositions have been achieved.* On the other hand, to get adequate data we require subjects to learn a task to a point of stable performance. With such intensive practice, it seems possible that the brain is sufficiently flexible that special-purpose routines would be developed that are specific to that task. Thus, an alternative plausible expectation is that at least some modular subprocesses are task-specific rather than task-general. In that sense, perhaps there is no "fundamental architecture of the mind," but rather a flexible architect who has some stylistic tendencies we can explore.

7.7.3 Quantitative versus Qualitative Task Changes

As shown by Ex. 7.3.2, the distinction between process decomposition (with its "quantitative" task changes) and task comparison (in which the task changes are "qualitative") can be subtle. In that example, the fMRI data suggest that the effect of the notation factor is probably better thought of as qualitative rather than quantitative—as replacing one encoding process by another, rather than influencing the settings or parameters of the "same" encoding process. Nonetheless, that example provides evidence for modular processes. In general, qualitative task changes should be avoided because

---

* For speeded tasks in which processes are arranged as stages, Sanders (1998, Chapter 3) has amassed some suggestive evidence for a small set of mental modules.

they reduce the likelihood of discovering modules. However, evidence is required to assert qualitative task invariance. One kind of evidence is the pattern of factor effects: for each factor, each change in level should influence the same operations and leave the other operations invariant. The usefulness of such evidence is one of several reasons for using factors with more than two levels. Unfortunately, few studies have done so.

7.7.4   Specialized Processors and Modular Processes

Does the existence of a localized neural processor that implements a particular process imply the modularity of that process? To address this question, consider one kind of evidence used to establish the existence of the processor: T_a and T_b are two classes of tasks such that brain region R_α is activated during T_a but not during T_b (or is activated *more* during one than the other), and such that we are willing to assume that all tasks T_a require a particular process α to be carried out, whereas none of tasks T_b do. Although it may seem plausible, such task specificity of R_α does not imply that the process α it implements in a given task is a modular subprocess in the sense of being modifiable separately from other subprocesses in that task. Suppose, for example, that α provides a motivational or attentional resource that is required by one or more other processes γ that differ across tasks T_a. A change in α would then induce a change in γ, so they would not be separately modifiable.

7.7.5   Relation between Mental and Neural Modules

Consider modular mental processes in a task, supported by behavioral evidence, and modular neural processes in that task, supported by brain measurements. Does either of these imply the other? On which psychophysical-physiological "linking propositions" (Teller, 1984) does the answer to this question depend? It would be helpful to have more studies (such as examples 7.2.1, 7.3.2, and 7.3.3) in which both brain and behavioral measures are taken, both directed at process decomposition. One starting point would be to take cases for which behavioral data already exist that persuasively favor a modular decomposition, as was done in examples 7.3.2 and 7.3.3, and ask whether there is a corresponding decomposition based on brain data into modular neural processes that are influenced by the same factors, and invariant with respect to the same other factors.

7.7.6   Separate Modifiability as a Criterion for Modularity

Is separate modifiability too strong or too weak to be a useful criterion for partitioning a process? What are the relative merits of alternative criteria for modularity, and alternative approaches to module identification? Is the weaker differential modifiability* more useful than separate modifiability?

7.7.7 Implications of Brain Metabolism Constraints

The metabolic requirements of brain activity are large relative to the available energy supply, with the implication that given the spike rates of active neurons, no more than about 1% of the neurons in the brain can be concurrently active (Lennie, 2003). This seems to be consistent with "sparse coding" and to argue against the idea of "massive parallelism." Other implications of these severe metabolic limitations for the plausibility and possibility of alternative processing architectures, and for the modularity of processors, have still to be worked out.

7.8 Further Reading

For more extensive discussion of the inferential logic associated with the process decomposition method, and detailed discussion of the six examples in Chapter 6, and examples 7.2.1 and 7.2.3 of the present chapter, see the text and appendices of Sternberg (2001), and references therein. For further discussion of examples 7.3.2 and 7.4.1, see Sternberg (2004). For the method of additive factors and numerous examples of its application, see Roberts and Sternberg (1993), Sanders (1998), Sternberg (1998a), Sternberg (2001, Sections 16, A.16.2-3 and references therein). For Hadley's defense of the existence and plausibility of mental modules against attacks by Fodor (2000), Kosslyn (2001), and Uttal (2001), see Hadley (2003). For other discussion of the properties that Fodor (1983) ascribed to modular processes, see Coltheart (1999) and Jacobs (1997). For discussion of double dissociation of tasks, as in the task-comparison method, a good place to start is with Dunn and Kirsner (2003).

Acknowledgments

For providing unpublished details of their data I thank Stanislas Dehaene, John Kounios, Lotfi Merabet, Alvaro Pascual-Leone, Philippe Pinel, Eric Schumacher, and Fren Smulders. For helpful comments on the manuscript I thank Stanislas Dehaene, Silke Goebel, Allen Osman, Eric Schumacher,

---

* If differential modifiability obtains, one can find factors F and G such that both factors influence both processes **A** and **B**, but for **A** (**B**) the effect of F (G) is the larger.

Fren Smulders, and Vincent Walsh. For numerous enlightening discussions I thank Seth Roberts.

References

Ashbridge, E., Walsh, V., and Cowey, A. 1997. Temporal aspects of visual search studied by transcranial magnetic stimulation. *Neuropsychologia, 35*, 1121–1131.

Biederman, I. and Kaplan, R. 1970. Stimulus discriminability and stimulus response Compatibility: Evidence for independent effects on choice reaction time. *J. Exp. Psychol., 86*, 434–439.

Chatterjee, A. 2005. A madness to the methods in cognitive neuroscience? *J. Cognit. Neurosci.* 27: 847–849.

Coltheart, M. 1999. Modularity and cognition. *Trends Cognit. Sci.* 3: 115–120.

Dehaene, S. 1996. The organization of brain activations in number comparison: Event-related potentials and the additive-factors method. *J. Cognit. Neurosci.,* 8: 47–68.

Dunn, J. C. and Kirsner, K., What can we infer from double dissociations? *Cortex, 39,* 1–7, 2003.

Fodor, J. 1983. *The Modularity of Mind: An Essay on Faculty Psychology.* Cambridge, MA: MIT Press.

Fodor, J. 2000. *The Mind Doesn't Work That Way: The Scope and Limits of Computational Psychology.* Cambridge, MA: MIT Press.

Frowein, H. W. and Sanders, A. F. 1978. Effects of visual stimulus degradation, S-R compatibility and fore period duration on choice reaction time and movement time. *Bull. Psychonomic Soc.* 12: 106–108.

Goebel, S., Walsh, V., and Rushworth, M. F. S. 2001. The mental number line and the human angular gyrus. *NeuroImage,* 14: 1278–1289.

Hadley, R. F. 2003. A defense of functional modularity. *Connect. Sci.,* 15: 95–116.

Haxby, J. V. 2004. Analysis of topographically organized patterns of response in fMRI data: Distributed representations of objects in ventral temporal cortex. In *Attention and Performance XX: Functional Neuroimaging of Visual Cognition,* Kanwisher, N. and Duncan, J. (Eds.), 83–97. Oxford: Oxford University Press.

Heeger, D. J. et al. 2000. Spikes versus BOLD: What does neuroimaging tell us about neuronal activity? *Nat. Neurosci.* 3: 631–633.

Heeger, D. J. and Ress, D. 2002. What does fMRI tell us about neuronal activity? *Nat. Rev. Neurosci.* 3: 142–151.

Jacobs, R. A. 1997. Nature, nurture, and the development of functional specializations: A computational approach. *Psychonomic Bull. and Rev.* 4: 299–309.

Kosslyn, S. M. 2001. The strategic eye: another look. *Minds and Machines,* 11: 287–291.

Kounios, J. 2007. Functional modularity of semantic memory revealed by event-related brain potentials. In *Neural Basis of Semantic Memory,* Hart, J., Jr. and Kraut, M. A. (Eds.), 65–104. Cambridge: Cambridge University Press.

Lennie, P. 2003. The cost of cortical computation. *Curr. Biol.,* 13: 493–497.

Lie, C.-H., Specht, K., Marshall, J. C., and Fink, G. R. 2006. Using fMRI to decompose the neural processes underlying the Wisconsin Card Sorting Test. *NeuroImage,* 30: 1038–1049.

Merabet, L., Thut, G., Murray, B., Andrews, J., Hsiao, S., and Pascual-Leone, A. 2004. Feeling by sight or seeing by touch? *Neuron*, 42: 173–179.

Osman, A., Bashore, T. R., Coles, M. G. H., Donchin, E., and Meyer, D. E. 1992. On the transmission of partial information: Inferences from movement-related brain potentials. *J. Exp. Psychol. Hum. Percept. Perform.* 18: 217–232.

Petersen, S. E., Fox, P. T., Posner, M. I., Minton, M., and Raichle, M. E. 1988. Positron emission tomographic studies of the cortical anatomy of single-word processing. *Nature*, 331: 585–589.

Pinel, P., Dehaene, S., Rivière, D., and LeBihan, D. 2001. Modulation of parietal activation by semantic distance in a number comparison task. *NeuroImage*, 14: 1013–1026.

Roberts, S. and Sternberg, S. 1993. The meaning of additive reaction-time effects: Tests of three alternatives. In *Attention and Performance XIV: Synergies in Experimental Psychology, Artificial Intelligence, and Cognitive Neuroscience*, Meyer, D. E. and Kornblum, S., (Eds.), 611–653. Cambridge, MA: MIT Press.

Rumelhart, D. E., McClelland, J. L., and PDP Research Group, *Parallel Distributed Processing: Explorations in the Microstructure of Cognition, Volume 1: Foundations*, MIT Press, Cambridge, MA, 1986.

Sanders, A. F. 1977. Structural and functional aspects of the reaction process. In *Attention and Performance VI*, Dornic, S. (Ed.), 3–25. Mahwah, NJ: Erlbaum.

Sanders, A. F. 1980. Some effects of instructed muscle tension on choice reaction time and movement time. In *Attention and Performance VIII*, Nickerson, R. S. (Ed.), 59–74. Mahwah, NJ: Erlbaum.

Sanders, A. F., Wijnen, J. L. C., and Van Arkel, A. E. 1982. An additive factor analysis of the effects of sleep loss on reaction processes. *Acta Psychol.* 51: 41–59.

Sanders, A. F.1998. *Elements of Human Performance: Reaction Processes and Attention in Human Skill*. Mahwah, NJ: Erlbaum.

Schumacher, E. H. and D'Esposito, M. 2002. Neural implementation of response selection in humans as revealed by localized effects of stimulus-response compatibility on brain activation. *Hum. Brain Mapp.* 17: 193–201.

Schwartz, S. P., Pomeraqntz, J. R., and Egeth, H. E. 1977. State and process limitations in information processing: An additive factors analysis. *J. Exp. Psychol.: Human Perception and Performance*, 3, 402–410.

Shallice, T. 1988. *From Neuropsychology to Mental Structure*. Cambridge, MA: Cambridge University Press.

Simon, H. A. 1962. The architecture of complexity. *Proc. Am. Phil. Soc.* 106: 467–482.

Simon, H. A. 2005. The structure of complexity in an evolving world: The role of near decomposability. In *Modularity: Understanding the Development and Evolution of Natural Complex Systems*, Callebaut, W. and Rasskin-Gutman, D. (Eds.), ix–xiii. Cambridge, MA: MIT Press.

Smulders, F. T. Y., Kok, A., Kenemans, J. L., and Bashore, T. R. 1995. The temporal selectivity of additive factor effects on the reaction process revealed in ERP component latencies. *Acta Psychol.* 90: 97–109.

Sternberg, S. 1998. Discovering mental processing stages: The method of additive factors. In *An Invitation to Cognitive Science*, Vol. 4: *Methods, Models, and Conceptual Issues*, Scarborough, D. and Sternberg, S. (Eds.), 703–863. Cambridge, MA: MIT Press.

Sternberg, S. 2001. Separate modifiability, mental modules, and the use of pure and composite measures to reveal them. *Acta Psychol.* 106: 147–246.

Sternberg, S. 2004. Separate modifiability and the search for processing modules. In *Attention and Performance XX: Functional Neuroimaging of Visual Cognition*, Kanwisher, N. and Duncan, J. (Eds.), 125–139. Oxford: Oxford University Press.

Teller, D. 1984. Linking propositions. *Vis. Res.* 24: 1233–1246.

Uttal, W. R. 2001. *The New Phrenology: The Limits of Localizing Cognitive Processes in the Brain.* Cambridge, MA: MIT Press.

Walsh, V. and Pascual-Leone, A. 2003. *Transcranial Magnetic Stimulation: A Neurochronometrics of Mind.* Cambridge, MA: MIT Press.

Section III

Psychophysiology of Work

8

The New Interface of Brain, Mind, and Machine: Will the Emergent Whole Be Greater than the Sum of the Parts?

Chris Berka, Daniel J. Levendowski, Gene Davis, Vladimir T. Zivkovic, Milenko M. Cvetinovic, and Richard E. Olmstead

CONTENTS

8.1 The Intersection of Brain, Mind, and Machine 167
8.2 The Venerable History of the EEG... 169
8.3 Detection of Drowsiness 169
8.4 The Differing Susceptibility of Individuals to the effects of Sleep
Deprivation 174
8.5 EEG-Based Feedback Alarms and Driving Performance 175
8.6 Neuroassays for Alertness and Memory.. 176
8.7 Neuropharmacoassays: Quantification of Drug Use and
Withdrawal 178
8.8 The New Science of Neuroergonomics: Alternative Approaches
to Evaluating New Products and Guiding Future Designs................. 179
8.9 Augmenting Human Capacity: Seamless Integration of Brain,
Mind, and Machine 179
8.10 The Future of Education: Integrating Brain Monitoring into
Training Systems.. 180
8.11 NeuroTeam: Exploring Group Dynamics 182
References 184

8.1 The Intersection of Brain, Mind, and Machine

The investigation and understanding of the human mind are being transformed by the convergence of ubiquitous computing, consumer electronics, and advances in neurotechnology. For better or worse, humans and machines are inextricably linked at work, at school, at play, in their automobiles, and on the battlefield as they connect to an increasingly complex web of local and globally networked communities. Contemporary humans are aided by laptops, hand-helds, and global satellite positioning and enchanted by engaging video games. Neural prosthetics such as cochlear implants and neurostimulation devices for chronic pain and depression are now common.

The promise of expanding mental capacity and enhancing performance is driving cross-disciplinary teams of engineers, neuroscientists, cognitive psychologists, and biophysicists to participate in this inevitable merging of man and machine. The decade of the brain contributed an explosion of technology for actively monitoring brain activity and relating it to cognition and behavior, including functional magnetic resonance imaging (fMRI), positron emission (PET) scans, magnetoencephalography (MEG), and dense array electroencephalography (EEG). These windows on the mind offer the possibility of transforming the ways humans interface with technology and with the world. One approach to expanding the capacity of human information processing is to radically rethink the design of human–machine system interfaces to optimize the flow and exchange of data between humans and machines. Given the pace and sophistication of evolving technology, it is surprising that the primary human–computer interfaces continue to be the arcane keyboard or the laborious and still imperfect speech recognition systems. This may change in the near future because the feasibility of a variety of direct brain-to-computer interfaces has recently been established (Sellers and Donchin, 2006; Wolpaw et al., 2003; Pfurtscheller et al., 2004; Allison and Pineda, 2006).

This chapter reviews recent progress in developing systems for monitoring and analyzing the human electroencephalogram (EEG) and the prospects for current and future technologies integrating EEG into the human–computer interface. It is now possible to routinely apply brain monitoring in fields outside the clinical laboratory, including education and training, human factors evaluations, military operations, and market research. To enhance cognitive capabilities and extend the limits of the human information processing system, a novel approach to the design and implementation of human–machine interfaces is required. A new field of investigation termed *neuroergonomics* has taken on this challenge with interdisciplinary research and practice that integrates understanding of the neural bases of cognition and behavior with the design, development, and implementation of technology (Marek and Pokorski, 2004; Parasuraman, 2005; Kramer and Parasuraman, 2005). The vision of neuroergonomics is to use knowledge of brain–behavior relationships to optimize the design of safer, more efficient work and home environments that increase motivation and productivity.

Another evolving new field, that of social neuroscience, further reflects the new frontier of interdisciplinary explorations on understanding the brain–mind interactions of individuals and groups within the context of social networks (Cacioppo et al., 2000). Social neuroscience brings together anthropologists, neuroscientists, social and cognitive psychologists, and philosophers with the goal of understanding how the mind works. In addition to exploring perception, attention, and memory, their comprehensive approach is designed to understand the mind–brain connections involved in far more complex human behaviors such as attraction, altruism, attachment, attitudes, cooperation, competition, empathy, persuasion, obedience, and morality. Social neuroscience also seeks to investigate the effects of the barrage of multisensory stimulation on the developing brain. It is clear that with each new generation, the pace of technology adaptation and integration is accelerating. This acceleration in sensory input, in combination with the fact that social networks are no longer bounded by geography, may be the most significant influences on human mental evolution in the new millennium.

Expansion of human capacity with technology is not without risk. The average human information processing system is already in a constant state of information overload. The human processor can also be seriously compromised by fatigue, stress, boredom, illness, and other factors. One of the benefits of actively monitoring brain activity is the potential for identifying these states and ensuring that the delivery of information is appropriately matched to the preparedness of the receiver. Brain monitoring also provides an assessment of the impact of new information or the style of presenting information to determine whether the user is engaged, bored, or overloaded.

8.2 The Venerable History of the EEG

Investigations of human mental activity have employed EEG recordings for nearly a century. The EEG has been regarded as a window on the mind since the first recordings were made by Hans Berger in 1929 (Berger, 1929). It is no surprise that the EEG has intrigued many investigators, particularly those interested in an objective method for assessing mental states. An untrained eye can readily observe the transitions in the EEG that occur as a result of closing the eyes or the distinctive EEG signals associated with vigilance, relaxation, and falling asleep.

Today, EEG is routinely used for overnight sleep studies in the laboratory and in neurology to characterize epilepsy and neurological disorders, but the great leaps in EEG research can be largely attributed to the wealth of information generated by psychologists and neuroscientists using EEG to investigate brain, mind, and behavior (see Fabiani, Gratton, and Coles, 2000 for a review). Although the relationships between specific mental states and EEG are just beginning to be understood, the foundation of work in detecting global state changes is sufficient to begin developing practical applications.

8.3 Detection of Drowsiness

There are many applications for a drowsiness detection system. An estimated 20% of the general population suffers from sleep disorders, and 5% of the population experiences chronic daytime drowsiness that results in deleterious effects on normal daily activities (Bixler et al., 1979; Lavie, 1981; Coleman et al., 1982; Dement and Vaughan, 1999). The public health consequences and economic impact of impaired alertness, including increased morbidity and risk for accidents have been well documented. Impaired vigilance is now believed to be a primary contributor to transportation and industrial accidents (Berka et al., 2005; Brookhuis and de Waard, 1993; Brookings, Wilson, and Swain, 1996; Gevins and Smith, 2005; Gevins et al., 1997, 1998; Kramer and Parasuraman, 2005; Kramer, Trejo, and Humphrey, 1996; Levendowski et al., 1999, 2001, 2002; Makeig, 1993; Makeig and Jung, 1995; Marek and Pokorski, 2004). The effects of even small amounts of sleep loss each night accumulate over time, resulting in a "sleep debt." As sleep debt increases, alertness, memory, and decision-making are increasingly impaired (Levendowski et al., 1999, 2001, 2002; Makeig, 1993; Makeig and Jung, 1995). Individuals have been shown to become accustomed to this chronic accumulation of fatigue and are often unaware of the impact on their performance. A study conducted by the Automobile Association of America found that 50% of people tested following sleep deprivation were unable to predict whether they would fall asleep within the next two minutes (Itoi et al., 1993).

As automation replaces manual labor and the role of the operator is simply to monitor, maintaining vigilance becomes more difficult with performance decrements increasing with time-on-task (Parasuraman, Molloy, and Singh, 1993; Singh, Molloy, and Parasuraman, 1993). The integration of EEG monitoring for real-time assessment of operator status offers the possibility of allocating tasks between machines and humans based on the operator status. Intelligent feedback or "closed-loop" systems can facilitate active intervention by the operator or through a third party (man or machine), increasing safety and productivity (Parasuraman et al., 1992; Parasuraman, Molloy, and Singh, 1993; Parasuraman, Mouloua, and Molloy, 1996; Singh, Molloy, and Parasuraman, 1993).

The EEG is the physiological "gold standard" for the assessment of alertness. Subtle shifts from vigilance to drowsiness can be identified by quantifying changes in EEG waveforms (Akerstedt and Kecklund, 1991; Akerstedt, Kecklund, and Knutsson, 1991; Akerstedt et al., 1993; O'Hanlon and Beatty, 1977). The measurement of EEG indices of alertness-drowsiness in the laboratory setting has resulted in highly sensitive and reliable correlations with performance, including the ability to predict performance on a second-by-second basis (Makeig, 1993; Makeig and Jung, 1995, 1996). Torsvall and Akerstedt (Torsvall and Akerstedt, 1987) identified EEG patterns highly predictive of sleep onset. In their research, eye closures or slow eye movements occurred too late in the behavioral chain of events to be useful in providing an early warning drowsiness detection system.

Recording and analysis of EEG has traditionally been confined to laboratory settings because of the technical obstacles of recording high-quality data and the computational demands of real-time analysis. Advances in electronics and data processing set the stage for ambulatory EEG applications. A

FIGURE 8.1
EEG headset: front, side, and back views.

recently developed wireless EEG sensor headset facilitates easy acquisition of high-quality EEG combining battery-powered hardware with a sensor placement system to provide a lightweight, easy-to-apply method to acquire and analyze six channels of high-quality EEG (Figure 8.1).

The EEG sensor headset requires no scalp preparation and provides a comfortable and secure sensor–scalp interface for 12 to 24 h of continuous use. The headset was designed with fixed sensor locations for three sizes (e.g., small, medium, and large). Standardized sensor placements include locations over frontal, central, parietal, and occipital regions. Amplification, digitization, and radio frequency (RF) transmission of the signals are accomplished with miniaturized electronics in a portable unit worn on the head. The combination of amplification and digitization of the EEG close to the sensors and wireless transmission of the data facilitates the acquisition of high-quality signals even in high-electromagnetic-interference environments. Data are sampled at 256 samples/second with a bandpass from 0.5 to 65 Hz (at 3 dB attenuation) obtained digitally with Sigma-Delta A/D converters.

The system provides impedance monitoring of the sensors to identify poor sensor connections and to inform the user if sensors need to be replaced. The battery power of the headset is also monitored so that an alarm can be delivered if the power is low.

Quantification of the EEG in real time, referred to as the B-Alert® system, is achieved using signal analysis techniques to identify and decontaminate fast and slow eye blinks, and identify and reject data points contaminated with excessive muscle activity, amplifier saturation, and excursions due to movement artifacts. Decontaminated EEG is then segmented into overlapping 256 data-point windows called *overlays*. An epoch consists of three consecutive overlays. Fast-Fourier transform is applied to each overlay of the decontaminated EEG signal multiplied by the Kaiser window ($\alpha = 6.0$) to compute the power spectral densities (PSDs). The PSD values are adjusted to take into account zero values inserted for artifact-contaminated data points.

Wavelet analyses are applied to detect excessive muscle activity (EMG) and to identify and decontaminate eye blinks. Once the artifacts are identified in the time-domain data, the EEG signal is decomposed using a wavelets transformation. The wavelets eye blink identification routine uses a two-step discriminant function analysis. The DFA classifies each data point as a control, eye blink, or theta activity. Multiple data points that are classified as eye blinks are then linked, and the eye blink detection region is established. Decontamination of eye blinks is accomplished by computing mean wavelet coefficients for the 0–2, 2–4, and 4–8 Hz bins from nearby noncontaminated regions and replacing the contaminated data points. The EEG signal is then reconstructed from the wavelets bins ranging from 0.5 to 64 Hz. Zero values are inserted into the reconstructed EEG signal at zero crossing before and after spikes, excursions, and saturations. EEG absolute and relative power spectral density (PSD) variables for each 1-s epoch using a 50% overlapping window are then computed.

A single 30-min baseline EEG test session is required for each participant to adjust the software to accommodate individual differences in the EEG. The output of the B-Alert software includes EEG metrics (values ranging from 0.1 to 1.0) for alertness/drowsiness calculated for each 1-s epoch of EEG, using quadratic and linear discriminant function analyses of model-selected EEG variables derived from power spectral analysis of the 1-Hz bins from 1 to 40 Hz. These metrics have proven utility in tracking both phasic and tonic changes in cognitive states, in predicting errors that result from either fatigue or overload, and in identifying the transition from novice to expert during skill acquisition (Berka et al., 2004; Berka, Levendowski, Westbrook et al., 2005).

Independent validation of the automated analysis was conducted with EEG and neurocognitive measures simultaneously acquired to quantify alertness from 24 participants during 44 h of sleep deprivation. Performance on a three-choice vigilance task (3C-VT), paired-associate learning/memory task (PAL), and modified Maintenance of Wakefulness Test (MWT), and sleep-technician-observed drowsiness (eye-closures, head-nods, EEG slowing) was quantified. B-Alert classifications were significantly correlated with technician-observations, visually scored EEG, and performance measures (Figure 8.2). B-Alert classifications during 3C-VT, and technician observations and performance during the 3C-VT and PAL evidenced progressively increasing drowsiness as a result of sleep deprivation, with a stabilizing effect observed at the batteries occurring between 0600 and 1100 that suggests a possible circadian effect similar to those reported in previous sleep deprivation studies (Mitler et al., 2002; Berka, Levendowski, Westbrook et al., 2005). Participants were given an opportunity to take a 40-min nap approximately 24 h into the sleep deprivation portion of the study (i.e., 7 PM on Saturday). The nap was followed by a transient period of increased alertness. Approximately 8 h after the nap, behavioral and physiological measures of drowsiness returned to levels prior to the nap.

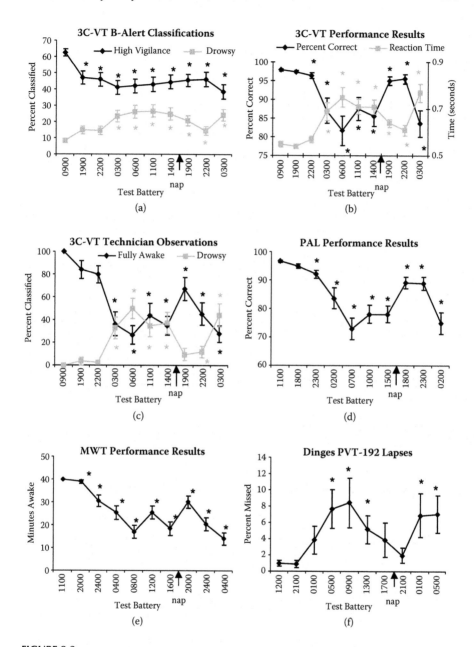

FIGURE 8.2
Repeated measures across 10 test batteries (mean + SE): (a) B-Alert EEG % high vigilance and % drowsy classifications, (b) % correct and reaction time during 3C-VT, (c) technician observations as fully awake or drowsy, (d) % correct during PAL, (e) minutes awake during MWT, (f) % missed during Dinges PVT-192.

8.4   The Differing Susceptibility of Individuals to the effects of Sleep Deprivation

Until recently, sleep deprivation was believed to exert effects on human performance in a dose-dependent manner with each hour of sleep debt resulting in an equivalent and predictable amount of performance deficit (Belenky, 2001; Rogers, 2002). A growing number of studies have now shown that individuals differ in their vulnerability to the effects of sleep deprivation (Drummond, 2003; Leproult et al., 2002; Baynard, 2003; Belenky, 2003; Balkin, 2001; Doran, Van Dongen, and Dinges, 2001; Mitler et al., 2002; Habeck et al., 2004). A series of sleep deprivation studies applying the simultaneous analysis of the B-Alert indices and performance measures suggested the possible existence of three groups of subjects in the population: those extremely vulnerable to the effects of sleep deprivation, a moderately vulnerable group, and a group relatively invulnerable to sleep deprivation. Other investigators have confirmed the existence of three groups (Doran, Van Dongen, and Dinges, 2001; Habeck et al., 2004); however, this stratification is based on small sample sizes, and the possibility exists that the level of vulnerability is normally distributed across the population. Whether these differences are stable over time ("traits") and are associated with demographic (e.g., age, sex, race) or other (e.g., education, I.Q., socioeconomic status) characteristics remains to be determined. A genetic basis for the susceptibility to sleep deprivation may ultimately be identified.

To assess the individual differences in the 44-h sleep deprivation study, cluster analysis was used to stratify individuals into three groups based on their level of impairment as a result of sleep deprivation. The combination of the B-Alert EEG indices and neurobehavioral measures identified individuals whose performance was most susceptible to sleep deprivation

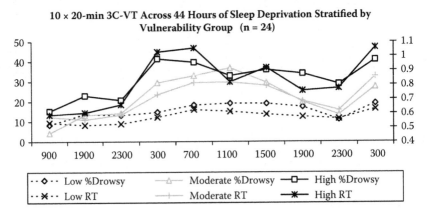

FIGURE 8.3
Mean 3C-VT RT and % drowsy for three groups stratified based on vulnerability to sleep deprivation.

(Figure 8.3). These objective measures could be applied in an operational setting to provide a "biobehavioral assay" to determine vulnerability to sleep deprivation.

8.5 EEG-Based Feedback Alarms and Driving Performance

An additional study was conducted to explore the feasibility of an integrated approach that combined real-time quantification of EEG indices and audio feedback alarms to assist 14 healthy participants in overcoming performance deficits on neurocognitive tests and in a driving simulator task during a sleep deprivation session. As expected, sleep deprivation significantly increased drowsiness as measured by B-Alert EEG classifications and impairments in neurocognitive tests and driving simulator performance (Figure 8.4). Timely administration of feedback resulted in increased alertness as measured by changes in EEG indices and performance, particularly during the driving simulator task. Most participants reported that the feedback alarms were beneficial in helping them maintain alertness. This suggests that a closed-loop EEG-based system combined with intelligent feedback can improve performance and decrease operator errors resulting from fatigue.

The effectiveness of the feedback alarms was somewhat overshadowed by individual differences in susceptibility to the effects of sleep deprivation. A subgroup of participants never achieved sufficient levels of drowsiness during the sleep deprivation session to evidence changes in the EEG or performance or to fully realize the beneficial effects of the feedback alarms. On the opposite end of the continuum, some participants showed signs of extreme drowsiness very early in the experimental session. These data suggest that mathematical models of fatigue will be effective only if they are able to account for individual differences.

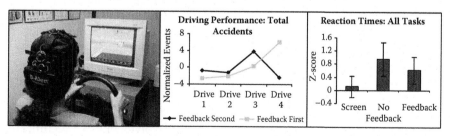

FIGURE 8.4
EEG indices of drowsiness triggered audio feedback alarms to sleep-deprived subjects and resulted in improved reaction time and performance in a driving simulation study.

8.6   Neuroassays for Alertness and Memory

Alterations in neurocognitive function play a major role in numerous sleep, neurological, and psychiatric disorders and can significantly impact quality of life, treatment efficacy, rehabilitation, and ability to function at work and carry out normal daily activities. The methods currently available for neurocognitive investigations including traditional neuropsychological assessment (e.g., Halstead-Reitan, Wechsler), neurophysiological measures (e.g., laboratory electroencephalogram (EEG), Multiple Sleep Latency Test), and behavioral tests (e.g., Continuous Performance Test, Psychomotor Vigilance Test) are time consuming, expensive, and require substantial technical expertise to administer and interpret.

An Alertness and Memory Profiling system (AMP) was designed to simultaneously assess EEG and neurobehavioral function to assess patients with sleep and neurological disorders. The AMP offers a multivariate approach, integrating neurophysiological, neuropsychological, behavioral, and subjective measures of alertness, attention, and memory into an easy-to-administer protocol.

AMP quantifies EEG and performance during the performance of the following tests:

- 20-min 3-Choice Vigilance Test
- 20-min Perceptual Acuity Test
- Image Learning and Recognition
- Image Learning with Distractors
- Paired Associate Learning (PAL)
  - Verbal
  - Image-number
- Sternberg Verbal Memory Scan
- Epworth and Visual Analog Scales for self-reported sleepiness

Although these evaluations have traditionally been conducted separately, acquisition and analysis of data can require days or weeks with conventional laboratory methods. The goal was to design a multimodal system to provide quantitative assessments of neurocognitive functions including alertness, attention, learning, and memory for accurate diagnosis and treatment outcome assessment of sleep and neurological disorders. The system also provides epidemiologists with a simple, inexpensive method for acquiring multiple parameters from large populations. The AMP (Figure 8.5) can be easily administered in a clinic or in the workplace using portable hardware and software technologies that minimize the involvement of a technician by providing self-paced instructions, monitoring patient responses during

FIGURE 8.5
AMP Neurocognitive Test Battery with photo of participant completing an AMP session.

train-to-criterion sessions, and conducting routine monitoring of EEG signal quality.

Studies are planned that will use the AMP to investigate depression, Alzheimer's, drug addiction, and other sleep and neurological disorders. Applications include diagnostic and treatment outcome evaluations, pharmaceutical investigations, ergonomics and sports medicine, military/forensic research, and educational/occupational studies.

Initial investigations suggested that AMP could be useful in characterizing the changes in alertness, learning, and memory associated with both partial (one-half normal sleep duration) and full (up to 44 total hours) sleep deprivation in healthy subjects (Mitler et al., 2002; Levendowski et al. 1999; Levendowski et al. 2000, 2001, 2002; Berka et al., 2004). AMP data have been acquired from a group of more than 200 healthy subjects (age range: 21–70), during fully rested and sleep-deprived conditions, providing the foundation for a normative database. The AMP was proven useful in characterizing and quantifying the neurocognitive effects of treating obstructive sleep apnea (OSA) patients with continuous positive airway pressure (CPAP), providing quantitative evidence of amelioration of some cognitive impairments following treatment with CPAP (Westbrook et al., 2004).

Four neurocognitive factors were derived from the AMP data: sustained attention, processing speed, verbal memory, and visuospatial memory. After 3 months of treatment with CPAP, the majority of OSA patients evidenced levels of sustained attention and processing speed within the normal range, but continued to show verbal memory impairments in comparison to the healthy subjects. The levels of hypoxemia observed in the overnight sleep studies of the OSA patients were predictive of the level of impairment in verbal memory and visuospatial memory, but not predictive of sustained attention or processing speed (Berka et al., 2007; Westbrook et al., 2004).

8.7 Neuropharmacoassays: Quantification of Drug Use and Withdrawal

Despite the widely publicized health risks of smoking, an estimated 25% of the U.S. population continues to smoke. Nicotine is a powerful stimulant that affects mood and cognition through the brain's dopamine system in a manner similar to that of other abused drugs such as cocaine. Nicotine is frequently used to counteract fatigue by people suffering from sleep deprivation as a result of lifestyle or undiagnosed sleep disorders. Although nicotine can temporarily sustain wakefulness and can improve performance in attention and memory tasks, abrupt withdrawal of the drug can cause deleterious effects on performance that can impact productivity and safety in military and industrial settings.

A recent study evaluated the effects of nicotine administration and withdrawal on learning and memory performance, including monitoring EEG indices of levels of task engagement, distraction, drowsiness, and mental workload. Data were acquired unobtrusively and quantified during performance of cognitive tests. EEG was acquired with the wireless sensor headset from 20 participants during three sessions (baseline-smoking permitted, 14 mg transdermal nicotine administration, and nicotine withdrawal-placebo patch) (Berka et al., 2006). Participants completed Verbal Memory Scan (VMS), Image Learning and Recognition (ILR), and ILR with Interference (ILR-I) tests at each session. VMS, a Sternberg serial-probe recognition memory task, measures speed and accuracy of verbal working memory, presenting lists of five words (memory sets) followed by single-word probes. The ILR requires encoding of 20 images from a specific category (e.g., animals, food). The 20 encoding images are presented twice and then randomly interspersed with 80 additional images during recognition, when subjects indicate whether or not the image was in the encoding set. The ILR is followed by an Interference (ILR-I) test where a set of 20 new images now becomes the encoding set and must be distinguished from the initial ILR set of encoding images during the recognition period.

EEG metrics for engagement and cognitive workload were calculated for each 1 s of EEG using quadratic and linear discriminant function analyses of model-selected variables derived from EEG power spectra (1-Hz bins, 1–40 Hz). Across tasks and subjects, nicotine significantly increased EEG-engagement and EEG-workload levels and improved performance in VMS and ILR-I as measured by accuracy and reaction time (RT). Nicotine withdrawal resulted in abnormally high levels of drowsiness and distraction coupled with slow RT and decreased accuracy. When compared to VMS and ILR-I data acquired from nonsmokers, EEG engagement and EEG workload during nicotine administration appeared inappropriately elevated relative to task demands. These data suggest that nicotine may induce inappropriate allocation of cognitive resources and that withdrawal reflects diminished resources with concomitant deleterious effects on performance.

Neuropharmacoassays have numerous potential applications: identifying underlying comorbidities (e.g., sleep disorders) that may contribute to drug use, characterizing neurocognitive changes following drug ingestion, making predictions regarding an individual's likelihood of success in rehabilitation, monitoring changes during rehabilitation, and identifying potential biomarkers for predisposition to drug use that may reflect genetic influences.

8.8 The New Science of Neuroergonomics: Alternative Approaches to Evaluating New Products and Guiding Future Designs

Potentially rich sources of data are underutilized because they cannot be sorted rapidly and organized efficiently enough to accommodate the capacity of the human information processing system. Interfaces need to be transformed such that information is presented when needed, in a modality that is readily perceivable, and in a form that is readily interpretable. The vision of neuroergonomics is to radically rethink the design of human–machine system interfaces to optimize the flow and exchange of data between humans and machines. This interdisciplinary area of research and practice integrates knowledge of brain–behavior relationships to optimize the design of work, home, and recreational environments. A complementary result of this endeavor is to better inform neuroscience regarding real-world human performance.

One promising avenue of research in neuroergonomics involves integrating the capability to continuously monitor operators' EEG to quantify levels of fatigue, attention, task engagement, and mental workload in real-world usability studies (Kramer, Trejo, and Humphrey, 1996; Gevins and Smith, 2005; Smith and Gevins, 2005; Wilson, 2005; Sterman, 1993; Sterman and Mann, 1995; Murata, 2005; Berka et al., 2004; Berka, Levendowski, Davis et al., 2005). Monitoring EEG allows developers—of products ranging from consumer electronics to heavy machinery to Web sites—direct access to the mental state of the user to provide near instantaneous feedback on the appeal, comfort, and usability of the design. The use of brain monitoring could have a significant impact on the design engineers of the future.

8.9 Augmenting Human Capacity: Seamless Integration of Brain, Mind, and Machine

Another new multidisciplinary endeavor, "Augmented Cognition," or Aug-Cog, unites neuroscience and computer engineering with an emphasis on detecting the cognitive state of the operator and integrating that information into the task environment. One goal of Aug-Cog is to create closed loop computational systems that will sense human cognitive state changes and adapt to them in real time (Schmorrow et al., 2005). The outcome is a closed-loop system that dynamically regulates and optimizes human–system interaction in real time with the goal of maintaining information load within the limits of the human information processor. If computational systems can learn from and adapt to their human operators, a fundamental shift will be achieved in the way humans interact with technology.

The feasibility of integration of real-time analysis of EEG indices of workload, task engagement, and attention into the human–computer loop was recently demonstrated in several complex task environments (Berka, Levendowski, Westbrook et al., 2005; Schmorrow and McBride, 2004; Morizio, Thomas, and Tremoulet, 2005; Ververs et al., 2005) including an Aegis radar operator simulation, a Tactical Tomahawk Weapons simulation environment, and in mobile warfighter operational scenarios (Berka, Levendowski, Davis et al., 2005; Berka, Levendowski, Westbrook et al., 2005; Ververs et al., 2005). In these preliminary demonstrations, EEG and other physiological indicators of user state were used to adjust interaction to suit a user's engagement level. If the system identified that a user was overloaded, the indices triggered information to be off-loaded or withheld. When a user was inoperative or had additional capacity, the indices triggered greater information dissemination or task reallocation. By actively providing assistance in managing information flow, the systems enhance human performance of complex tasks (Berka, Levendowski, Westbrook et al., 2005; Ververs et al., 2005; Dickson, 2005; Morizio, Thomas, and Tremoulet, 2005).

8.10 The Future of Education: Integrating Brain Monitoring into Training Systems

With the U.S. facing a decline in science, math, and engineering skills, educators in science and mathematics are pioneering many novel approaches to science education. Machine learning tools are beginning to provide refined models of student skill acquisition and learning behaviors in science and mathematics (Beal, 2004; Walles, Beal, and Arroyo, 2005; Arroyo, Beck, and Beal, 2003; Beck, Woolf, and Beal, 2000). Computer-based models have been designed to monitor actions including speed of processing, correct and incorrect answers or requests for help. More sophisticated models have been designed to assess learning styles and the use of optimal strategic approaches to problem solving (Stevens et al., 2004). Although these learner models are capable of forecasting student difficulties, or identifying when students may

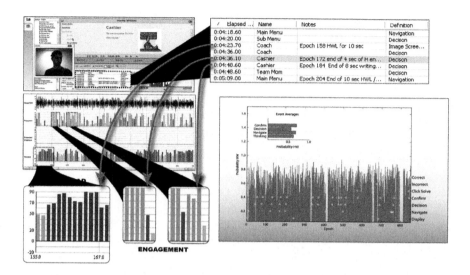

FIGURE 8.6

Quantification of EEG workload (WL), distraction (DIST), and task engagement (ENG) for a high-school student performing an automated interactive chemistry experiment.

require an educational intervention, they still must rely on relatively limited input, including learner actions that can be detected by the tutoring system (e.g., menu choices, mouse clicks, answers entered, requests to view multimedia help) and calculations of the time required for the student to complete various tasks and activities.

The next generation of technology designed to build models of student learning and progress will include EEG monitors to provide nonintrusive and nondistractive assessment of cognitive processes including attention, working memory, mental workload, decision making, and problem solving. These systems allow collection of indicators from students without distracting prompts such as verbal protocols, written logs, or self-reported behavioral indicators. These data can then be combined with real-time computational models of the tasks, and associated outcomes allow direct identification of conditions under which learning gains of students are impeded by the task complexity and the need for the integration of diverse skills.

In a collaborative study conducted by UCLA and Advanced Brain Monitoring, real-time readouts of workload, distraction, and task engagement were obtained at 1 s intervals and interleaved with timeline representations of the problem-solving process (Stevens, Galloway, and Berka, 2006) (Figure 8.6). The dynamics of workload (WL) showed significant elevation immediately after the selection of each new piece of problem-solving evidence and showed high/low fluctuations while the evidence was being interpreted and a decision was being made regarding the next test selection. Elevated distraction (DT) was mainly observed when a piece of evidence was misinterpreted, and often this was accompanied by high DT levels on subsequent test selec-

FIGURE 8.7
Future EEG use in the classroom, objectively measuring teacher/professor/course material effectiveness.

tions. Engagement (ENG) levels rose and were maintained until several seconds before the selection of the next test. These results indicate that real-time monitoring of EEG can begin to contribute an important dynamic dimension to classroom problem solving and could help design approaches for real-time feedback to improve learning (Figure 8.7).

Using EEG to monitor attention, task engagement, and mental workload on each of these tasks can facilitate the development of performance models that will complement and extend current probabilistic and individualized models of learning, resulting in radical transformations in technology-based learning systems.

8.11   NeuroTeam: Exploring Group Dynamics

The NeuroTeam concept involves simultaneous EEG monitoring of groups of people to identify trends, patterns, and relationships of EEG parameters

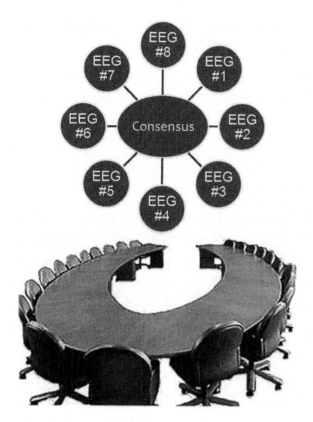

FIGURE 8.8
Future EEG application in corporations, objectively measuring communication efficiency, team consensus/discord, and employee neurocognitive factors.

during group interactions. Visual symbolic representations of the EEG signatures related to group interactions can be displayed to provide feedback to the team leaders or the entire group (Figure 8.8). Although NeuroTeaming is a new program in the early stages of development, multiple application scenarios have been proposed and are currently under investigation. The systems will facilitate investigations of areas relevant to social neuroscience, including exploration of the neural correlates of attitude formation, cooperation and competition among teams, as well as the capacity of leaders to empathize with and motivate their colleagues.

In a teaching environment, it will allow teachers to monitor their students' levels of fatigue, engagement, and mental workload in real time during the delivery of a lecture or to "listen in" on the mental activities of students working alone or in groups. Alternatively, NeuroTeaming can be applied in any team environment to explore the dynamics of group interactions and decision making and to assess the EEG parameters that are associated with reaching consensus. NeuroTeaming data can be displayed and provide feedback to the team to use as guidance in the group environment. NeuroTeam can also be used in a therapeutic environment to allow patients and therapist to achieve concordant physiology.

References

Akerstedt, T., Hume, K., Minors, D., and Waterhouse, J. 1993. Regulation of sleep and naps on an irregular schedule. *Sleep,* 16(8): 736–43.

Akerstedt, T. and Kecklund, G. 1991. Stability of day and night sleep—a two-year follow-up of EEG parameters in three-shift workers. *Sleep,* 14(6): 507–10.

Akerstedt, T., Kecklund, G., and Knutsson, A. 1991. Manifest sleepiness and the spectral content of the EEG during shift work. *Sleep,* 14(3): 221–5.

Allison, B. Z. and Pineda, J. A. 2006. Effects of SOA and flash pattern manipulations on ERPs, performance, and preference: implications for a BCI system. *Int. J. Psychophysiol.,* 59(2): 127–40.

Arroyo, I., Beck, J., and Beal, C. 2003. Learning within the zone of proximal development with the AnimalWatch intelligent tutoring system. Paper read at the American Educational Research Association Annual Meeting, at Chicago.

Balkin, T. J. 2001. Sleep Deprivation Research at WRAIR. Paper read at DARPA Workshop, August 21–23, at Las Vegas, NV.

Baynard, M. D., M, G., Moest, E. I., Ballas, C., Dinges, D. F., Van Dongen, H. 2003. Interindividual differences in psychomotor vigilance performance deficits during repeated exposure to sleep deprivation. *Sleep,* Vol. 26 (Abstract Supplement).

Beal, C. 2004. Adaptive user displays for intelligent tutoring software. *CyberPsychol. Behav.* 7(6): 689–692.

Beck, J, Woolf, B., and Beal, C., 2000. Learning to Teach: a machine learning architecture for intelligent tutor construction. Paper read at 18th National Conference on Artificial Intelligence (AAAI–2000).

Belenky, G. 2001. Neurobiological basis of effective operational performance. Paper read at Continuous Assisted Performance Teaming Workshop, DARPA, August 21–23, 2001, at Las Vegas, NV.

Belenky, G. L., B. P., Wesensten, N. J., Balkin, T. J. 2003. Variation in sensitivity to sleep restriction as a function of age revealed by growth modeling analysis. *Sleep,* Vol. 26 (Abstract Supplement).

Berger, H. 1929. Uber das Elektroenzephalorgam des Menschen. *Arch. Psychiatr. Nervenk.,* 90: 527–570.

Berka, C., Levendowski, D., Cvetinovic, M., Petrovic, M., Davis, G., Lumicao, M., Zivkovic, V., Popovic, M., and Olmstead, R. 2004. Real-time analysis of EEG indices of alertness, cognition, and memory acquired with a wireless EEG headset. *Int. J. Human-Computer Interaction,* 17(2): 151–170.

Berka, C., Levendowski, D., Davis, G., Lumicao, M., Ramsey, C., Stanney, K., Reeves, L., Harkness, S., and Tremoulet., P. 2005. EEG indices distinguish spatial and verbal working memory processing: implications for real-time monitoring in a closed-loop tactical tomahawk weapons simulation. Paper read at 1st International Conference on Augmented Cognition, July 22–27, at Las Vegas, NV.

Berka, C., Levendowski, D., Davis, G., Yau, A., Whitmoyer, M., Fatch, R., Zivkovic, V., and Olmstead., R. 2006. Nicotine administration and withdrawal effects on EEG metrics of attention, memory and workload: implications for cognitive resource allocation. In *Augmented Cognition: Past, Present and Future*, Schmorrow, D., Stanney. K., and Reeves. L., Arlington, V. A., (Eds.): Strategic Analysis, Inc.

Berka, C., Levendowski, D., Lumicao, M., Yau, A., Davis, G., Zivkovic, V., Olmstead, R., Tremoulet, P., and Craven, P. L. 2007. EEG Correlates of task engagement and mental workload in vigilance, learning and memory tasks—in press. *Aviation Space and Environmental Medicine*, 78(5, Section II, Supplement).

Berka, C., Levendowski, D., Westbrook, P., Davis, G., Lumicao, M., Ramsey, C., Petrovic, M., Zivkovic, V., and Olmstead., R. 2005. Implementation of a closed-loop real-time EEG-based drowsiness detection system: effects of feedback alarms on performance in a driving simulator. Paper read at 1st International Conference on Augmented Cognition, July at Las Vegas, NV.

Berka, C., Westbrook, P., Levendowski, D., Lumicao, M., Ramsey, C. K., Zavora, T., and Offner, T. 2005 Implementation model for identifying and treating obstructive sleep apnea in commercial drivers. Paper read at International Conference on Fatigue Management in Transportation Operations, September 12–14, at Seattle, WA.

Bixler, E. O., Kales, A., Soldatos, C. R., Kales, J. D., and Healey, S. 1979. Prevalence of sleep disorders in the Los Angeles metropolitan area. *Am. J. Psychiatry*, 136(10): 1257–62.

Brookhuis, K. A. and de Waard., D. 1993. The use of psychophysiology to assess driver status. *Ergonomics*, 36(9): 1099–110.

Brookings, J. B., Wilson, G. F., and Swain, C. R. 1996. Psychophysiological responses to changes in workload during simulated air traffic control. *Biol. Psychol.*, 42(3): 361–77.

Cacioppo, J. T., Berntson, G. G., Sheridan, J. F., and McClintock, M. K. 2000. Multilevel integrative analyses of human behavior: social neuroscience and the complementing nature of social and biological approaches. *Psychol. Bul.*, 126(6): 829–43.

Coleman, R. M., Roffwarg, H. P., Kennedy, S. J., Guilleminault, C., Cinque, J., Cohn, M. A., Karacan, I., Kupfer, D. J., Lemmi, H., Miles, L. E., Orr, W. C., Phillips, E. R., Roth, T., Sassin, J. F., Schmidt, H. S., Weitzman, E. D., and Dement, W. C. 1982. Sleep-wake disorders based on a polysomnographic diagnosis. A national cooperative study. *JAMA* 247(7): 997–1003.

Dement, W. C. and Vaughan, C. C. 1999. *The promise of sleep: A pioneer in Sleep Medicine Explores the Vital Connection between Health, Happiness, and a Good Night's Sleep*. House, R. (Ed.), New York: Delacorte Press.

Dickson, B. T. 2005. The cognitive cockpit—a test-bed for augmented cognition. Paper read at 1st International Conference on Augmented Cognition, July 22–27, at Las Vegas, NV.

Doran, S. M., Van Dongen, H. P., and Dinges, D. F. 2001. Sustained attention performance during sleep deprivation: evidence of state instability. *Archives Italiennes de Biologie* 139 (3): 253–267.

Drummond, S., S. J., Brown, G. G., Dinges, D. F., Gillin, J. C. 2003. Brain regions underlying differential PVT performance. *Sleep*, Vol. 26 (Abstract Supplement).

Fabiani, M., Gratton, G., and Coles, M. G. 2000. Event-related brain potentials. In *Handbook of Psychophysiology*, Cacioppo, J. T., Tassinary, L. G., and Berntson, G. G. (Eds.), Cambridge: Cambridge University Press.

Gevins, A. and Smith, M. E. 2005. Assessing fitness-for-duty and predicting performance with cognitive neurophysiological measures. Paper read at Biomonitoring for Physiological and Cognitive Performance during Military Operations, March 31–April 1, at Orlando, FL.

Gevins, A., Smith, M. E., Leong, H., McEvoy, L., Whitfield, S., Du, R., and Rush, G. 1998. Monitoring working memory load during computer-based tasks with EEG pattern recognition methods. *Human Factors*, 40 (1): 79–91.

Gevins, A., Smith, M. E., McEvoy, L., and Yu, D. 1997. High-resolution EEG mapping of cortical activation related to working memory: effects of task difficulty, type of processing, and practice. *Cerebral Cortex*, 7 (4): 374–85.

Habeck, C., Rakitin, B. C., Moeller, J., Scarmeas, N., Zarahn, E., Brown, T., and Stern, Y. 2004. An event-related fMRI study of the neurobehavioral impact of sleep deprivation on performance of a delayed-match-to-sample task. *Cognitive Brain Res.* 18: 306–321.

Itoi, A., Cilveti, R., Voth, M., Dantz, B., Hyde, P., Gupta, A., and Dement, W. M. D. 1993. Relationship between awareness of sleepiness and ability to predict sleep onset. Stanford, CA: AAA Foundation for Traffic Safety.

Kramer, A. and Parasuraman, R. 2005 Neuroergonomics: application of neuroscience to human factors. In *Handbook of Psychophysiology*, Caccioppo, J. T., Tassinary, L. G., and Berntson, G. G. (Eds.), New York: Cambridge University Press.

Kramer, H. C., Trejo, L. J., and Humphrey, D. G. 1996. *Psychophysiological Measures of Workload: Potential Applications to Adaptively Automated Systems*. Mouloua, P. M. (Ed.), *Automation and Human Performance*. NJ: Lawrence Erlbaum Associates.

Lavie, P. 1981. Sleep habits and sleep disturbances in industrial workers in Israel: main findings and some characteristics of workers complaining of excessive daytime sleepiness. *Sleep* 4(2): 147–58.

Leproult, R., Colecchia, E. F., Berardi, A. M., Stickgold, R. Kosslyn, S. M., and Van Cauter, E. 2002. Individual differences in subjective and objective alertness during sleep deprivation are stable and unrelated. *Am. J. Physiol. Regul. Integr. Comp. Physiol.* 284 (2): R280–90.

Levendowski, D., Berka, C., Olmstead, R., and Jarvik, M. 1999. Correlations between EEG indices of alertness measures of performance and self-reported states while operating a driving simulator. Paper read at 29th Annual Meeting, Society for Neuroscience, October 25, at Miami Beach, FL.

Levendowski, D., Berka, C., Olmstead, R., Konstantinovic, Z. R., Davis, G., Lumicao, M., and Westbrook, P. 2001. Electroencephalographic indices predict future vulnerability to fatigue induced by sleep deprivation. *Sleep*, 24 (Abstract Supplement): A243–A244.

Levendowski, D., Olmstead, R., Konstantinovic, Z. R., Berka, C., and Westbrook., P. 2000. Detection of electroencephalographic indices of drowsiness in real-time using a multi-level discriminant function analysis. *Sleep*, 23 (Abstract Supplement #2): A243–A244.

Levendowski, D., Westbrook, P., Berka, C., Popovic, M. V., Pineda, J. A., Zavora, T., Lumicao, M., and Zivkovic, V. T. 2002. Event-related potentials during a test of working memory differentiate sleep apnea patients from healthy subjects. *Sleep*, 25 (Abstract Supplement): A460–A461.

Makeig, S. 1993. Auditory event-related dynamics of the EEG spectrum and effects of exposure to tones. *Electroencephal. Clin. Neurophysiol.*, 86 (4): 283–293.

Makeig, S. and Jung, T. P. 1995. Changes in alertness are a principal component of variance in the EEG spectrum. *Neuroreport,* 7 (1): 213–6.

Makeig, S. and Jung, T. P. 1996. Tonic, phasic, and transient EEG correlates of auditory awareness in drowsiness. *Brain Res.: Cognitive Brain Res.,* 4 (1): 15–25.

Marek, T. and Pokorski, J. 2004. Quo vadis, ergonomia?—25 years on. *Ergonomia,* 26: 13–18.

Mitler, M. M., Westbrook, P., Levendowski, D. J., Ensign, W. Y., Olmstead, R. E., Berka, C., Davis, G., Lumicao, M. N. Cvetinovic, M., and Petrovic, M. M. 2002. Validation of automated EEG quantification of alertness: methods for early identification of individuals most susceptible to sleep deprivation. *Sleep,* 25 (Abstract Supplement): A147–A148.

Morizio, N., Thomas, M., and Tremoulet, P. 2005. Performance augmentation through cognitive enhancement (PACE). Paper read at 1st International Conference on Augmented Cognition, July, at Las Vegas, NV.

Murata, A. 2005. An attempt to evaluate mental workload using wavelet transform of EEG. Paper read at Human Factors, Fall 2005.

O'Hanlon, J. F. and Beatty, J. 1977. Concurrence of electroencephalographic and performance changes during a simulated radar watch and some implications for the arousal theory of vigilance. pp. 189–201. Mackie, R. R. (Ed.). *Vigilance: Theory, Operational Performance, and Physiological Correlates.* New York: Plenum Press.

Parasuraman, R. 2005. *Neuroergonomics: The Brain at Work.* New York: Oxford University Press.

Parasuraman, R., Bahri, T., Deaton, J. E., Morrison, J. G., and Barnes, M. 1992. *Theory and Design of Adaptive Automation in Adaptive Systems.* Warminster, PA: Naval Air Warfare Center, Aircraft Division.

Parasuraman, R., Molloy, R., and Singh, I. L 1993. Performance consequences of automation induced complacency. *Int. J. Aviation Psychol.,* 3: 1–23.

Parasuraman, R., Mouloua, M., and Molloy, R. 1996. Effects of adaptive task allocation on monitoring of automated systems. *Hum. Factors,* 38 (4): 665–79.

Pfurtscheller, G., Graimann, B., Huggins, J. E., and Levine, S. P. 2004. Brain-computer communication based on the dynamics of brain oscillations. *Suppl. Clin. Neurophysiol.* 57: 583–91.

Rogers, N. L., V. D. H, Powell, I. V., JW, Carlin, M. M., Szuba, M. P., Maislin, G., and Dinges, D. F. 2002. Neurobehavioural functioning during chronic sleep restriction at an adverse circadian phase. *Sleep,* Vol. 25 (Abstract Supplement).

Schmorrow, D. and McBride, D. 2004. Overview of the DARPA Augmented Cognition. *Int. J. Human-Computer Interaction,* 17 (2).

Schmorrow, D., Stanney, K. M., Wilson, G. F., and Young, P. 2005. Augmented cognition in human-system interaction. In *Handbook of Human Factors and Ergonomics,* (3rd ed.), Salvendy, G. (Ed.), New York: John Wiley.

Sellers, E. W. and Donchin, E. 2006. A P300-based brain-computer interface: initial tests by ALS patients. *Clin. Neurophysiol.,* 117 (3): 538–48.

Singh, I. L., Molloy, R., and Parasuraman, R. 1993. Automation-induced complacency: Development of the complacency-potential rating scale. *Int. J. Aviation Psychol.,* 3: 111–121.

Smith, M. E. and Gevins, A. 2005. Neurophysiologic monitoring of mental workload and fatigue and during operation of a flight simulator. Paper read at Biomonitoring for Physiological and Cognitive Performance during Military Operations, March 31–April 1, at Orlando, FL.

Sterman, M. B. 1993. Application of quantitative EEG analysis to workload assessment in an advanced aircraft simulator. Paper read at Proceeding of Human Factors Society, at Seattle, WA.

Sterman, M. B. and Mann, C. A. 1995. Concepts and applications of EEG analysis in aviation performance evaluation. *Biol. Psychol.*, 40 (1–2): 115–30.

Stevens, R, T Galloway, and C Berka. 2006. Integrating EEG models of cognitive load with machine learning models of scientific problem solving. In *Augmented Cognition: Past, Present and Future*, edited by D. Schmorrow, K. Stanney and L. Reeves. Arlington, VA: Strategic Analysis.

Stevens, R., Soller, A., Cooper, M., and Sprang, M. 2004. Modeling the development of problem-solving skills in chemistry with a web-based tutor. *Lecture notes in Computer Science*, 3220: 580–591.

Torsvall, L. and Akerstedt, T. 1987. Sleepiness on the job: continuously measured EEG changes in train drivers. *Electroencephalogr. Clin. Neurophysiol.*, 66 (6): 502–11.

Ververs, P. M., Whitlow, S., Dorneich, M., and Mathan, S. 2005. Building Honeywell's Adaptive System for the Augmented Cognition Program. Paper read at 1st International Conference on Augmented Cognition, July 22–27, at Las Vegas, NV.

Walles, R., Beal, C., and Arroyo, I. 2005. Cognitive Predictors of response to web-based tutoring in SAT-Math. Paper read at Biennial Meeting of the Society for Research in Child Development, at Atlanta, GA.

Westbrook, P., Berka, C., Levendowski, D., Lumicao, M., Davis, G., Olmstead, R., Petrovic, M., Yuksel, Y., Ramsey, C., and Zivkovic., V. 2004. Quantification of alertness, memory and neurophysiological changes in sleep apnea patients following treatment with nCPAP. *Sleep*, 27: A223.

Wilson, G. F. 2005. Operator functional state assessment for adaptive automation implementation. Paper read at Biomonitoring for Physiological and Cognitive Performance during Military Operations, March 31-April 1, at Orlando, FL.

Wolpaw, J. R., McFarland, D. J., Vaughan, T. M., and Schalk, G. 2003. The Wadsworth Center brain-computer interface (BCI) research and development program. *IEEE Trans. Neural. Syst. Rehabil. Eng.*, 11 (2): 204–7.

9

The Interaction of Sleep and Memory

Jeffrey M. Ellenbogen

CONTENTS

9.1 Introduction.. 189
9.2 Position 1: Sleep Does Not Facilitate Memory 191
9.3 Position 2: Sleep Facilitates Memory 193
9.4 Conclusion... 198
Appendix—Conventional Sleep Terminology for Human Sleep 198
References 199

9.1 Introduction

We sleep in order to be less sleepy. This tautological assertion is as flawed as suggesting that we eat in order to be less hungry.

We do, of course. But the behavioral drive to eat merely motivates our consumption of foods that, in turn, bestow nourishments to key physiologic functions. But if sleep is like food—supplying essential elements to the body—what physiologic processes does sleep provide? Among many (non-exclusive) candidates, increasing evidence asserts that sleep is essential for several critical brain functions, including certain kinds cognitive processing such as memory consolidation* and memory integration.†

Whether sleep leads to benefits in memory is not just a contemporary question. In fact, it has been debated for more than 100 years. In one of the first experimental studies of memory, Ebbinghaus described the patterns of memory decay over time (1885). He noted a curious observation in his

---

* Although there are many uses of the term *consolidation*, in this chapter I restrict myself to the definition provided by Dudai: the "progressive postacquisition stabilization of memory" (Dudai, 2004). Meaning that consolidation is a process that makes memories resistant to loss.

† Memory integration, otherwise known as relational memory, refers to the ability to take existing memories and stitch (bind) them together. Transitive inference is one example: if one learns that A > B and that B > C, one can infer that A > C without ever directly learning that particular pairing. A very brief discussion of experimental evidence for the relationship of sleep and memory integration is mentioned in Figure 9.2.

data—the rate of forgetting was slowest between 9 and 24 h. Although Ebbinghaus acknowledged that sleep was occurring during this time period, he dismissed the possibility that sleep was making any meaningful contribution, referring to sleep as an implausible cause of his findings. Instead, he referred to them as resulting from "accident influences"; that is, noise in his data. Capitalizing on this intriguing finding (and reaffirming the principle that one scientist's noise is another's insight), Jenkins and Dallenbach (1924) hypothesized that sleep was, in fact, responsible for the period of slower forgetting seen in Ebbinghaus's data.

In their study, Jenkins and Dallenbach performed one of the first hypothesis-driven, empirical investigations of the relationship between sleep and memory. They examined the rate of forgetting nonsense syllables across periods of sleep during the night, compared to wakefulness during the day. Their results demonstrated a steeper rate of forgetting during wakefulness, as compared to sleep; and went on to conclude that the worse performance in the wake group was due to the negative influence of interference that normally occurs during wakefulness. They gave no consideration to the alternative possibility that sleep actively strengthens memories, causing the sleep group to perform better.

Modern refinements in the experimental study of sleep and of memory are bringing the fascinating relationship of sleep and memory into sharper focus. A long-standing position—put forth first by Jenkins and Dallenbach, and suffusing into contemporary thinking—contends that sleep passively protects memories by temporarily sheltering them from interference, thus providing precious little benefit for memory. However, recent evidence is unmasking a more substantial and long-lasting benefit of sleep for memory. Although the precise causal mechanisms within sleep that result in memory consolidation remain elusive, clues are emerging, and recent data provide strong evidence that neurobiological processes within sleep actively enhance memories, including different kinds of memory systems.

First, periods of sleep, compared to those comprised of wakefulness, appear to actively improve human performance of recently acquired skills (Stickgold, 2005; Gais and Born, 2004; Smith, 2001; Ficca and Salzarulo, 2004). Examples include the learning of motor sequences akin to playing a few notes on a piano (Walker et al., 2002; Cohen et al., 2005; Fischer et al., 2005); the ability to discriminate rapidly presented visual objects (Stickgold, James, and Hobson, 2000); and other motor learning such as the serial-reaction-time task (Robertson, Pascual-Leone, and Press, 2004; Peigneux et al., 2003; Maquet et al., 2000; Fischer et al., 2006; Robertson, Press, and Pascual-Leone, 2005) and auditory discrimination of a synthetic language (Fenn, Nusbaum, and Margoliash, 2003).

Joining these so-called nondeclarative (implicit) processes, recent studies support the beneficial effects of sleep for memories of facts and events in time—those that depend heavily on the small but critical structure deep in the brain's temporal lobe: the hippocampus.

In this chapter, I provide a framework for understanding this complex and controversial topic—whether sleep supports memory—by broadly dividing the debate into two readily distinguishable positions that argue for, or against, the notion that sleep plays a key role in memory processing.

Those who argue that sleep makes no meaningful contribution to memory have two main stances. One, succinctly stated, is that sleep offers nothing for memory. The second, more nuanced position, asserts that sleep temporarily shields memory from the negative effects of interference, implying, in short, that sleep provides a passive and transient benefit for memory and nothing more.

However, proponents of the view that sleep plays a critical role in memory point to emerging data that support their assertion: that biological properties of sleep do, in fact, have memory-enhancing characteristics. I provide examples, discuss distinctions between, assumptions within, as well as strengths and weaknesses of each of these two opposing perspectives.

9.2   Position 1: Sleep Does Not Facilitate Memory

Several lines of reasoning have been employed to refute the view that sleep plays a role in memory consolidation. First, some have pointed out that those with diminished amounts of REM sleep* (either from consuming REM-suppressing agents, such as certain types of antidepressants, or more rarely, through brainstem damage) have continued to lead normal productive lives (Vertes, 2004; Vertes and Eastman, 2000; Vertes and Siegel, 2005; Siegel, 2001). Certainly, if sleep were critical for memory, the argument goes, then those with such impairments in REM sleep would have apparent deficits in memory.

Yet, these same critics acknowledge that the cognitive capacities of those individuals with diminished REM were not systematically examined. Furthermore, little study of tasks on which performance reportedly depends on sleep has ever been conducted in these individuals.

And it is intriguing to think about what one might conclude, even if such studies of REM impairment and memory were performed. They certainly could clarify our understanding of REM's particular role in sleep-enhanced memory consolidation. However, would that generalize to assumptions about sleep, in total? Some would argue no. Among them would likely be those that hypothesize that NREM sleep (pronounced "non-REM," referring to sleep stages N1, N2, and N3; see appendix) is more important for memory consolidation than REM sleep. If that were true, impaired REM sleep might even enhance sleep-dependent memory consolidation, if it were to increase the relative amount of NREM sleep that takes the place of REM (Wixted, 2004). However, to date, few empirical studies exist to guide this thinking.

---

* For an introduction to sleep terminology, see the appendix, "Sleep Terminology," at the end of the text.

An additional argument, put forth against the notion that sleep benefits memory, states that improvements of memory overnight can be explained by the mere passage of time, rather than be attributed to a function of sleep per se. (Vertes, 2004). The problem with this perspective is that many studies looking at sleep and memory also employ waking control groups (e.g., Ellenbogen et al., 2006; Tucker et al., 2006; Gais and Born, 2004). Therefore, the wake and sleep groups have equal amounts of time, yet the participants in the sleep group perform better. It would seem, then, that when sleep is compared to wakefulness and performance is better after sleep, that some benefit of sleep for memory ought to be acknowledged.

Yet there are those who maintain the position that sleep does not provide a meaningful enhancement of memory. They argue that sleep transiently shelters memory from interference. Harping back to Jenkins and Dallenbach, this position simultaneously acknowledges that memory recall should be better after sleep than after wakefulness, but only because of a passive—and transient—protection that sleep affords memory from interference.

This more nuanced variation within position one (that sleep plays no meaningful role in memory) exists to accommodate empirical evidence showing relative improvements of memory recall after sleep. Sleep only transiently sustains memories, by protecting them from interference while asleep, but does not consolidate them. Thus, recall is better in the morning immediately after sleep, compared to after a day awake, but only until exposure to interference in the subsequent day. However, because sleep does not consolidate memories, they will be once again rendered vulnerable to interference in the waking day to come, as vulnerable as they would be had the person not slept at all.

Viewed from this passive-protection perspective, any study that demonstrates superior recall performance immediately after periods of sleep, compared to those of wakefulness, is not showing that sleep improves memory; rather, it demonstrates the negative effects of waking mental experience on memory. Therefore, sleep is a temporary shelter—a respite for memory—from the inevitable negative effects of interfering mental activity during wakefulness. Much like a document that cannot be edited or deleted while a computer is in "sleep" mode, memories remain unchanged across a night of sleep; neither for the better, nor worse. In short, sleep adds nothing to memories.

Two recent studies call this hypothesis into question, by directly examining verbal recall, interference and sleep. One of these studies manipulated when time of training, time of testing, and sleep occurred (Gais, Lucas, and Born, 2006): sleep was found to improve memory recall independent of the amount of time awake, providing evidence that sleep does more than passively protect the previously formed declarative memories.

Another study promotes the benefit of sleep for memory, by experimentally manipulating interference (Ellenbogen et al., 2006). This study directly challenged the assumption that sleep is merely a passive protection of memories. Rather than simply test subjects on memory performance after sleep, the authors unmasked the extent of sleep's benefit for memory by introducing interference *after* periods of wakefulness or sleep. Using a well-established

AB–AC interference paradigm (Barnes and Underwood, 1959), subjects first learned paired associates, designated A_iB_i. After sleep at night, or wakefulness during the day, half of the subjects in each group learned new paired associates, A_iC_i, before being tested for recall of the original (A_iB_i) list. Their results demonstrated that compared to wakefulness, sleep provided modest protection against memory deterioration, even in the absence of interference training (Figure 9.1a), and that a large protection was seen against *post*-sleep interference (Figure 9.1b). In other words, sleep made memories resistant to interference after sleep, suggesting that the memory had been strengthened during sleep.

In this study, two questions remained to be answered before it could be concluded that sleep actively enhanced memory. First, did the time of day of testing confound the study (i.e., all sleep groups were trained at night and tested in the morning, whereas the wake groups were trained in the evening and tested in the morning)? Regarding training, the authors point out that the number of training trials to criterion were no different in the sleep or wake group, arguing that the encoding process was similar in both groups, whether training took place in the morning or evening. To answer whether time of day of testing confounded the study, the authors needed to employ an additional control group: 24-h between training and testing, where training took place at night, followed by sleep, and then testing took place the following night (rather than the next morning). This 24-h group showed that the benefit of sleep was not dependent on time of day that testing took place. Further, it demonstrated that the benefit of sleep for memory was sustained throughout the subsequent waking day (Figure 9.1).

Thus, this study showed that memories tested after a night of sleep were highly resistant to interference, and remained resistant across the subsequent day. Differently, memories after a day of wakefulness were highly susceptible to interference. Collectively, these findings provide evidence that consolidation must also occur during sleep for the memories to become resistant to interference the following day.

It is important to distinguish this study by Ellenbogen et al. from that of Jenkins and Dallenbach and their contemporary supporters. Jenkins and Dallenbach claimed that sleep transiently protects memories from interference during sleep, whereas the study from Ellenbogen et al. demonstrates that sleep leads to consolidation of verbal memories, making them resistant to interference after a night of sleep, and thus provides evidence that sleep makes memories stronger.

9.3 Position 2: Sleep Facilitates Memory

Among those that acknowledge sleep's contribution to memory consolidation, there are two main hypotheses. The first hypothesis is "permissive

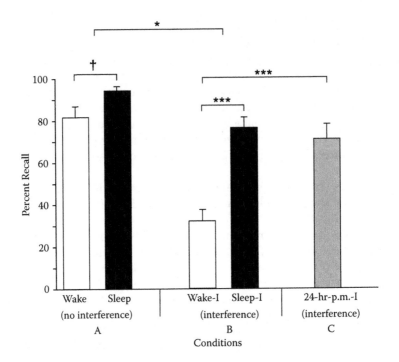

FIGURE 9.1
Shown here is the performance data from a study of verbal memory (from Ellenbogen et al., *Curr. Biol.* 16(13): 1290–4, 2006). All subjects learned a list of 20 pairs and recalled them after periods containing sleep, wakefulness, or both. In the no-interference conditions (A), participants recalled the learned words after 12-h periods. The sleep group was marginally better. In the interference conditions (B and C), participants learned an associative interference list immediately prior to being asked to recall the original list. This new list was meant to disrupt the memory of the original list. Shown in the figure, the sleep group was highly resistant to the negative effects of interference in the subsequent day; they were able to retain the original word list, despite attempts at interference implying that sleep stabilized the memory (consolidation). In the 24-h group (C), learning took place in the evening and testing the following evening, with sleep intervening between learning and testing. This group shows that the benefit of sleep persists throughout the subsequent waking day. (Importantly, no portion of this study involved sleep deprivation: wake groups were awake during the daytime.) Error bars are standard error of the mean. $\dagger\, p = .06$; $*\, p < 0.05$; $***\, p < 0.001$.

consolidation": sleep creates conditions conducive to memory consolidation, and the second hypothesis is "active consolidation": that unique properties of sleep directly participate in memory consolidation. I will discuss each of these, in turn.

The permissive consolidation hypothesis asserts that sleep indirectly contributes to consolidation. It incorporates the aforementioned hypothesis of Jenkins and Dallenbach—that less interference occurs during sleep—but, differently, extends this view to include consolidation. The permissive consolidation hypothesis achieves this seeming contradiction by assuming that sleep is a state of reduced interference and this setting allows an already available consolidation mechanism to function most effectively (Wixted, 2004). During the day, recently encoded memories are bombarded by interference from waking cerebral activity, which not only weakens them, but also impedes their effective consolidation. In contrast, memories during sleep—unbridled by waking interference—have a privileged opportunity to consolidate.

Behavioral measures alone cannot parse whether sleep leads to the circumstance conducive to consolidation, or whether unique biological properties of sleep consolidate memories. The distinction rests on complete knowledge of the precise physiologic markers of consolidation, interference, and their relationship to sleep, all of which are subjects of ongoing investigation.

To unequivocally claim that sleep actively consolidates memories, one must understand the specific properties of sleep physiology. By showing that a specific form of memory consolidation critically depends on a brain property unique to sleep, one can conclusively validate the position that sleep actively enhances memory.

Several recent studies extended the works of Barrett et al. (Barrett and Ekstrand, 1972) and Fowler (Fowler, Sullivan, and Ekstrand, 1973), emphasizing the role of slow wave sleep (SWS) in memory consolidation (Plihal and Born, 1997; Gais and Born, 2004). (SWS is otherwise known as N3 sleep; see appendix.) They showed superior memory recall among subjects that slept across the first few hours of the night (so-called early sleep, a portion of sleep with relatively large percentage of SWS) compared to the same time period awake. The same benefit was not seen when the conditions were switched to the second half of the night, a portion of sleep containing relatively little SWS. And in an additional group, the investigators intentionally elevated brain levels of acetylcholine by administering the drug physostigmine to participants, resulting in a reduced benefit of sleep for memory during the early sleep. Taken together, these findings support the claim that the naturally occurring nadir of acetylcholine in the first half of a night's sleep interacts with SWS to consolidate memories.

Other studies also emphasize SWS as an important physiologic process that contributes to memory consolidation (Takashima et al., 2006). In vivo, intracellular recordings demonstrate that neocortical neurons spontaneously reactivate during SWS (Steriade and Timofeev, 2003); it is conceivable that reactivation leads to long-term potentiation (i.e., strengthening) of memory traces.

Several animal studies also support the hypothetical role of sleep in memory consolidation. They demonstrate that recently acquired, hippocampus-based memories are "replayed" during sleep (e.g., Wilson and McNaughton, 1994; Pavlides and Winson, 1989; Nadasdy et al., 1999; O'Neill, Senior, and Csicsvari, 2006). Interestingly, this reactivation of hippocampus-dependent memories after spatial navigation has also been seen in wakefulness; but while awake, these memories are chronologically replayed backward ("reverse replay"; Foster and Wilson, 2006), unlike the sleeping patterns that are replayed forward. One interpretation of this distinction hypothesizes that these differing patterns reflect different roles: whereas initial learning relies on reverse replay, consolidation relies on forward replay (Suzuki, 2006).

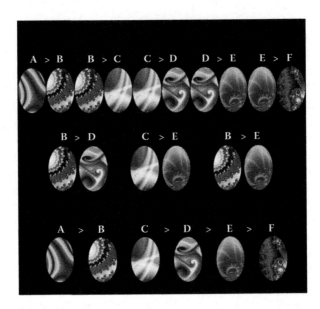

FIGURE 9.2

Shown are the items used in a behavioral study examining the role of sleep in relational memory by using a test of transitive inference (from Ellenbogen et al., *Proc. Natl. Acad. Sci. U.S.A.* 104(18): 7723–8, 2007). Participants learned which item was correct when paired with another item (top row), one pair at a time. They were then tested in their ability to make inference judgments (inference items seen in the middle row). The group of participants that slept was more likely to correctly identify the most distant inference pair (represented schematically by B-E), providing evidence that sleep leads to the ability to relate one memory to another, allowing the flexible use of existing memories. Knowing the inference pairs (middle row in figure) strongly implied knowledge of the hierarchy (bottom row). (The actual study used a color version of these items.)

Further supporting the notion that memories are replayed during sleep, neuroimaging findings in humans further demonstrate increased hippocampal activity during sleep following spatial learning, an increase that is proportional to the degree of overnight behavioral improvement (Peigneux et al., 2004).

Collectively, these studies suggest that hippocampus-dependent memories are reactivated during sleep, and that this reactivation leads to strengthened memory traces.

Although SWS might be sufficient to consolidate memories, independent of other sleep stages, an alternative hypothesis states that all stages of sleep are important for memory consolidation. A strength of this perspective notes that each stage of normal human sleep (including SWS and REM) occurs in succession several times throughout the night. (Each of these so-called ultradian cycles lasts approximately 1 h.) Perhaps the interplay of sleep stages within an ultradian cycle—repeating over several iterations throughout the night—orchestrates a complex process of feed-forward and feedback mechanisms between the hippocampus and neocortex, resulting in enhanced memory consolidation by repeatedly shuffling information back and forth.

One study examined this theoretical hippocampal-neocortical dialogue (Buzsaki, 1996), and its putative role for memory, by disrupting the ultradian cycle (Ficca et al., 2000). Results demonstrated that ultradian cycles throughout the night are essential for memory consolidation, arguing for a combined role of both NREM and REM.

Other candidate mechanisms exist that propose that sleep directly enhances memory, including sleep spindles (Gais et al., 2002; see appendix). Animal models demonstrate that these distinct electrophysiological phenomena simultaneously occur with activity in the hippocampus during sleep (Siapas and Wilson, 1998). Perhaps this represents a reorganization mechanism that transfers information between the hippocampus and neocortex during sleep, resulting in consolidation of memories (Steriade, 2000).

One concern is that some studies of sleep spindles and memory note that spindle quantity correlates with intelligence (Bodizs et al., 2005; Schabus et al., 2006); thus, the conclusion that spindle quantity results in enhanced performance after sleep might be confounded by intelligence. One effective strategy to deal with this possible experimental-design problem is to examine within-subject changes between a baseline night to the experimental night, to account for individual differences in spindle quantity resulting from intelligence. Without this measure, it can be difficult to discern whether high spindle content correlates with overnight improvement in memory; or whether the overnight improvement is simply a function of higher IQ alone; or an interaction between the two (i.e., those with higher IQ have more spindles, and those with more spindles have a more pronounced overnight improvement); or that those with more spindles have a higher IQ because of more pronounced overnight sleep consolidation processes.

Limitations aside, a reported 34% within-subject increase in spindle density early in the night following task training (Gais et al., 2002) and increased spindle density after learning difficult lists of words (Schmidt et al., 2006) argue in favor of a role for sleep spindles in memory processing. Similarly, a second study found a strong correlation between spindle density and overnight verbal memory retention but no correlation between memory for face recognition and spindles, arguing against a general intelligence effect (Clemens, Fabo, and Halasz, 2005). A more recent study looking at naps and motor memory demonstrated regions of the brain that had enhanced sleep spindle activity which correlated with improved motor skill abilities: left-handed motor training led to enhancement of spindle quantity in the right motor cortex of the brain, which, in turn, resulted in improved performance, suggesting a key interaction between learning, spindles of discrete brain regions, and improvements in performance.

From a different perspective, several researchers examined whether sleep can help repair damaged memories. Fenn et al. (Fenn, Nusbaum, and Margoliash, 2003) showed that sleep can restore damaged memories in perceptual learning of a spoken language. Related to this, Norman et al. (Norman, Newman, and Perotte, 2005) employed a computational neural network (the complementary learning system model; McClelland, McNaughton, and

O'Reilly, 1995; O'Reilly and Rudy, 2001) to examine the effects of sleep on learning. Knowing that learning new information can sometimes disrupt existing knowledge (the stability-plasticity problem), they looked at whether preexisting memories, impaired by recently acquired knowledge, could be restored by REM-like models of sleep. Results from their computational model demonstrated that REM sleep can "repair damaged memories."

Lastly, two independent studies examined the role of sleep for the consolidation of emotional memory. One study demonstrated that sleep in the latter half of the night, a period with relatively large quantities of REM sleep physiology, lead to improved recall (Wagner, Gais, and Born, 2001). The second study demonstrated that an entire night of sleep led to enhanced consolidation of arousing emotional stimuli (Hu, Stylos-Allen, and Walker; in press). Taken together, these studies argue that sleep consolidates emotionally arousing memories by employing mechanisms in REM sleep.

9.4   Conclusion

In summary, future technological advances in measures of sleep physiology, coupled with refinements in behavioral memory tasks employed in these studies, will undoubtedly be instrumental in teasing apart the precise contributions of sleep for a wide range of memory and related cognitive systems. Many questions remain to be answered. Nonetheless, the available evidence converges on the notion that neurobiological processes within sleep are directly responsible for part of memory consolidation in humans.

Appendix—Conventional Sleep Terminology for Human Sleep

The standard electrophysiologic measurements of sleep, collectively called *polysomnography* (PSG), include elements from a constellation of electrodes placed on the surface of the scalp and body that measure electromagnetic signals emitted by (1) the brain, (2) eye movements, and (3) muscle activity of the body. The most widely held convention (Iber et al., 2007) divides the characteristic waves seen in sleep into named stages by using readily discernable electrophysiologic signatures, including frequency of signals, their amplitude, and wave morphology (characteristic shape).

Emerging from this convention, two main categories exist: rapid eye movement sleep (REM) and non-rapid-eye-movement sleep (NREM, pronounced "non-REM").

REM sleep is defined by three surface electrophysiologic signatures: low muscle tone, mixed (relatively fast) frequency brain waves, and occasional rapid eye movements. NREM, on the other hand, is divided into stages N1–N3, with stage N3 often called slow-wave sleep.

Stage N1 (NREM 1) is the transition period between wakefulness and sleep. It is a state of drowsiness, where the waking brain waves begin to slow down in frequency.

Stage N2 (NREM 2) is a slightly deeper stage, where the frequency of brain waves tend to slow down even further, and particular wave patterns are seen: sleep spindles and K-complexes. Spindles are brief, high frequency (11–16 Hz) bursts of brain activity. K-complex refers to a single, characteristic-appearing sharp wave that stands out from the background waveforms.

Stage N3 (NREM 3) is often referred to slow-wave sleep. As the name implies, the brain waves seen in this stage are the slowest (at least 20% of the waves are less than 2 Hz). The slow waves are organized and coordinated, reflecting synchronized activity of multiple brain regions, including the thalamus and cerebral cortex, and multiple regions of the cortex.

(In this new classification terminology, there is no longer a "stage 4." Stage N3 refers to both NREM stage 3 and NREM stage 4 of the old convention.)

References

Barnes, J. M. and Benton, J. Underwood. 1959. Fate of first-list association in transfer theory. *J. Exp. Psychol.* 58 (2): 97–105.

Barrett, T. R. and Ekstrand, B. R. 1972. Effect of sleep on memory: III. Controlling for time-of-day effects. *J. Exp. Psychol.* 96 (2): 321–327.

Bodizs, R., Kis, T., Lazar, A. S., Havran, L., Rigo, P., Clemens, Z., and Halasz, P. 2005. Prediction of general mental ability based on neural oscillation measures of sleep. *J. Sleep Res.* 14 (3): 285–92.

Buzsaki, Gyorgy. 1996. The hippocampo-neocortical dialogue. *Cereb. Cortex* 6 (2): 81–92.

Clemens, Z., Fabo, D., and Halasz, P., 2005. Overnight verbal memory retention correlates with the number of sleep spindles. *Neuroscience.* 132 (2): 529–35.

Cohen, D. A., Pascual-Leone, A., Press, D. Z., and Robertson, E. M. 2005. Off-line learning of motor skill memory: A double dissociation of goal and movement. *Proc. Natl. Acad. Sci. U.S.A.* 102 (50): 18237–41.

Dudai, Y. 2004. The neurobiology of consolidations, or, how stable is the engram? *Annu. Rev. Psychol.* 55: 51–86.

Ebbinghaus, H. 1885. *Ueber das Gedactnis: Untersuchungen zur experimentellen Psychologie.* Translated by Ruger, H. A. and Bussenius, C. E.: Teachers' College, NY. Original edition, 1885.

Ellenbogen, J. M., Hu, P. T., Payne, J. D., Titone, D., and Walker, M. P. 2007. Human relational memory requires time and sleep. *Proc. Natl. Acad. Sci. U.S.A.* 104(18): 7723–8.

Ellenbogen, J. M., Hulbert, J. C., Stickgold, R., Dinges, D. F., and Thompson-Schill, S. L. 2006. Interfering with theories of sleep and memory: sleep, declarative memory, and associative interference. *Curr. Biol.* 16(13): 1290–4.

Fenn, K. M., Nusbaum, H. C., and Margoliash, D. 2003. Consolidation during sleep of perceptual learning of spoken language. *Nature* 425 (6958): 614–616.

Ficca, G. and Salzarulo, P. 2004. What in sleep is for memory. *Sleep Med.* 5(3): 225–30.

Ficca, G., Lombardo, P., Rossi, L., and Salzarulo, P. 2000. Morning recall of verbal material depends on prior sleep organization. *Behav. Brain Res.* 112(1–2): 159–163.

Fischer, S., Drosopoulos, S., Tsen, J., and Born, J. 2006. Implicit learning—explicit knowing: a role for sleep in memory system interaction. *J. Cogn. Neurosci.* 18(3): 311–9.

Fischer, S., Nitschke, M. F., Melchert, U. H., Erdmann, C., and Born. J. 2005. Motor memory consolidation in sleep shapes more effective neuronal representations. *J. Neurosci.* 25(49): 11248–55.

Foster, D. J. and Wilson, M. A. 2006. Reverse replay of behavioural sequences in hippocampal place cells during the awake state. *Nature* 440(7084): 680–3.

Fowler, M. J., Sullivan, M. J., and Ekstrand, B. R. 1973. Sleep and memory. *Science* 179(70): 302–4.

Gais, S., and Born, J. 2004. Declarative memory consolidation: mechanisms acting during human sleep. *Learn Mem.* 11(6): 679–85.

Gais, S., and Born, J. 2004. Low acetylcholine during slow-wave sleep is critical for declarative memory consolidation. *Proc. Natl. Acad. Sci. U.S.A.* 101(7): 2140–4.

Gais, S., Lucas, B., and Born, J. 2006. Sleep after learning aids memory recall. *Learn Mem.* 13(3): 259–62.

Gais, S., Molle, M., Helms, K., and Born, J. 2002. Learning-dependent increases in sleep spindle density. *J. Neurosci.* 22(15): 6830–4.

Hu, P., Stylos-Allen, M., and Walker, M. P. In press. Sleep facilitates consolidation of emotionally arousing declarative memory. *Psychol. Sci.*

Iber, C., Ancoli-Israel, S., Chesson, A., and Quan, S. F. 2007. *The AASM Manual for the Scoring of Sleep and Associated Events: Rules, Terminology and Technical Specifications.* 1st ed. Westchester, Illinois: American Academy of Sleep Medicine.

Jenkins, J. G. and Dallenbach, K. M. 1924. Obliviscence during sleep and waking. *Am. J. Psychol.* 35: 605–612.

Maquet, P., Laureys, S., Peigneux, P., Fuchs, S., Petiau, C., Phillips, C., Aerts, J., Del Fiore, G., Degueldre, C., Meulemans, T., Luxen, A., Franck, G., Van Der Linden, M., Smith, C., and Cleeremans, A. 2000. Experience-dependent changes in cerebral activation during human REM sleep. *Nat. Neurosci.* 3 (8): 831–6.

McClelland, J. L., McNaughton, B. L., and O'Reilly, R. C. 1995. Why there are complementary learning systems in the hippocampus and neocortex: insights from the successes and failures of connectionist models of learning and memory. *Psychol. Rev.* 102 (3): 419–57.

Nadasdy, Z., Hirase, H., Czurko, A., Csicsvari, J., and Buzsaki, G. 1999. Replay and time compression of recurring spike sequences in the hippocampus. *J. Neurosci.* 19 (21): 9497–507.

Norman, K. A., Newman, E. L., and Perotte, A. J. 2005. Methods for reducing interference in the Complementary Learning Systems model: oscillating inhibition and autonomous memory rehearsal. *Neural Netw.* 18 (9): 1212–28.

O'Neill, J., Senior, T., and Csicsvari, J. 2006. Place-selective firing of CA1 pyramidal cells during sharp wave/ripple network patterns in exploratory behavior. *Neuron* 49 (1): 143–55.

O'Reilly, R. C. and Rudy, J. W. 2001. Conjunctive representations in learning and memory: principles of cortical and hippocampal function. *Psychol. Rev.* 108 (2): 311–45.

Pavlides, C. and Winson, J. 1989. Influences of hippocampal place cell firing in the awake state on the activity of these cells during subsequent sleep episodes. *J. Neurosci.* 9 (8): 2907–18.

Peigneux, P., Laureys, S., Fuchs, S., Collette, F., Perrin, F., Reggers, J., Phillips, C., Degueldre, C., Del Fiore, G., Aerts, J., Luxen, A., and Maquet, P. 2004. Are spatial memories strengthened in the human hippocampus during slow wave sleep? *Neuron* 44 (3): 535–45.

Peigneux, P., Laureys, S., Fuchs, S., Destrebecqz, A., Collette, F., Delbeuck, X., Phillips, C., Aerts, J., Del Fiore, G., Degueldre, C., Luxen, A., Cleeremans, A., and Maquet, P. 2003. Learned material content and acquisition level modulate cerebral reactivation during posttraining rapid-eye-movements sleep. *Neuroimage* 20 (1): 125–34.

Plihal, W. and Born, J. 1997. Effects of Early and Late Nocturnal Sleep on Declarative and Procedural Memory. *J. Cognitive Neurosci.* 9 (4): 534–547.

Robertson, E. M., Pascual-Leone, A., and Press, D. Z. 2004. Awareness modifies the skill-learning benefits of sleep. *Curr. Biol.* 14 (3): 208–12.

Robertson, E. M., Press, D. Z., and Pascual-Leone, A. 2005. Off-line learning and the primary motor cortex. *J. Neurosci.* 25 (27): 6372–8.

Schabus, M., Hodlmoser, K., Gruber, G., Sauter, C., Anderer, P., Klosch, G., Parapatics, S., Saletu, B., Klimesch, W., and Zeitlhofer, J. 2006. Sleep spindle-related activity in the human EEG and its relation to general cognitive and learning abilities. *Eur. J. Neurosci.* 23 (7): 1738–46.

Schmidt, C., Peigneux, P., Muto, V., Schenkel, M., Knoblauch, V., Munch, M., de Quervain, D. J., Wirz-Justice, A., and Cajochen, C. 2006. Encoding difficulty promotes postlearning changes in sleep spindle activity during napping. *J. Neurosci.* 26 (35): 8976–82.

Siapas, A. G. and Wilson, M. A. 1998. Coordinated interactions between hippocampal ripples and cortical spindles during slow-wave sleep. *Neuron* 21 (5): 1123–8.

Siegel, Jerome M. 2001. The R.E.M. sleep-memory consolidation hypothesis. *Science* 294 (5544): 1058–1063.

Smith, C. 2001. Sleep states and memory processes in humans: procedural versus declarative memory systems. *Sleep Med. Rev.* 5 (6): 491–506.

Steriade, M. 2000. Corticothalamic resonance, states of vigilance and mentation. *Neurosci.* 101 (2): 243–76.

Steriade, M. and Timofeev, I. 2003. Neuronal plasticity in thalamocortical networks during sleep and waking oscillations. *Neuron* 37 (4): 563–76.

Stickgold, R. 2005. Sleep-dependent memory consolidation. *Nature* 437 (7063): 1272–8.

Stickgold, R., James, L., and Hobson, J. A. 2000. Visual discrimination learning requires sleep after training. *Nat. Neurosci.* 3(12): 1237–8.

Suzuki, W. A. 2006. Encoding new episodes and making them stick. *Neuron* 50 (1): 19–21.

Takashima, A., Petersson, K. M., Rutters, F., Tendolkar, I., Jensen, O., Zwarts, M. J., McNaughton, B. L., and Fernandez, G. 2006. Declarative memory consolidation in humans: a prospective functional magnetic resonance imaging study. *Proc. Natl. Acad. Sci. U.S.A.* 103 (3): 756–61.

Tucker, M. A., Hirota, Y., Wamsley, E. J., Lau, H., Chaklader, A., and Fishbein, W. 2006. A daytime nap containing solely non-REM sleep enhances declarative but not procedural memory. *Neurobiol. Learn Mem.* 86 (2): 241–7.

Vertes, R. P. 2004. Memory consolidation in sleep; dream or reality. *Neuron* 44 (1): 135–48.

Vertes, R. P. and Eastman, K. E. 2000. The case against memory consolidation in REM sleep. *Behav. Brain Sci.* 23 (6): 867–76; discussion 904–1121.

Vertes, R. P. and Siegel, J. M. 2005. Time for the sleep community to take a critical look at the purported role of sleep in memory processing. *Sleep* 28(10): 1228–9; discussion 1230–3.

Wagner, U., Gais, S., and Born, J. 2001. Emotional memory formation is enhanced across sleep intervals with high amounts of rapid eye movement sleep. *Learn Mem.* 8 (2): 112–9.

Walker, M. P., Brakefield, T., Morgan, A., Hobson, J. A, and Stickgold, R. 2002. Practice with sleep makes perfect: sleep-dependent motor skill learning. *Neuron* 35 (1): 205–11.

Wilson, M. A. and McNaughton, B. L. 1994. Reactivation of hippocampal ensemble memories during sleep. *Science* 265: 676–679.

Wixted, J. T. 2004. The psychology and neuroscience of forgetting. *Annu. Rev. Psychol.* 55: 235–69.

10

Attention, Selection for Action, Error Processing, and Safety

Magdalena Fafrowicz and Tadeusz Marek

CONTENTS

Abstract...203
10.1 Technological Progress, Attention, and Errors..................................204
10.2 Attention Networks..205
10.3 Attention Networks and Errors...206
10.4 Executive Network and Anterior Cingulate Cortex..........................207
    10.4.1 EEG Studies: Error-Related Negativity, Feedback
        Error-Related Negativity, and Error Positivity.......................210
    10.4.2 Human Error, ERN, PE, Feedback ERN, and
        Neuroadaptive Interfaces..213
Acknowledgments..214
References..214

Abstract

Advances in cognitive neuroscience are leading to a clearer understanding of the attention networks of the human brain. Positron emission tomography (PET) and functional magnetic resonance imaging (fMRI) have enabled us to identify brain mechanisms underlying attention processes. The progress in this domain shows attention processes in an entirely different perspective as compared to classical attention concepts. The attention system governs all cognitive processes, from perception to decision and execution. The system is based on neural networks with different locations for the orienting, executive, and vigilance subsystems. Disturbance occurring in the interaction of attention subsystems is one of the main sources of errors appearing in the work process; the demands of dynamic changes in circumstances exceed the regulating capabilities of attention. In some critical situations, the system is not able to meet the borderline demands. To understand the way in which disturbances occur in the coordination of attention subsystems, interconnections between them are analyzed. The efficiency of attention networks plays a basic role at every moment of work processes. There are different configurations of attention network activity under different task demands, and excessive task demands can decrease dramatically the effectiveness of the system. In recent times, the efficiency of attention networks has become a crucial point in ergonomic design (especially in advanced technologies). From the neuroscience point of view, there are several dimensions related to attention subsystems that have to be taken into account in analyzing safety at work. This chapter deals with these problems. Particular attention will be given to recent discoveries on the role of the anterior cingulate cortex in the selection of targets for action, and to error processing.

Key words: neuronal attention system, saccadic eye movements, task demands, error processing, safety.

10.1   Technological Progress, Attention, and Errors

As a result of the rapid development of modern technologies imposing ever-higher requirements concerning the efficiency of cognitive processes on workers, this area is becoming essential for research mainly from the point of view of risk prevention (Marek 2003).

Maladjustment of cognitive requirements created by technological progress to human abilities can produce serious consequences, especially in the form of errors that often lead to accidents. Researchers and practitioners specializing in the field of safety at work are convinced that a significant increase in the number of errors attributed to people rather than technology is actually due to the growing incompatibility between the individual and modern technology. The absence of compatibility can be regarded as a "hidden risk," defined by Chapanis as a danger existing in the system, which is not directly perceptible or self-evident (Chapanis 1979; Marek and Pokorski 2004).

In advanced technologies, which impose sophisticated requirements, designing of risk prevention programs based on behaviorally oriented job analysis is not enough to reduce incompatibility and, consequently, minimize the hidden risk. Work activity analysis performed on a behavioral level ought to be replaced by analyses of neural mechanisms underlying this activity.

The efficiency of attention plays a crucial role in every moment of work. Attention determines how our brain processes information at sensory, motor, and other areas. New technologies impose specially high demands on the attention system. On the one side, variable conditions might expose this system to entirely different, and excessive, demands and create errors. On the other hand, error detection and monitoring is one of the most fundamental processes regarding the adaptive functions of attention and the possibility of realizing one's goals and needs. Therefore, the error monitoring process has to be automatic to enable this adaptation to be the most effective. Here, an important role is played by early, effective access, which involves the immediate evaluation and definition of events and stimuli. This access precedes subsequent attention—cognitive recognition and interpretation. Error detection is related to this first automatic process.

Thus, the efficiency of attention becomes a crucial point in risk prevention and in the effectiveness of a human–technology system.

10.2 Attention Networks

The attention system governs all cognitive processes, from perception to decision and execution. According to present-day knowledge, the attention processes are regarded as a system, which is anatomically and functionally independent.

The functioning of the attention system is determined by the activity of various structures cooperating with one another. Attention is not managed by a single center, nor is it the function of the whole brain. Particular functions of attention are managed by various functional–anatomic neural networks of the brain. Three such attention networks can be distinguished: (1) the network responsible for maintaining the state of alertness, which is involved in both the overall readiness for action and specific activities directed at the goal, (2) the network that manages the orienting and monitoring processes, and (3) the executive network that controls information recognition and processing.

The basic task of the alert attention network is maintaining a sustained state of alertness. There are two areas of the brain that are active when subjects are required to maintain a state of alertness: the right frontal and right parietal lobes.

The orienting attention network consists of three neural structures. Each structure performs a different function required to orient attention. The parietal lobe acts to disengage attention from its current focus. The superior colliculus moves the spotlight of attention from its current engagement to the area of the cue. Finally, the pulvinar lobe selects the contents of the attended area or attended object and enhances those contents so that they are given priority for processing.

Even while attention is engaged in a new location, the executive attention network comes into play. The anterior cingulate cortex (ACC) provides a connection between widely different dimensions of attention (attention to objects, visual locations, and semantic contexts and actions, including working memory). The network has the task of bringing an object or given area into conscious awareness. It also includes recognition of the identity of the object or area, and the realization that they fulfill a sought-after goal. There is a kind of execution of an instruction. The network has the responsibility of ensuring that the instruction is followed. The basic function of the network is to select a target from among many alternatives. It detects the target and executes control over the action. According to Posner and Raichle, "the order of the computations is determined by the degree to which the computations are primed by the executive attention network." (Posner and Raichle 1994).

The networks perform, by interacting with each other, complex functions that determine human cognitive activity. For example, the alerting network has a strong effect on the orienting as well as executive networks. During alert states, the orienting attention network is tuned, and the executive attention network's activity is inhibited. The alerting network favors the right cerebral hemisphere. Similarly, in the case of the orienting attention network, the right parietal lobe handles attention shifts in both visual fields (Posner and Raichle 1994).

10.3   Attention Networks and Errors

Excessive task demands can decrease dramatically the effectiveness of the attention systems. There are different configurations of attention network activities under different task demands. The demands of dynamic changes of circumstances exceed the regulating capabilities of the attention system. Disturbances occurring in the interaction of attention networks are the main source of errors appearing in the work process (Marek, Fafrowicz, and Pokorski 2004). They result mainly from the rapid and dynamic changes of task demands. The time parameter plays the fundamental role here and is in most cases responsible for the inefficiency of the system. In some critical situations, the system is not able to meet the borderline demands. The attention networks routinely go about their business through an orchestration of facilitatory and inhibitory processes. Each attention operation is likely to be associated with the activation of some structures and inhibition of others. During activity of the alerting attention network, response time is reduced, and error rates increase (the executive attention network is inhibited). The alerting and orienting attention systems inhibit the activity of the executive attention network, which shows stronger activation while mental load and level of task complexity increase. Response time increases, and error rates decrease. The executive attention network inhibits orienting and alerting activity.

To understand the way in which disturbances occur in the coordination of attention networks, leading subsequently to wrong decisions and actions or blockage of decisions and actions, the interconnections between attention subsystems have to be explained. The activities of the orienting and executive networks in visual perception could be identified with the peripheral and central parts of the visual field, respectively. There is a body of evidence that the extent of the central visual field (physiologically determined by the dimensions of the retinal area, characterized by the highest visual acuity—up to 2°) might change according to various factors. Its range determines the functional range of attention, depending on the complexity of the perceived area, experience, or level of arousal (Marek and Fafrowicz 1995). The functional range of attention is first of all linked to actual (definite) fixation, allowing into the region of sharp vision the elements of the perception field, to be recognized and analyzed in detail. The fixation time is determined by the time required for this recognition. The functional range of attention could be different in consecutive fixations, bigger or smaller (wider or narrower).

The process of attention shifting from one field of perception to another is one of the basic processes lying at the foundation of scanning. The so-called covered shift of attention constitutes a kind of preparation for the operation of programming and performing a saccadic eye movement, which leads to a change of the fixation point. The direction and amplitude of a saccadic eye movement are determined on the basis of the difference between the primary fixation point and the new point of attention. The process of scanning and its dynamism are determined by two contradictory mechanisms. One of them is responsible for the depth and the other for the flexibility—the span—of scanning. These mechanisms could be identified with local- and global-oriented attention, respectively. Keeping functional balance between the two mechanisms seems to be a basic condition for effective cognitive processing. The remaining mechanisms are determined mainly depending on the type and structure of the task.

The range of attention is closely related to its concentration or, as it is sometimes defined, the depth of processing of the information in the attention field. Enlargement of the range of attention causes the processing to be more shallow. On the other hand, the narrowing of the field of attention leads to deeper processing.

One of the principal factors causing widening or narrowing of the attention field is the state of the executive, alerting, and orienting networks. Narrowing of the attention field is predominantly caused by difficult and stressful situations. To adopt a simplified model, it can be stated that, with the narrowing of the functional field of attention, the processing of information of the executive type increases, whereas, with the widening of the attention field, detective information processing starts. Under stress, significant narrowing of the attention field occurs with simultaneous increase in the depth and efficiency of information processing. Such a state is described as *tunnel vision* (Marek, Fafrowicz, and Pokorski 2004). In extreme cases, severe narrowing of the field of attention might be associated with the switching off of the orienting subsystem responsible for peripheral vision. It is a well-known fact that at the heart of stress is loss of control. Therefore, it seems to be logical that, in stressful situations, the executive subsystem is activated to regain control and bring the system back to balance. After 20 years of research on neuronal attention networks, it is clear that a crucial role in this process is played by the executive network.

10.4   Executive Network and Anterior Cingulate Cortex

Various terms are used for executive attention, such as supervisory, focused, selective, or conflict resolution attention. According to Norman and Shallice's (1986) model of higher-level executive attention, there are five types of situations in which the executive network is required:

1. Situations involving planning or decision making
2. Involving error correction
3. Situations in which the response is novel and not well learned
4. Those judged to be difficult or dangerous
5. Those that require overcoming of habitual responses

The executive attention system comprises the mechanisms for monitoring and resolving conflict between responses, thoughts, and feelings (Raz 2004). This network is related also to the feeling of mental effort (Fernandez-Duque, Baird, and Posner 2000).

Anatomical, clinical, electrophysiological, and hemodynamical data provide support for the existence, in the area of the cingulate cortex, of two specialized subdivisions—the anterior cingulate cortex (ACC) and the posterior cingulate cortex (PCC). The former functions as an executive part, and the latter an evaluative part. As mentioned previously, ACC is recognized as a nodal element of the executive attention network. ACC's strong neuronal connections to the limbic, association, and motor cortex explain how this structure's activations influence complex reactions such as selective attention, motivation, or goal-directed behavior. ACC integrates inputs from various sources including motivation, evaluation, and representations from cognitive and emotional networks, and influences activities in other brain regions modulating cognitive, motor, endocrine, and visceral responses (Bush, Luu, and Posner 2000).

ACC is functionally subdivided into a dorsal part and a rostral-ventral part (Devinsky, Morrell, and Vogt 1995; Bush, Luu, and Posner 2000). The dorsal and ventral parts of ACC are interconnected with each other (Etkin et al. 2006).

Dorsal ACC has strong reciprocal connections with the lateral prefrontal cortex, and parietal cortex, premotor and motor areas (Devinsky, Morrell, and Vogt 1995), whereas the rostral maintains interconnections with the amygdala, periaqueductal gray, nucleus accumbens, hypothalamus, anterior insula, hippocampus, and the orbitofrontal cortex. The rostral part has efferent connections with the autonomic, visceromotor, and endocrine systems (Bush, Luu, and Posner 2000).

It has been proved that rostral ACC is inextricably connected with conditioned emotional learning, vocalizations associated with expressing internal states, assessments of motivational content and assigning emotional valence to internal and external stimuli, maternal–infant interactions (Devinsky, Morell, and Vogt 1995), regulation of emotional responses and evaluation of emotional and motivational information (Bush, Luu, and Posner 2000), and resolution of emotional conflict (Etkin et al. 2006).

Dorsal ACC is perceived as responsible for different functions, including motor control (Turken and Swick 1999), response selection and cognitively demanding information processing, motivation, novelty and error detection, monitoring competition, anticipation, working memory, and reward-based decision making (Bush, Luu, and Posner 2000).

ACC plays a significant role in cognition. Dorsal ACC is activated by cognitively demanding tasks (e.g., color Stroop, divided-attention tasks, verbal- and motor-response selection tasks, and working memory tasks), whereas rostral ACC is stimulated by affect-related tasks (e.g., studies of emotional processing, inducing sadness, symptom provocation studies in psychiatric disorders such as schizophrenia and depression). Moreover, cognitively demanding tasks, in addition to activating dorsal ACC, deactivated rostral ACC, and emotional tasks activated rostral ACC with deactivation of the dorsal ACC (Bush 2004). Generally, dorsal ACC is activated in cognitive conflict tasks, and the rostral ACC after an error occurs (Luu and Pederson 2004; Raz and Buhle 2006).

While performing cognitively demanding tasks, many components of the limbic system were suppressed (anterior cingulate rostral division, orbitofrontal cortex, amygdala, and insular cortex), which indicates the reciprocal moderate inhibiting influence of these two subdivisions (dorsal ACC and rostral ACC).

Analyses of brain mapping studies support functional differences of specific areas of ACC, delineating cognitive (dorsal) and affective (rostral) subdivisions. There is evidence of ACC as part of a circuit of a form of attention that regulates both cognitive and emotional processing.

According to brain activation studies, ACC is responsible for error processing and responds specifically to the occurrence of conflict and error detection.

Studies providing evidence of ACC activation following conflict occurrence and error detection may be divided into two groups. The first is driven by hemodynamic studies (PET and fMRI studies), whereas the second is based on event-related potentials in EEG recordings.

The first type of studies included tasks requiring the participants to override automatic and inappropriate responses. The classic Stroop-like tasks were the most popular in this sort of research. Increased ACC activation in Stroop tasks were widely observed (Bush et al. 1998, 1999; Bush, Luu, and Posner 2000; Luu, Tucker, and Makeig 2004; Whalen et al. 1998), which suggests this structure's link to conflict occurrence. Hemodynamic studies also providing evidence of ACC engagement in conflict detection required participants to choose one of many equal responses. Most of the experiments conducted in these paradigm studies were based on verbal tasks (Botvinick et al. 2001) and provided evidence of ACC activation under conditions of undetermined responding.

The second group of studies relates to the links of ACC activation to the commission of errors. Recently, there has been a tendency to propose association of errors with conflict due to the possible overlap of pathways of correct and wrong answers (Botvinick et al. 2001). This interference of pathways may be treated as a conflict.

Considering issues of conflict processing, the differentiation between conflict-monitoring and conflict-resolving processes has to be emphasized. In line with new neuroimaging studies (Etkin et al. 2006), the activity found in the dorsal cingulate and the dorsomedial prefrontal cortex was linked to conflict monitoring, whereas the activity in the lateral prefrontal cortices was associated with conflict resolution.

In the opinion of Carter and coworkers (Carter, Botvinick, and Cohen 1999), ACC plays a main role in conflict monitoring, rather than in conflict resolution.

However, several studies have shown activation of the anterior cingulate gyrus and supplementary motor areas, the orbitofrontal cortex, the dorsolateral prefrontal cortex, the basal ganglia, and the thalamus during effortful cognitive processing, conflict resolution, error detection, and emotional control (Bush, Luu, and Posner 2000; Fernandez-Duque, Baird, and Posner 2000; Posner and Fan 2004). Moreover, the anterior cingulate and lateral frontal cortices are areas of action for the dopamine receptors system (Posner and Fan 2004). This suggests that these structures are involved in learning processes and, thereby, in conflict resolution.

Detailed analyses of the "monitoring versus resolution" dilemma were provided by dense-array EEG studies.

10.4.1 EEG Studies: Error-Related Negativity, Feedback Error-Related Negativity, and Error Positivity

Event-related potential (ERP) technology is valued for its direct reflection of neuronal functions, unlike fMRI and PET methodologies, which rely on indirect measures of neuronal activity. The sluggishness of the hemodynamic signal does not permit fMRI to provide firm evidence regarding when ACC is active relative to the stimulus and the response. Thus, ERP recordings may provide further insight into the temporal dynamics of ACC activity during performance monitoring.

An ERP component called error negativity (Ne), or error-related negativity (ERN), was first reported by Falkenstein and his colleagues (Falkenstein et al. 1991) and Gehring (Gehring et al. 1993).

ERN is a large negative-polarity peak in the event-related brain potential waveform (ERP) that occurs when subjects make errors in reaction-time tasks. It begins at the moment of the error and reaches a maximum about 50–100 ms later (Gehring et al. 1993; Holroyd, Dien, and Coles 1998). ERN has a frontocentral distribution over the scalp, and it is largest at the top of the head.

According to Coles and coworkers, the production of the ERN seems to be tied to slips in unwilled actions but not mistakes in willed actions (Coles, Scheffers, and Fournier 1995). The amplitude of this electrophysiological marker reflects the degree of mismatch between the two representations of the erroneous response and the correct response, or the degree of error detected by the participants. The amplitude of ERN increases with the importance of errors (Falkenstein, Hohnsbein, and Hoormann 1995; Gehring et al. 1993). ERN is identified as a response-locked component of ERP and is unrelated to inhibiting or correcting the erroneous response. The key ERN-induced factor is the discrepancy between what is expected and what is executed.

Another component of ERP, the ERN-like deflection, is recognized as a neural response related to performance-monitoring activities. There is a systematic relationship between the participants' subjective perception of their response accuracy and the ERN amplitude, with increasingly larger ERN associated with incorrectly perceived responses. This phenomenon has been called the "feedback ERN." It appears when an error feedback is presented. The ERN-like deflection is stimulus locked (to distinguish it from the response-locked ERN deflection) and has midfrontal distribution and peaks between 250 and 350 ms after feedback onset (e.g., Luu and Tucker 2004). The discovery of feedback ERN seemed to indicate that ERN production depends rather on the detection of the error or learning from the error, and is not only associated with error commission. Taken together, these results motivated the theory that ERNs are elicited by a process of error detection.

ERN has been assumed to originate in ACC (Dehaene, Posner, and Tucker 1994). This is supported by converging lines of evidence from FMRI, PET, magnetoencephalography, and single-cell recordings (Luu, Tucker, and Makeig 2004). Imaging studies have found that the dorsal and rostral regions of ACC are activated when errors occur (Polli et al. 2005), but it should be noticed that the results of the research of van Veen and coworkers (van Veen et al. 2004) suggest that error feedback negativity may be generated not by ACC but by other elements of the systems evaluating performance and feedback. It is also possible that feedback ERN is a more complex process, and the contribution of ACC to this process might be more paradigm dependent. However, Luu and Tucker (2002, 2004) found that ERN is made of two components. The first is located in rostral ACC and is locked to the response, whereas the second is located in dorsal ACC and is locked to the feedback (response-locked measures).

Tucker and colleagues (Tucker et al. 1999) have observed that medial frontal negativity (MFN) response discriminated between two types of feedback (good and bad feedback signals), and between good and bad targets. The authors argue that the signal activity (350 ms) reflects the evaluation of feedback on the subject's performance as well as the initial target evaluation. Luu and coworkers (Luu et al. 2003) found that an MFN is made up of dorsal ACC source. They concluded that the stimulus-locked ERN (or MFN) and response-locked ERN are not the same. Kiehl, Liddle, and Hopfinger (2000), using a go-no-go paradigm, found that conflict activates the dorsal part of ACC, whereas error activates rostral ACC. Menon and colleagues (Menon et al. 2001) also found error-specific activation of the rostroventral region of ACC, whereas high-response conflict activated caudal ACC.

Luu and Tucker (2002) and Luu and Pederson (2004) suggest that dorsal ACC monitors the context of action (the target, response deadline, feedback value), whereas rostral ACC monitors the response. This point of view is confirmed by PET study. Eliot and Dolan (1988) found that dorsal ACC is activated in the phase of hypothesis generation (hypothesis concerning correctness of response), whereas ventral ACC remains active when a choice is made. According to the concept of ACC as a form of attention that is committed to regulating actions in contexts, ACC participates in three processes of action regulation: (1) monitoring context violation, (2) monitoring response relative to the context, and (3) evaluating the motivational or affective consequence of expectancy violations (Luu and Pederson 2004).

According to Luu and Tucker (2002), ACC deals with rapid actions and highly novel situations related to fight–flight conditions, in contrast to PCC, which deals with routine actions. The authors argue that ACC, in association with amygdala, plays a crucial role in emergency situations, whereas PCC, together with the hippocampal network, acts in the more gradual context-updating mode of regulation.

In this perspective, special attention is given to the relationship between ACC and amygdala. The inhibitory effects of rostral ACC activity on amygdala have been shown in fMRI studies (Etkin et al. 2006).

Bush, Luu, and Posner (2000) examined the relationship between the amplitude of ERN and the dimension known as negative emotionality. ERN related to ACC seems to reflect an affective response to error. The idea is that an affective evaluation occurs during error detection, and this evaluation varies along a continuum related to the distress of making an error. It was found that a high level of negative emotionality is correlated with low activity of the rostral ACC. In that case, the inhibitory effect of rostral ACC activity on amygdala is weaker (Hull 2002; Shin et al. 2005). Similar regularity was found in the case of depression (Kumari et al. 2003; Etkin et al. 2005).

The amplitude of ERN can be modulated by different factors. It increases with the importance of the error (Falkenstein et al. 1995). ERN decreases as the quality of performance is affected by fatigue (Scheffers et al. 1996). It depends on task difficulty (West and Alain 1999), stimulus and response uncertainty (Pailing and Segalowitz 2004), aging (Falkenstein et al. 2001; Mathewson et al. 2005), and alcohol intake (Ridderinkhof et al. 2002).

The next error-related component named error positivity (Pe) is observed between 200 and 400 ms following an incorrect response (Falkenstein et al. 2000; Nieuvenhuis et al. 2001). Pe is a positive deflection at the centroparietal sites and reflects an error-related process that is independent of ERN. Error positivity occurs only after a conscious perception of the error (in contrast to the ERN) (Leudhold and Sommer 1999; Falkenstein 2004). Van Veen and Carter (2002) found that one of the Pe sources is located in the rostral part of ACC and is related to emotional error processing.

Hsieh and coworkers found that sleep deprivation influenced the error correction ability (Pe) as well as the efficiency of the error monitoring process (ERN) (Hsieh et al. 2007).

10.4.2 Human Error, ERN, PE, Feedback ERN, and Neuroadaptive Interfaces

The key task of neuroadaptive technologies is to promote a safer and highly effective human–machine system performance and to design neuroadaptive interfaces, characteristics that change from variations in the state of neural networks. The variations of neuroadaptive interface states are indexed by the appropriate neural networks' activities, which control the functionally adaptive modulation of the system. Hettinger and coworkers pointed out, "While fully functional adaptive interfaces … do not currently exist, there are promising steps being taken toward their development, and great potential value in doing so—value that corresponds directly to and benefits from a neuroergonomic approach to system development." (Hettinger et al. 2003).

A growing number of research projects and developments related to neuroadaptive technologies has appeared in the literature in recent years. Let us recall a short review presented at the Eleventh International Conference on Human Aspects of Advanced Manufacturing: Agility and Hybrid Automation (Marek et al. 2007). This includes research devoted to different types of neural indices of cognitive workload (Marek and Fafrowicz 1993; Baldwin 2003; Just, Carpenter, and Miyake 2003; Kerns et al. 2004); research dealing with neural indices in vigilance (Fafrowicz, Marek, and Noworol 1993; Warm and Parasuraman 2007); Donchin and coworkers' research on EEG P300-based brain–computer interface (Donchin, Spencer, and Wijesinghe 2000); and Pfurtscheller and Neuper's research devoted to motor imagery and direct brain–computer communication (Pfurtscheller, Scherer, and Neuper 2007). Scerbo and coworkers (Scerbo, Freeman, and Mikulka 2003) analyzed EEG signals as brain-based systems for adaptive automation; Hettinger and his colleagues described brain–computer interfaces (Hettinger et al. 2003). Kramer and McCarley focused on oculomotor behavior and described how knowledge concerning neural mechanisms of this type of behavior might be applied in neuroergonomic study and design (Kramer and McCarley 2003). Automation cuing modulated cerebral blood flow and vigilance in a simulated air-traffic control task. This was the topic by research of Hitchcock and colleagues (Hitchcock et al. 2003); Gevins and Smith reviewed a long-term program of research aimed at developing cognitive workload–monitoring methods based on EEG measures (Gevins and Smith 2003). Fafrowicz, in her research, analyzed the operation of attention disengagement, measured by oculomotor indices, and its diurnal variability from the point of view of human error (Fafrowicz 2006); human brain activity in motor control task was the research topic of Karwowski and his coworkers (Karwowski et al. 2007); and finally, three excellent papers deal with the problem of handicapped people and neuroadaptive technologies (Pfurtscheller, Scherer, and Neuper 2007; Poggel, Merabet, and Rizzo 2007; Riener 2007).

The elimination of human error seems to be the crucial problem within the scope of neuroadaptive technologies. From this point of view, the most interesting research is on executive neural networks (Fu and Parasuraman

2007; Grafman 2007). What is important from the perspective of neuroadaptive interfaces is that the network is activated by error appearance, error monitoring, and correction.

The very first candidates considered as excellent indices of error processing (monitoring, correction, and correction) are ERN, Pe, and feedback ERN.

As Fu and Parasuraman (2007) state, the relevance of ERN to neuroergonomic research and applications is straightforward. ERN allows identification, prediction, and perhaps prevention of operator errors in real time. A study by Fiehler and coworkers (2004) indicated that rostral ACC is a common neuronal substrate for error detection and correction. Thereby, ERN and Pe could be used to identify the human operator's tendency to commit or recognize, or to correct, an error. It could be the basis of an adaptive interface with the ERN and Pe detector to notify the operator or machine regarding the error committed, or corrected.

Acknowledgments

This work was supported by a grant from the Polish Ministry of Science and Higher Education (N106 034 31/3110) (2006–2009).

References

Baldwin, C. L. (2003). Neuroergonomics of mental workload: new insights from the convergence of brain and behaviour in ergonomics research. *Theoretical Issues in Ergonomics Science*, 4, 1–2, 132–141.

Botvinick, M. M., Braver, T. S., Barch, D. M., Carter, C. S., and Cohen, J. D. (2001). Conflict monitoring and cognitive control. *Psychological Review*, 108(3), 624–652.

Bush, G. (2004). Multimodal studies of cingulate cortex. In Posner, M. I. (Ed.), *Cognitive Neuroscience of Attention*, New York: Guilford Press, 207–218.

Bush, G., Frazier J. A., Rauch, S. L., Seidman, L. J., Whalen, P. J., Jenike, M.A., Rosen, B.R., and Biederman, J. (1999). Anterior cingulate cortex dysfunction in attention deficit/hyperactivity disorder revealed by fMRI and the Counting Stroop. *Biological Psychiatry*, 45, 1542–1552.

Bush, G., Luu, P., and Posner, M. I. (2000). Cognitive and emotional influences in the anterior cingulate cortex. *Trends in Cognitive Sciences*, 4, 215–222.

Bush, G., Whalen, P. J., Rosen, B. R., Jenike, M. A., McInerney, S. C., and Rauch, S. L. (1998). The Counting Stroop: An interference task specialized for functional neuroimaging—validation study with functional MRI. *Human Brain Mapping*, 6, 270–282.

Carter, C. S., Botvinick, M. M., and Cohen, J. D. (1999). The contribution of the anterior cingulate cortex to executive processes in cognition. *Reviews in Neuroscience*, 10, 49–57.

Chapanis, A. (1979). Quo Vadis, Ergonomia? *Ergonomia*, 2, 109–122.

Coles, M. G. H., Scheffers, M. K., and Fournier, L. (1995). Where did you go wrong? Errors, partial errors and the nature of human information processing. *Acta Psychologica*, 90, 129–144.

Dehaene, S., Posner, M. I., and Tucker, D. M. (1994). Localization of a neural system for error detection and compensation. *Psychological Science*, 5, 303–305.

Devinsky, O., Morrell, M. J., and Vogt, B. A. (1995). Contributions of anterior cingulate cortex to behaviour. *Brain*, 118, 279–306.

Donchin, E., Spencer, K. M., and Wijesinghe, R. (2000). The mental prosthesis: Assessing the speed of a P300-based brain-computer interface. *IEEE Transactions on Rehabilitation Engineering*, 8, 174–179.

Eliot, R. and Dolan, R. J. (1988). Activation of different anterior cingulate foci in association with hypothesis testing and response selection. *Neuroimage*, 8, 17–29.

Etkin, A., Egner, T., Peraza, D. M., Kandel, E. R., and Hirsch, J. (2006). Resolving emotional conflict: a role for the rostral anterior cingulate cortex in modulating activity in the amygdala. *Neuron*, 51, 1–12, 2596–2607.

Etkin, A., Pittenger, C., Polan, H. J., and Kandel, E. R. (2005). Toward a neurobiology of psychotherapy: basic science and clinical applications. *Journal of Neuropsychiatry and Clinical Neurosciences*, 17, 45–158.

Fafrowicz, M., Marek, T., and Noworol, C. (1993). Changes in attention disengagement process under repetitive visual discrete tracking task measured by occulographical index. In Maras, W. S., Karwowski, W., Smith, J. S., and Pacholski, L. (Eds.), *The Ergonomics of Manual Work*. London: Taylor and Francis, 433–436.

Fafrowicz, M. (2006). Operation of attention disengagement and its diurnal variability, *Ergonomia IJE&HF*, 28 (1), 13–31.

Falkenstein, M. (2004). ERP correlates of erroneous performance. In Ullsperger, M. and Falkenstein, M. (Eds.), *Errors, Conflicts, and the Brain Current Opinions on Performance Monitoring*. Max Plank Institut fuer Kognitions und Neurowissenschaften, Leipzig.

Falkenstein, M., Hielscher, H., Dziobek, I., Schwarzenau, P., Hoormann, J., and Sundermann, B. et al. (2001). Action monitoring, error detection, and the basal ganglia: an ERP study. *Neuroreport*, 12, 157–161.

Falkenstein, M., Hohnsbein, J., Hoormann, J., and Blanke, L. (1991). Effects of crossmodal divided attention on late ERP components. II. Error processing in choice reaction tasks. *Electoencephalography and Clinical Neurophysiology*, 78, 447–455.

Falkenstein, M., Hohnsbein, J., and Hoormann, J. (1995). Event-related potential correlates of errors in reaction tasks. *Electroencephalography and Clinical Neurophysiology Supplement*, 44, 287–296.

Falkenstein, M., Hoormann, J., Christ, S., and Hohnsbein, J. (2000). ERP components on reaction errors and their functional significance: a tutorial. *Biological Psychology*, 51, 87–107.

Fernandez-Duque, D., Baird, J. A., and Posner, M. I. (2000). Executive attention and metacognitive regulation. *Conscious Cognition*, 9, 288–307.

Fiehler, K., Ullsperger, M., and von Cramon, D. Y. (2004). Neural correlates of error detection and error correction: is there a common neuroatomical substrate? *European Journal of Neuroscience*, 19, 3081–3087.

Fu, S. and Parasuraman, R. (2007). Event-related potentials (ERPs) in neuroergonomics. In Parasuraman, R. and Rizzo, M. (Eds.), *Neuroergonomics. The Brain at Work*. New York: Oxford University Press, 32–50.

Gehring, W. J., Goss, B., Coles, M. G. H., Meyer, D. E., and Donchin, E. (1993). A neural system for error detection and compensation. *Psychological Science*, 4, 385–390.

Gevins, A. and Smith, M. E. (2003). Neurophysiological measures of cognitive workload during human-computer interaction. *Theoretical Issues in Ergonomics Science*, 4, 1–2, 113–131.

Grafman, J. (2007). Executive functions. In Parasuraman, R. and Rizzo, M. (Eds.), *Neuroergonomics. The Brain at Work*. New York: Oxford University Press, 159–177.

Hettinger, L. J., Branco, P., Encarnacao, L. M., and Bonato, P. (2003). Neuroadaptive technologies: applying neuroergonomics to the design of advanced interface. *Theoretical Issues in Ergonomics Science*, 4, 1–2, 220–237.

Hitchcock, E. M., Warm, J. S., Matthews, G., Dember, W. N., Shear, P. K., Tripp, L. D., Mayleben, D. W., and Parasuraman, R. (2003). Automation cueing modulates cerebral blood flow and vigilance in a simulated air traffic control task. *Theoretical Issues in Ergonomics Science*, 4, 1–2, 89–112.

Holroyd, C. B., Dien, J., and Coles, M. G. H. (1998). Error-related scalp potentials elicited by hand and foot movements: Evidence for an output-independent error-processing system in humans. *Neuroscience Letters*, 242, 65–68.

Hull, A. M. (2002). Neuroimaging findings in post-traumatic stress disorder. Systematic review. *The British Journal of Psychiatry*, 181, 102–110.

Hsieh, S., Cheng, I-Ch., and Tai, L-L. (2007). Immediate error correction process following sleep deprivation. *Journal of Sleep Research*, 16, 137–147.

Just, M. A., Carpenter, P. A., and Miyake, A. (2003). Neuroindices of cognitive workload: Neuroimaging, pupillometric and event-related potential studies of brain work. *Theoretical Issues in Ergonomics Science*, 4, 1–2, 56–88.

Karwowski, W., Sherehiy, B., Siemionow, W., and Gielo-Perczak, K. (2007). Physical neuroergonomics. In Parasuraman, R. and Rizzo, M. (Eds.), *Neuroergonomics. The Brain at Work*. New York: Oxford University Press, 221–238.

Kerns, J. G., Cohen, J. D., MacDonald, A. W. III, Cho, R. Y., Stenger, V. A., and Carter, C. S. (2004). Anterior cingulate conflict monitoring and adjustments in control. *Science*, 303, 1023–1026.

Kramer, A. F. and McCarley, J. S. (2003). Oculomotor behaviour as a reflection of attention and memory processes: neural mechanisms and applications to human factors. *Theoretical Issues in Ergonomics Science*. 4, 1–2, 21–55.

Kiehl, K. A., Liddle, P. F., and Hopfinger, J. B. (2000). Error processing and the rostral anterior cingulate: an event-related fMRI study. *Psychophysiology*, 37, 216–223.

Kumari, V., Mitterschiffthaler, M. T., Teasdale, J. D., Malhi, G. S., Brown, R. G., Giampietro, V., Brammer, M. J., Poon, L., Simmons, A., and Williams, S. C. et al. (2003). Neural abnormalities during cognitive generation of affect in treatment-resistant depression. *Biological Psychiatry*, 54, 777–791.

Leudhold, H. and Sommer, W. (1999). ERP correlates of error processing in spatial S-R compatibility tasks. *Clinical Neurophysiology*, 110, 342–357.

Luu, P. and Pederson, S. M. (2004). The anterior cingulate cortex. Regulation actions in context. In Posner, M. I. (Ed.), *Cognitive Neuroscience of Attention*. New York: Guilford Press, 232–242.

Luu, P. and Tucker, D. M. (2002). Self-regulation and the executive functions: electrophysiological clues. In Zani, A and Preverbio, A. M. (Eds.), *The cognitive electrophysiology of mind and brain*. San Diego: Academic Press, 199–223.

Luu, P. and Tucker, D. M. (2004). Self-regulation by the medial frontal cortex: limbic representation of motive set-points'. In Beauregard, M. (Ed.) *Consciousness, Emotional Self-Regulation and the Brain*. Amsterdam: John Benjamin, 123–161.

Luu, P., Tucker, D. M., Derryberry, D., Reed, M., and Poulsen, C. (2003). Activity in human medial frontal cortex in emotional evaluation and error monitoring. *Psychological Science*, 14, 47–53.

Luu, P., Tucker, D. M., and Makeig, S. (2004). Frontal midline theta and the error-related negativity: neurophysiological mechanisms of action regulation. *Clinical Neurophysiology*, 115, 1821–1835.

Marek, T. (2003). Attention—neuroergonomics point of view. In Min K. Chung (Ed.), *Ergonomics in the Digital Age*, 1–4.

Marek, T. and Fáfrowicz, M. (1993). Mental effort under repetitive visual discrete tracking task. In Maras, W. S. Karwowski, W., Smith, J. S., and Pacholski, L. (Eds.), *The Ergonomics of Manual Work*. London: Taylor and Francis, 399–402.

Marek, T. and Fáfrowicz, M. (1995). The basis of creative visual perception. In Maruszewski, T. and Nosal, Cz. (Eds.), *Creative Information Processing—Cognitive Models*. Delft: Eburon Publisher, 105–114.

Marek, T., Fafrowicz, M., Golonka, K., Mojsa-Kaja, J., Oginska, H., and Tucholska, K. (2007). Neuroergonomics, neuroadaptive technologies, human error, and executive neuronal network. In Pacholski, L. M. and Trzcieliński, S. (Eds.) *Ergonomics in Contemporary Enterprise*. Madison, WI: IEA Press.

Marek, T., Fafrowicz, M., and Pokorski, J. (2004). Mechanisms of visual attention and driver error. *Ergonomia IJE&HF*, 26(3), 201–208.

Marek, T. and Pokorski, J. (2004). Quo vadis, Ergonomia?—25 years on. *Ergonomia IJE&HF*, 26 (1), 13–18.

Mathewson, K. J., Dywan, J., and Segalovitz S. J. (2005). Brain bases of error-related ERPs as influences by age and task. *Biological Psychology*, 70, 88–104.

Menon, V., Adleman, N. E., White, C. D., Glover, G. H., and Reiss, A. L. (2001). Error-related brain-activation during go/nogo response inhibition task. *Human Brain Mapping*, 12, 131–143.

Nieuvenhius, S., Ridderinkof, K. R., Blom, J., Band, G. P., and Kok, A. (2001). Error-related brain potentials are differentially related to awareness of response errors: evidence from an antisaccade task. *Psychophysiology*, 38, 752–760.

Norman, D. A. and Shallice, T. (1986). Attention to action: Willed and automatic control of behaviour. In Davidson, R. J. Schwartz, G. E., and Shapiro, D. (Eds.), *Consciousness and Self-Regulation*. New York: Plenum, 1–18.

Pailing, P. E. and Segalowitz, S. J. (2004). The error-related negativity (ERN/Ne) as a state and trait measure: motivation, personality and ERPs in response to errors. *Psychophysiology*, 41, 84–95.

Pfurtscheller, G., Scherer, R., and Neuper, Ch. (2007). EEG based brain-computer interface. In Parasuraman, R. and Rizzo, M. (Eds.), *Neuroergonomics. The Brain at Work*. New York: Oxford University Press, 315–328.

Poggel, D. A., Merabet, L. B., and Rizzo, J. F. (2007). Artificial vision. In Parasuraman, R. and Rizzo, M. (Eds.), *Neuroergonomics. The Brain at Work*. New York: Oxford University Press, 329–359.

Polli, F. E., Barton, J. J. S., Cain, M. S., Thakkar, K. N., Rauch, S. L., and Manoach, D. S. (2005). Rostral and dorsal anterior cingulate cortex make dissociable contributions during antisaccade error commission. *PNAS*, 102(42), 15700–15705.

Posner, M. I. and Fan, J. (2004). Attention as an organ system. In Pomerantz J. R. and Crair, M. C. (Eds.), *Topics in Integrative Neuroscience: From Cells to Cognition*. Cambridge, UK: Cambridge University Press.

Posner, M. I. and Raichle, M. E. (1994). *Images of Mind*. New York: HPHLP.

Raz, A. (2004). Anatomy of attentional networks. *The Anatomical Record*. 281B, 21–36.

Raz, A. and Buhle, J. (2006). Typologies of attentional networks. *Nature Reviews Neuroscience, 7*, 367–379.

Ridderinkhof, R., De Vlugt, Y., Bramlage, A., Spaan, M., Elton, M., Snel, J., and Band, G. P. H. (2002). Alcohol consumption impairs detection of performance errors in mediafrontal cortex. *Science, 298*, 2209–2211.

Riener, R. (2007). Neurorehabilitation robotics and neuroprosthetics. In Parasuraman, R. and Rizzo, M. (Eds.), *Neuroergonomics. The Brain at Work.* New York: Oxford University Press, 346–357.

Scerbo, M. W., Freeman, F. G., and Mikulka, P. J. (2003). A brain-based system for adaptive automation, *Theoretical Issues in Ergonomics Science, 4*, 200–219.

Scheffers, M. K., Coles, M. G. H., Bernstein, P., Gehring, W. R., and Donchin, E. (1996). Event-related brain potentials and error-related processing: an analysis of incorrect responses to go and no-go stimuli. *Psychophysiology, 33*, 42–53.

Shin, L. M., Wright, C. I., Cannistraro, P. A., Wedig, M. M., McMullin, K., Martis, B., Macklin, M. L., Lasko, N. B., Cavanagh, S. R., and Krangel, T. S. et al. (2005). A functional magnetic resonance imaging study of amygdala and medial prefrontal cortex responses to overtly presented fearful faces in posttraumatic stress disorder. *Archives of General Psychiatry, 62*, 273–281.

Tucker, D. M., Hartry-Speiser, A., McDougal, L., Luu, P., and deGrandpre, D. (1999). Mood and spatial memory: Emotion and the right hemisphere contribution to spatial cognition. *Biological Psychology, 50*, 103–125.

Turken, A. U. and Swick, D. (1999). Response selection in the human anterior cingulate cortex. *Nature Neuroscience, 2*, 920–924.

Van Veen, V. and Carter, C. S. (2002). The timing of action monitoring processes in the anterior cingulate cortex. *Journal of Cognitive Neuroscience, 14*(4), 593–602.

van Veen, V., Holroyd, C. B., Cohen, J. D., Stenger, A., and Carter, C. S. (2004). Errors without conflict: Implications for performance monitoring theories of anterior cingulate cortex. *Brain and Cognition, 56*, 267–276.

Warm, J. S. and Parasuraman, R. (2007). Cerebral hemodynamics and vigilance. In Parasuraman, R. and Rizzo, M. (Eds.), *Neuroergonomics. The Brain at Work.* New York: Oxford University Press, 146–158.

Whalen, P. J., Bush, G., McNally, R. J., Wilhelm, S., McInerney S. C., Jenike M. A., and Rauch S. L. (1998). The emotional counting Stroop paradigm: a functional magnetic resonance imaging probe of the anterior cingulate affective division. *Biological Psychiatry, 44*, 1219–1228.

West, R. and Alain, C. (1999). Event-related neural activity associated with the Stroop task. Brain Research. *Cognitive Brain Research, 8*, 157–164.

Section IV

Activity Theory and Ecological Psychology and Their Application

11

Activity Theory: Comparative Analysis of Eastern and Western Approaches

Waldemar Karwowski, Gregory Z. Bedny, and Olexiy Y. Chebykin

CONTENTS

11.1 Introduction 221
11.2 General Activity Theory 222
11.3 AT in the West 227
11.4 Systemic-Structural Activity Theory (SSAT) 230
11.5 Learning in AT 238
11.6 Action Theory 240
11.7 Conclusion 242
References 243

11.1 Introduction

Activity theory (AT) is a branch of psychology that studies internal mental processes in an integrated manner with human behavior. A central concept in AT is "activity," which refers to a collection of internal (cognitive) and external (behavioral) processes that are guided by a conscious goal (Petrovsky, 1986). AT originated in the former Soviet Union in the 1930s. However, since the early 1980s there has been a growing interest in the study of AT in the West. This is reflected in the rising number of publications on the topic in Western journals and in the formation of an International Association of Activity Theory. In contrast, there has been a decline in AT research in its birthplace—the former Soviet Union. This decline can be attributed in large part to sociopolitical events. Historically, AT incorporates Marxist philosophy, which was promulgated by the socialist government of the Soviet Union. With the fall of the socialist government, the academic community of Eastern Europe has largely abandoned Marxist philosophy. The new generation of Eastern European psychologists have lost interest in AT because of its historical association with Marxism.

Despite the negative role Marxist ideology played in Soviet science, Marxist philosophy engendered some unique perspectives within psychology. For example, more than other branches of psychology, AT emphasizes the social determination of the psyche and connects human activity with human work (Yaroshevsky, 1994). Furthermore, more so than in other branches of psychology, AT has a strong tradition of applied psychology research, particularly in the study of human work and learning (Anokhin, 1962,1969; Bernshtein, 1966, 1996; Bedny, 1987; Gordeeva and Zinchenko, 1982; Galactionov, 1978; Konopkin, 1980; Kotik, 1974; Landa, 1976; Oshanin, 1977; Platonov, 1982; Pushkin, 1978; Zavalova, Lomov, and Ponomarenko, 1986; Zarakovsky, 2004). Within AT, theoretical and applied studies are tightly interconnected. Applied studies are an important source for the development of theoretical aspects of activity, and the theoretical aspects shape applied research. These unique properties of AT make it a valuable influence for psychology as a whole.

To further the understanding of AT in the wider psychological community, we highlight some of its major developments and attempt to synthesize the Eastern European and Western perspectives on AT. We pay particular attention to the literature on AT published in the Soviet Union that is not widely known in the West. We hope that our synthesis will further the development of AT and its contribution to the field of psychology as whole.

11.2   General Activity Theory

The development of AT in the Soviet Union has its roots in the work of many scientists. However, Vygotsky, Rubenshtein, and Leont'ev played a particularly important role in founding AT (Vygotsky, 1962, 1978; Rubenshtein, 1957, 1959; Leont'ev, 1978). Currently, the work of Vygotsky and Leont'ev is well known in the English-speaking world. At the same time, the work of Rubenshtein is practically unknown. To fill this gap in the literature, this section will focus on the work of Rubenshtein while briefly considering the contributions of Vygotsky and Leont'ev.

Vigotsky's major contribution to AT was the idea that signs as mental tools are major factors in human mental development. According to Vigotsky, humans acquire a system of meanings common to all those who belong to the same culture through social interactions. The processes of acquiring common cultural meanings proceed through the internalization of commonly used sign systems, such as language. Through internalization, signs of the culture become mental tools. According to Vygotsky, mental tools are analogous to physical tools. Whereas physical tools change physical objects, mental tools transform the human psyche and are the main means of mental activity. According to Vygotsky, signs, and therefore mental tools, have a cultural-historical character. The concept of activity was perceived as an actualization of human culture in individual behavior. Vygotsky emphasized the influence of culture on the mind and deemphasized the individual, psychological features of mental tools.

In contrast to Vygotsky, Leont'ev (1978) emphasized the importance of work and play with physical objects in mental development (rather than social interactions). Leont'ev and his students claimed that Vygotsky's theory idealized the historical and social determination of the human mind, and was insufficiently materialistic. According to this point of view, Vygotsky erroneously described human culture as the major force in the development of the human mind. Rather, work and play were the major influences in the development of the human mind. Mental tools that developed based on sign systems such as language were deemphasized by Leont'ev and his students. Historically, the Soviet psychologist's concept of the relationship of the human mind with reality became more closely aligned with the object-oriented activity perspective proposed by Leont'ev. According to post-Vygotskian AT, it was incorrect to reduce mental development to social and historical factors. Practical activity (when the subject interacts with material objects) mediates between the subject and reality.

In summary, both Vygotsky and Leont'ev described human development as a process of internalizing reality. However, in Vygotsky's theory, social interaction plays the major role in the process of internalization. In Leont'ev's theory, in contrast, object-oriented, practical activity is primal in the development of the mind.

We now turn to the work of Rubinshtein. According to Brushlinksy (2001), Rubinshtein was the first to define the notion of object-oriented activity. In Rubinshtein's (1922/1986) article "The principle of creative activity," he outlined the following main characteristics of activity:

1. Activity is a uniquely human behavior, performed by one or several subjects acting as a group.
2. Activity is the interaction of a subject with object or objects, and therefore is object oriented.
3. Activity is not strictly dependent on the external environment, and includes creative elements.

Rubinshtein augmented the previous idea by noting that activity included "creative elements," thus asserting that mental development could not be reduced to the process of internalization. He further refined the notion of activity in his influential article "The problem of psychology in the works of Marx" (Rubinshtein, 1934). In this article he proposed the principle of unity of consciousness and behavior, suggesting that consciousness and behavior were inextricably linked and that consciousness is shaped by behavior. According to Rubinshtein, consciousness develops through the process of transforming objects of activity. This idea was extremely important at a time when psychology was considered either a science of the mind or a science of behavior.

In the same article, Rubinshtein suggested that social and individual aspects of mental development were not contradictory, as psychologists claimed at the time. He noted that much of labor was inherently social. Social labor changes the external world and at the same time transforms the subject and his or her consciousness. According to Rubinshtein's approach (known as the *subject-oriented activity approach*), the psyche is not self-enclosed or isolated. It is connected to the specific, object-oriented activity of a human being and the social world.

Rubinshtein pointed out that items in the external environment are not in and of themselves "objects" within AT. They become "objects" in the formal sense only through an interaction with the subject (Rubinshtein, 1957). Similarly, a human becomes a subject as the person becomes aware of himself or herself as an individual isolated from the external environment. Humans put themselves in contraposition to the environment, and the environment becomes a possible object of activity. Through activity, the subject experiences reality not as a system of stimuli but as a web of objects, and the person experiences others as subjects of activity.

The notion of the "subject" of activity was further developed by Rubinshtein's student Brushlinsky in the subject-oriented theory of activity (Brushlinsky, 1999). According to this perspective, activity is always carried out by an individual or group subject. During activity, the subject always interacts with an object, rather than operating on symbols. For example, in the process of mental activity the subject interacts with objects, the significance of which is revealed through words, concepts, and symbols. Brushlinsky notes that purely "symbol-oriented-activity" is possible only for a computer, which operates on symbols. In this respect, the activity of a computer is qualitatively different from that of a human. Following in the tradition of Rubinshtein, Brushlinsky argued that language and other symbols are not in and of themselves the building blocks of human mental development; rather, practical activity (e.g., play, learning, and work) of children and adults is interconnected with social interaction and forms the foundation of mental development (Brushlinsky, 2001).

Rubinshtein (1959) developed a system of units for the analysis of activity. One of the major units he coined was the "operation" that is directed toward an object and is carried out through motor and cognitive actions. In his first fundamental work, Rubinshtein (1935, p. 337) wrote: "Because practical action comes in direct contact with objective reality, penetrating inside this reality, and transforming it, action is an incredibly powerful tool in the formation of thought, which in turn, reflects objective reality. Thought is carried into objective reality on the penetrating blade of action." Thus, the products of activity are evaluated by the subject and are an important vehicle for human mental development. Every human act changes not only the situation but also develops the self.

Rubinshtein refined important concepts within AT such as motive, goal, and task. He argued that objects in the surrounding world are connected to the motives of the subject and become the goals of activity. The feelings of need become connected to the goals of activity and thus become motives (Rubinshtein, 1989). In contrast, Leont'ev (1978) considered objects themselves to be motives.

Rubinshtein emphasized the role of independent exploration and interaction with the objective world as the source of the objective world's reflection in the consciousness of a person. He therefore disagreed with the notion of internalization as the fundamental driving force in mental development. He argued that a person does not simply internalize ready-made standards. Rather, the external world acts on the mind through the mind's own internal conditions. Activity includes elements of creativity, independence, and the specificity of individual development. According to Rubinshtein, activity is always to some extent independent and creative. From this follows an important assertion regarding mental development: "External influences on mental development always act through internal conditions" (Rubinshtein, 1989).

Leont'ev, on the other hand, argued that external and internal activity have similar structure because external physical activity is internalized and becomes internal mental activity. However, Rubinshtein argued that internal and external activity do not have the same structure. Later, Pushkin (1978) and Tikhomirov (1984) studied the interrelationship between external and internal activity. They showed that internal activity has elements that cannot be directly connected to aspects of external activity. These findings validated Rubinshtein's argument regarding the partial independence of mental activity from external physical activity.

Rubinshtein's conceptualization of activity was also distinct from Vygotsky's. Vygotsky argued that an individual's mental activity shared characteristics with social interactions. Vygotsky emphasized the external properties of social interactions, rather than the individual's experience of these interactions. According to Rubinshtein, Vygotsky's theory thus lacked a true concept of an individual. The role of individual properties of the individual himself or herself in shaping activity were not considered sufficiently by Vygotsky (Al'bukhanova, 1973). In contrast to this, Rubinshtein emphasized the individual character of existence, the transforming and creative nature of human activity.

Rubinshtein conceptualized activity as composed of planned and situation components. Some aspects of activity are planned out in advance of performance. However, activity unfolds as a process and therefore cannot be planned in all of its details. Aspects of activity are situational, in that they are adjusted to the situation as activity unfolds. Thus, a person's activity cannot be entirely predicted in advance, because of its flexible and plastic character.

Rubinshtein was the first to conceive of activity as both a process and a system of action. He developed a distinction between unconscious mental processes from the conscious elements of activity such as actions and goals. According to this idea, mental processes are largely unconscious as they unfold in time. Nevertheless, mental activity is regulated in a largely conscious manner through the conscious evaluation of mental actions, operations, and their results (Al'bukhanova, 1973). Thus, continuous and unconscious mental processes give rise to conscious, object-oriented activity. For example, the process of thinking becomes a mental activity when the goal of this process becomes conscious (Rubinshtein, 1957). Thinking as activity is analyzed in terms of motives, goals, actions, and operations of thinking. When we analyze thinking as a process, we focus on determination of thinking and a way in which thinking unfolds over time. Thinking as a process is not determined in advance and has unpredictable, situated aspects. The process of thinking during awareness of its goal is transferred into thinking activity (Rubinshtein, 1959).

Our analysis of the works of Rubinshtein and Vygotsky allows us to outline major differences in their concepts of activity. Rubinshtein conceived of activity as individualist in character. He analyzed the relationship of external behavior and internal cognition through the prism of individuality. Thus, he argued that internal activity was built by the individual based on external object-oriented and social activity. According to him, social determination of consciousness occurs not from without but rather from within as a result of the sociocultural existence of the individual.

In contrast, Vygotsky conceptualized mental activity as an internalized version of external activity. Thus, individual-psychological aspects of activity were not sufficiently considered in the works of Vygotsky. Criticizing Vygotsky's sociocultural theory, Rubinshtein demonstrates that individual psychological characteristics of a person are not completely derived from the social environment. In the same social environment, different individuals act in varying ways and are impacted by the social environment in different ways. Thus, a different concept of mental development was revealed from that proposed by Vygotsky. Mental activity is shaped by external activity and by the individual characteristics of the subject. Activity is a creative process through which the mind develops. In Rubinshtein's version of AT, this idea has been termed *the personality principle*. The personality of the individual shapes the character of his or her mental activity during interaction with the objective world. Rubinshtein argued against the separation of "natural" and "cultural" functions.

Our comparative analysis of the works of Rubinstein and Leont'ev reveals both differences and similarities. Similar to Rubinshtein, Leont'ev argued for the importance of practical activity as the bases of mental development. He also had similar views regarding the structure of activity: according to both Leont'ev and Rubinshtein, activity included goals, motives, actions, and operations. However, they disagreed on how the human mind developed. Rubinshtein was against the idea of internalization, whereas Leont'ev argued for the idea that practical activity is internalized and becomes mental activity. There are also other differences between the theories of these authors that we have not considered here, such as their ideas about human abilities.

Stepping back and critically comparing the work of Rubinshtein and Leont'ev with the work of Vygotsky, we note that both Rubinshtein and Leont'ev did not sufficiently consider the semiotic aspects of human activity, such as interactions with signs, symbols, and artifacts. Vygotsky stressed that humans create these elements and are influenced by them. According to Vygotsky, signs and symbols were an important part of human development. They were internalized and played a critical role in human cognition. Because signs and symbols are provided for the individual by society, the social world was an important piece of Vygotsky's theory of mental development. At the same time, as compared to Rubinshtein's concept of activity, Vygotsky did not sufficiently consider individual, object-oriented activity. Development cannot be reduced to the internalization of social standards.

In this section, we focused on the work of Rubinshtein because his work is practically unknown in the English-speaking world. We also briefly considered the work of Leont'ev and Vygotsky. The comparative analysis of their work demonstrates that each focused on a different aspect of activity. Vygotsky focused on the sign-mediated nature of activity. For him, the sign is the major mediator of mental development. For Leont'ev and Rubinshtein, the primary tools for mental development were material objects, rather than the sign. In reality, all of these aspects of activity are interconnected. Signs do not exist without objects and social interactions. Rather than putting these approaches in contradiction to each other, it is important to integrate them (Bedny and Karwowski, 2006).

11.3 AT in the West

Discussions of AT in the West are largely based on the work of Vygotsky and Leont'ev. According to Engestrom (1999), one theoretical assumption is that AT is strongly opposed to the idea that people's mental processes can be studied outside their sociocultural context. Since the early 1980s, sociocultural psychology (which has been linked to AT in the West) has attracted a great deal of attention. Engestrom (1999, p. 29) wrote that Vygotsky's basic idea about the mediation of cognition by tools and signs breaks down the Cartesian wall that isolates the individual mind from culture and society. Humans control their own behavior not from the inside on the basis of biological urges but from the outside, using and creating artifacts (Engestrom, 1999). This interpretation of AT is largely based on Vygotsky's work and thus overemphasizes the importance of external influences. Rubinshtein's perspective (as well as that of many modern AT researchers) has been largely overlooked. In contrast to Vygotsky, Rubinshtein emphasized the interaction of internal and external elements of activity. Because the contribution of Rubinshtein is not well known in the West, AT in the West is quite different from modern AT in the former Soviet Union.

Scientists in the West also confront a number of difficulties in the translation and interpretation of basic concepts and terminology derived from AT (e.g., units of analysis, object and subject of study, object of activity, goal, actions, needs and motives, etc.) Very often the same terminology in the West has totally different meanings. Actions get confused with task, or with notion of action in action theory. Limbs of the body get confused with the tools, etc. For example, Nardi (1997) wrote that according to AT a task is something automatic, neat, pure, and ignores motivational forces. As a result Kaptelinin and Nardi (2006, p.259) attempted to introduce the term *engagement* instead of the concept *task*. However, in AT the concept of task includes problem-solving and motivational components (Bedny and Karwowski, 2006).

The concept of goal in AT of the former Soviet Union has a totally different meaning relative to the way in which it is used in the West. For example, Lee et al. (1989) discuss the notion of goal as undifferentiated from motive, whereas within AT motive and goal are distinct though highly interactive components that form the vector "motive–goal." Alternatively, according to Preece et al. (1994, p. 411), the goal is the final state of the system that the human wishes to achieve. From this example, it appears that a goal is an externally given ready-made standard provided to the subject. In contrast, in AT the goal is formed by the subject in stages: goal recognition, goal interpretation, goal reformulation, goal formation, goal acceptance, etc. Therefore, even in the simplest situation, a goal requires recognition and some aspects of interpretation and acceptance.

Leont'ev (1978) discussed situations when the result of activity satisfies the goal and motivation, simultaneously. In this instance, the motive and goal coincide; however, they are still different elements of activity. For example, if two people are hungry, they have a psychological drive to reduce their hunger. This drive should be distinguished from the cognitive representation of obtaining food as a goal. Two people who experience different degrees of hunger and have the same goal—to obtain food—have different degrees (and perhaps qualities) of motivation.

Kaptelinin and Nardi (2006), analyzing Leont'ev's work, considered such concepts as needs and motives, goal, and their interrelationships in AT. To improve upon Leont'ev's analysis, they suggest separating the motive from the object. However, many modern AT researchers of the former Soviet Union do not consider object and motive to be identical. For example, Rubinshtein interprets the motive as a kind of experienced need (see, for example, this discussion in English by Nosulenko et al., 2005). Thus, the separation of motive from the object is a long-standing and well-established development that predates this analysis. Objects are not considered motives but, rather, sources of motivation. Needs, on the other hand, can become motives. However, for transformation of needs into motives, not only needs and objects but also goals are required. When needs are associated with a goal, they are transferred into motives. The notions of motive and goal give rise to the vector motive–goal, which lends activity a goal-directed character (Lomov et al., 1977).

Kaptelinin and Nardi (2006, p. 148) reject the idea that activity has multiple motives. Following D. Leont'ev (1993), they suggest that that there are "several needs and one motive." However, within modern AT in the former Soviet Union, it was thought that there can be several interacting motives and one kind of motivation. Motivation is a broader concept than motive. Instead of the notion "one activity and one motive," one can utilize the expression "one activity and one state of motivation."

The concepts of subject and object are also not always distinguished within Western AT as in former Soviet AT. For example, Engestrom (2000) studied children's medical care from the perspective of Western AT. In this study he formulates the physician as the subject and patient as the object. However, within the rubric of former Soviet AT, the patient is the subject while discussing his or her problems with the doctor. The object of physician's activity is the health conditions of the patient and not the patient. During social interaction, the patient is the subject. During diagnosis of the patient's state, the health condition of the patient becomes the object of activity. Hence, in the physician's diagnostic task the subject–subject relationship can be transferred into subject–object relationship and vice versa.

In Western psychology, there is inconsistent use of the term "object." For example, Kaptelinin and Nardy wrote, "A way to understand objects of activity is to think of them as objectives that give meaning to what people do" (2006, p. 66). The authors apparently understand the object as objectives. Additionally, there is often a failure to distinguish between the object of study and object of activity. Object of activity is something that is modified according to the goal of activity. Objects may be either concrete or abstract. Abstract objects are, for example, signs, symbols, or images, which are transformed by the subject in accordance with the existing goal (Bedny and Karwowski, 2006, Bedny and Harris, 2005).

Additionally, the terms *activity* and *actions* in regard to AT are used differently in the West. Within AT, actions are categorically different from tasks. Actions can either be cognitive or behavioral (motor). Cognitive actions can be classified according to the dominant cognitive processes of which they are a part. For example, cognitive actions can be perceptual, mnemonic, decision-making actions at the sensory-perceptual level, decision-making actions at the verbal logical level, etc. Behavioral actions can be classified based on the objects of the actions and methods of their performance. (Bedny and Meister, 1997; Bedny and Karwowski, 2006). In contrast, the task is a logically organized system of cognitive and motor actions that are directed toward achieving a high-order goal. We return to the discussion of actions in the next section. Here, it is sufficient to note that the term *action* refers to a formalized unit of analysis in AT.

In the West, the term action has a common-sense meaning as referring to motor activity. For example, the cognitive psychologist Preece et al. (1994, p. 414) discuss action as being a task that involves no problem-solving or problem-structure components. Similarly, Western researchers working in AT do not differentiate between the concepts of task and action. For example, Kaptelinin and Nardi (2006, pp. 67–68) write, "making a hunting weapon is an action that entails, at lower level, finding suitable material and tools for manufacturing of the weapon. Therefore, the level of actions is itself hierarchically organized and can be decomposed into an arbitrary numbers of sublevels, higher level of actions to lower level of actions. ... There is an evident difference between higher-level actions and lower-level actions (or task)." From these examples, we can see that actions and tasks that are differentiated in former Soviet AT are not differentiated by some Western scientists.

The term action is also used informally and loosely in the context of applied work. Kuutti (1997) defines "building a house" as an activity and "fixing the roof" as an action. However, both examples are simply production processes. Similarly, in a study of pediatric medical care, physician's care is described as consisting of actions (Engestrom, 2000). However, actions referred to in this study are considered tasks in the framework of AT. For example, according to AT, examination and diagnosis of patients are not an action, but, rather, a diagnostic task.

11.4   Systemic-Structural Activity Theory (SSAT)

General AT is a philosophical rather than practical framework for studying human performance. Beginning in the 1960s and through to the present, there has been a move to modify AT and apply it to the practical study of human work. Zarakovsky (2004) called this direction of AT "an operational AT approach." There are several important directions within this line of research, including the study of pilots' operative image during flight (Zavalova et al., 1986), the microstructural analysis of cognitive and motor actions (Kochurova et al., 1981) and Systemic Structural Activity Theory (SSAT) (Bedny and Karwowski, 2006). This section focuses on and provides a brief overview of SSAT.

SSAT is theoretically based on the work of the Vygotsky, Leont'ev, Rubinshtein, and many other scientists. Yet it is a unique and independent approach, distinct from those that precede it. An important aspect of SSAT is that it views activity as a structurally organized, self-regulative system, rather than as an aggregation of responses to multiple stimuli. Furthermore, it views activity as a goal-directed rather than as a homeostatic self-regulative system. A system is considered goal-directed and self-regulative if it continues to pursue the same goal under changed environmental conditions. The system can also change its goal while functioning, but is nonetheless continuously driven by a goal. Activity is a goal-directed, self-regulated system that integrates cognitive, emotionally motivational, and behavioral components.

SSAT proposes that to be fully described, activity must be analyzed in multiple ways and on multiple levels. SSAT distinguishes between two classes of analytical approaches: parametric and systemic. An example of a parametric approach is cognitive analysis. Cognitive analysis focuses on decomposition of activity into separate cognitive processes. Systemic approaches include functional analysis and morphological analysis. Functional analysis focuses on activity as a self-regulating system, and its major units are functional micro and macro blocks. The units of morphological analysis are cognitive and behavioral actions, operations. During algorithmic analysis of activity,

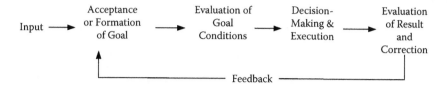

FIGURE 11.1
Simplified model of action as a one-loop system.

interdependent actions can be integrated into more complex units of analysis (Bedny and Karwowski, 2006).

Within SSAT, methods of studying activity are organized into different stages, such as qualitative analysis, algorithmic analysis, analysis of time structure of activity, and quantitative analysis. The qualitative stage of analysis includes parametric (measuring separate parameters and characteristics of activity) and systemic (functional analysis) methods. Algorithmic analysis and time structure description are morphological systemic stages of analysis. The quantitative stage of analysis focuses on quantitative evaluation of complexity and reliability of performance. Quantitative analysis is also related to the systemic analysis of activity. All these stages of analysis are interdependent and have a loop structure organization.

Each stage of analysis can be performed with different levels of decomposition. Macrostructural and microstructural analyses determine the level of analysis. In the context of practical application of SSAT to the analysis of performance, one must select the appropriate analysis levels for the present purpose. Some stages of analysis can be omitted depending on the goal of the study.

During morphological analysis SSAT decomposes activity into cognitive and behavior actions that unfold in time. The starting point of an action is the initiation of a conscious goal (goal acceptance or goal formation). An action is considered completed when the actual result of the action is evaluated in relation to its goal. The simplest model of action as a one-loop system is presented in Figure 11.1 (Bedny and Karwowski, 2006).

Actions are units of analysis or elements into which one can divide activity for the purpose of study. Depending on strategies of performance, a given task may contain different actions that have different logical organizations. Actions can be further subdivided into operations. For example, motor actions can be divided into motions.

Similarly, cognition is considered to be not only a system of cognitive processes but also a system of cognitive actions and operations. Sometimes, mental operations are called mental acts (Platonov, 1982). Further discussion and data regarding the analysis and description of cognitive and motor actions can be found in the works of Gordeeva, Zinchenko, (1982), Zaporozhets (1967), Zarakovsky and Pavlov (1987), Bedny, Seglin, and Meister (2000), Bedny and Harris (2005), etc. Cognitive and motor actions usually require shorter performance time (Bedny and Karwowski, 2006; Gal'sev, 1973; Gordeeva and Zinchenko, 1982; Kochurova, et al., 1981, Zarakovsky, 2004).

Activity is always composed of a set of individual actions, but identifying these actions can be complex, particularly in the analysis of cognitive activity. For example, during analysis of computer-based tasks, it may be necessary to use eye movement registration to extract and classify cognitive actions. SSAT develops normative requirements for units of analysis in the study of human activity. Standardized descriptions of cognitive and motor actions were suggested (Bedny and Karwowski, 2006).

Activity during task performance is the object of study. Within SSAT, activity is analyzed into units such as actions, operations, and function micro and macro blocks. These units of analysis are elements into which one divides the unified whole of activity for the purposes of activity description (Bedny and Karwowski, 2006).

At the micro level of analysis one can use mental and motor operations or functional micro blocks. The subdivision of activity into units is necessarily, to some degree, idealized. Activity is an extremely flexible and dynamic system. Therefore, extraction of individual actions and systematic description of activity can be performed with some degree of approximation. For example, a researcher will describe the more representative strategies of task performance despite other strategies that can be employed. Unexplained variability is a feature of many complex systems. For example, in mass production and manufacturing, each manufactured part is unique in its size and shape. However, if this variation of size and shape is within an established range of tolerance, all parts are considered to be of "the same" size and shape. Similarly, when actions or strategies of performance of activity fall inside an established range of tolerance, these actions or strategies are considered "the same." The notion of accepted and unexplained variability is an important principle of activity design. Therefore, for successful design, it must be acknowledged that real activity can only approach the designed models of activity.

Similarly, it is important to consider variability when we evaluate performance for accuracy and optimization. Activity can be compared with a symphony. Each time an orchestra performs the same symphony, its structure, which unfolds over time, is unique. However, this uniqueness does not contradict the fact that the symphony is the same. One should distinguish uniqueness of reproduced symphony from its incorrect, misrepresented performance. In the same way, repetition of the same activity can be viewed as a unique structure, which unfolds over time. If variation in strategy of performance is within the range of tolerance, then this is the same strategy of activity. The same activity can be performed employing different strategies. Each strategy has its own acceptable range of tolerance.

Analysis of activity during task performance requires one to differentiate between the goal of the task and the goal of individual actions. There are goals of individual actions and goals of the tasks as a whole. Similarly, there are objects of individual actions and objects of activity as a whole. For example, when the subject performs the action "reaches and grasps the computer mouse with the right hand," he performs a simple motor action, which includes two motions or operations (reach and grasp). The goal of this action is the mental image of the result "grasp the mouse." In this example, the object of the action is the mouse. When a subject performs the action "move the cursor to the required position and depress the left mouse button with the index finger for activation of a particular icon," this is an example of another motor action that includes two motions or operations (move and apply pressure). The mental representation of the possible consequence of the activation of a particular icon is the goal of action. In this motor action, the mouse is not an object but rather the tool of action, and the icon is the object. A pointer that the subject moves along the screen is also a tool of this action. The goal of a task can be the mental image of the final result, which is obtained during performance of all logically organized actions during task performance. An example of the object of a task can be the initial organization of interdependent elements on the screen and their final organization after performance of all required actions. If this final organization is not achieved, then the result of activity deviates from the required goal, and therefore, corrections are needed. The logical organization of actions can be described by utilizing algorithmic analysis of activity (an important method of the morphological stage of analysis). This method of analysis is entirely different from analysis of computer algorithms. The purpose of algorithmic analysis of activity is the discovery and description of the logical organization of human cognitive and behavioral actions during interaction with technology. Such algorithms can be described during analysis of work activity. Algorithmic analysis is of great practical importance because it permits the researcher to find and describe the most efficient strategies of activity performance. Task performance with technology requires the subject to know various rules, boundaries, and procedures. Therefore, such activity can be usefully described with the help of deterministic or probabilistic human algorithms, which should be distinguished from computer algorithms, mathematical algorithms, etc.

It must be acknowledged that creative tasks are somewhat difficult to describe in an algorithmic fashion. Nonetheless, this does not render algorithmic analysis unimportant, for two reasons: (1) Purely creative tasks are seldom encountered in the production process. Work frequently involves safety and quality requirements, and time limitations, which must be considered by the subject and render work not purely creative. (2) Most purely creative components of work can be improved with the aid of quasi-algorithmic descriptions. Here, one describes possible strategies and the algorithm allows for some degree of uncertainty. The quasi-algorithmic description can be useful in the study of creative activity. It helps one to better understand how subjects interact with technology in very difficult and unusual situations. Such a description does not guarantee success, but rather is a heuristic description useful for the analysis of unusual and creative tasks in the production environment. Algorithmic analysis subdivides activity into psychological units (members of the algorithm) and determines their logical organization and sequence. Members of an algorithm usually include one or several interdependent actions (motor, perceptual, decision-making actions, etc.). These actions are integrated through a supervening goal into the algorithm member. Individual members of an algorithm are hierarchically organized subsystems within the activity. Subjectively, individual members of an algorithm are perceived by a subject as elements of activity that have a logical completeness. Each member of an algorithm is limited to one to four actions, because of limits on the capacity of working memory, when actions are performed simultaneously or need to be ordered. Actions that comprise algorithm members can be extracted and classified based on developed systems of action classification (Bedny and Karwowski, 2006).

Human algorithms can be deterministic (when logical conditions have only two outputs) and probabilistic (when logical conditions have multiple outputs). In probabilistic algorithms, each output can have different probabilities. Probabilistic algorithms can be used to describe very complicated and flexible types of activity.

Algorithmic analysis requires decomposition of activity into cognitive and motor actions and operations and their combination into qualitatively different members of the algorithm. Consider an example of decomposition of activity into required units of analysis (see Table 11.1). For simplicity we do not consider distance for motor movements in this table.

According to SSAT there are two types of units of analysis. One type utilizes technological and the other uses psychological units of analysis. In this table, we utilize both types of units. Technological units of analysis are elements of the task that are extracted without any requirements regarding their psychological description and classification. For example, "decide to use the left or the right bin" is the technological unit of analysis for a cognitive element of activity. The same element of activity can be defined on the basis of standardized psychological descriptions of cognitive actions. In this case, we would utilize psychological units of analysis. For instance, "decision making at sensory–perceptual level" is a psychological unit of analysis. This description corresponds to the classification system for cognitive actions (see Bedny and Karwowski, 2006). It is possible to develop more precise classification systems for cognitive actions. One important characteristic for such units of analysis is performance time of standardized cognitive actions. Let us consider another example. "Move a part to an air-operated clamping device" is an example of a technological unit of analysis. The description of this element as "motion Move-M30C" by utilizing MTM-1 system is an example of a psychological unit of analysis. These symbols designate the following: M—move; 30—distance in inches; C—move an object to an exact location. Some additional data are used during the description of the "move" operation when the subject applies a force above 2.5 lb. Such a standardized description helps us to specify clearly what kind of movement is performed by a subject.

Technological units are typical elements of work, but they are not sufficiently precise. Psychological units of analysis, on the other hand, are standardized elements of activity. They are developed on the basis of taxonomic principles. Utilizing psychological units of activity helps us to describe any

TABLE 11.1

Example of Decomposition of Activity during Task Performance

Members of Algorithm	Actions	Operations (or Motions)
O^α_1—Look at one bulb, then another	**Simultaneous perceptual actions**	
	1. Look at bulb #1	1. The same
2. Move eyes to the second bulb		
	2. Look at bulb #2	1. The same
I_1—Decide which bin to choose	**Decision-making actions at the verbal-thinking level**	
	1. Decide to choose left bin (or see below)	1. The same
	2. Decide to choose right bin	1. The same
O^ε_2—Take part	**Discrete motor action**	
	1. Move arm and grasp part	1. Move right arm to part (motion Reach-R).
		2. Grasp part (motion Grasp-G)
$O^{-\varepsilon}_2$—Install part in air-operated clamping device	**Discrete motor action**	
	1. Move part to the clamping device and install it	1. Move part to air-operated clamping device (motion Move-M)
		2. Turn hand with part in required position (Motion Turn-T)
		3. Align part with clamps and insert part into fixed secure position (motion-Position- P)

type of activity in a standardized way. Psychological units of analysis help us to understand with a high level of precision what the subject really did. The algorithm member "O^α_1—Look at one bulb, then another" contains the symbol "O^α_1," which is the psychological, standardized description element of the activity at a more general level, and "Look at one bulb and another," the technological unit. The symbol O^α_1 demonstrates that it is a sensory–perceptual element of the activity. At the same time, "Look at one bulb and another" is a technological unit of description because we do not know the distance between bulbs, angle of vision, illumination, time for performance of these actions, etc. Hence, this member of the algorithm is described by utilizing a combination of psychological and technological units of analysis. At the later stage of description, technological units of analysis should be transferred into psychological units. For example, for quantitative evaluation of task complexity, technological units of analysis should be transferred into psychological units (Bedny and Karwowski, 2006). Performance times of activity elements are important characteristics of psychological units of analysis.

In Table 11.1, each member of the algorithm is decomposed into hierarchically organized units. However, it is critical to keep in mind that members of the algorithm have a logical rather than hierarchical organization. Each member of the algorithm can emerge as a relatively independent object of analysis. We can analyze different strategies of performance, utilizing algorithmic descriptions of activity. If necessary, we can augment algorithmic analysis with time structure analysis and quantitative analysis. For further discussion of these analysis types, consult Bedny and Karwowski (2006).

We now briefly turn to a discussion of functional analysis. Functional analysis is a qualitative systemic analysis of activity. It views an activity as a goal-directed, self-regulative system. As a self-regulative system, an activity can change its own structure and content on the basis of self-regulation. The major purpose of functional analysis is to discover and describe the more representative strategies of human performance. Functional analysis is conducted on a macro-level, and thus the functional macro-block is the major unit of analysis. In functional analysis, we describe the self-regulative system of activity as integrating cognitive, executive, evaluative, and emotional–evaluative components of activity. A subject develops different strategies of activity based on the principle of self-regulation. An activity is a situated system because it is adapted to situations on the basis of principles of self-regulation. This self-regulative system contains different functional mechanisms or function blocks. The functional mechanism or function block represents a coordinated system of subfunctions that serve a specific purpose in activity regulation. For example, the goal, evaluation of task difficulty and task significance, formation of the level of motivation, decision making, program of performance, etc., are specific mechanisms in activity regulation. During functional analysis we attempt to describe the activity not so much in terms of cognitive processes or cognitive and behavioral actions as in terms of more complicated integrative functional mechanisms. Every function block or mechanism can include the same cognitive processes. However, their functioning and integration can be performed in different ways, depending on the specificity of the task. The content of the function block can be changed, but its purpose in self-regulation should be the same. From this it follows that cognition may be studied from the perspective of its functional role in the regulation of activity.

Functional blocks can be thought of as different windows through which the scientist views the same object of study. For example, the study of the goal as a functional mechanism embodies a number of specific aspects of activity analysis. This analysis includes, for example, description of goal formation, goal acceptance, goal interpretation, relationships among verbally logical and imaginative components, significance of the goal, deviation of the subjectively accepted goal from the objectively required, etc. The general model of self-regulation contains more than 20 functional blocks, but not all

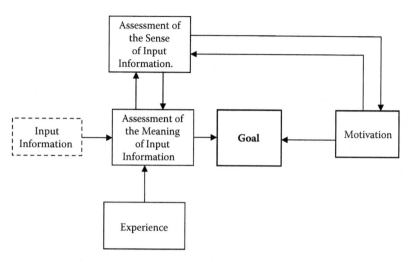

FIGURE 11.2
Functional model of the goal formation process.

functional blocks need to be utilized for a particular study. Only the most pertinent blocks should be chosen.

Functional blocks are connected to each other through feed-forward and feedback connections. Therefore, data must be interpreted with the interdependencies and mutual influences of functional blocks in mind. As an example of systemic qualitative analysis, we considered the simplest functional model of activity that describes the goal formation process only. In Figure 11.2, we present a functional model of the goal formation process.

The function block "assessment of the meaning of the input information" refers to the mechanism that is responsible for representing the state of reality in our consciousness. It is a person's interpretation of the different aspects of a situation. A meaningful interpretation of the situation is possible if the subject has adequate past experience and the ability to use required skills and knowledge. Depending on the subject's past experience and ability, he or she assigns a particular meaning to, and understanding of, the same situation. What is important at this stage of analysis is the difference between the objectively required interpretation and the actual subjective interpretation.

At this stage of analysis one can utilize a lot of data in AT about meaning and its interpretation (see, for example, Bedny and Karwowski, 2004). The function block "assessment of the sense of the input information" in this model is responsible for evaluation of personal significance of information. It is well known in AT that a factor of significance can influence the interpretation of a situation. Therefore, an objectively presented situation can be interpreted in different ways depending on the subjective significance of the situation. This is particularly obvious when we consider the political opinions of people in the United Sates. For example, if I am conservative and a Republican, I might ignore information presented by the Democratic Party and vice versa. The function block "sense" is related to the emotionally evaluative component of activity. The function block "motivation" is different from the function block "sense" because of its essential link with goal directness. The presence of the motivational functional block allows emotionally evaluative components of activity to be transformed into inducing components. This gives the self-regulative system of activity its goal-directed quality, which can be presented as the vector "motive → goal." The more significant the goal is, the more the person is motivated to reach the goal. Therefore, the vector "motive → goal" is the result of a self-regulative process.

The same activity can be studied from different perspectives and the data obtained combined on the basis of the analysis of interrelationships between functional blocks.

11.5   Learning in AT

Within cognitive psychology, cognition is considered a system of internal automated operations that are only loosely connected with the external object-related world. For example, in Anderson's (1987) concept, learning is reduced to purely mental processes, particularly the processes that take place in memory. Learning is described as a set of memorization mechanisms, and memory is treated as the agent of the learning process. The motivational aspects of learning and its goal-directed character are not considered. The focus is on automatic mental processes, which occur through the involuntary application of internal mental rules. These "production rules" are characterized by their automaticity. According to AT, the student, in Anderson's theory, cannot be considered a subject that interacts with the object in the learning activity. The student is considered to be a computer-like system. A great deal of attention in this theory is devoted to the involuntary aspects of learning. In contrast, AT focuses mainly on goal-directed voluntary processes.

AT treats learning largely as a set of consciously regulated processes. Although there are unconscious mental processes and acts, these are not the primary focus of AT. For example, in SSAT, learning is treated as a recursive process that has a goal-directed character and is governed by control, corrections, and analysis of feedback. A self-regulative process comprises conscious and unconscious levels. These levels of self-regulation are interdependent and can be transformed into each other. At the same time, the conscious level of self-regulation is more important in the learning process. Associated learning is an example of the unconscious level of self-regulation. Through the process of self-regulation, the learner acquires different strategies of activity performance. The more complex the learning process, the more numerous are the intermittent strategies utilized by the learner (Bedny and Karwowski, 2006).

The constructivist approach developed by Piaget (1952) shares some important features with AT. Within the constructivist approach, psychological development and learning involve the active interaction of the subject with the environment. The subject is viewed as an organism that attempts to reach equilibrium with the environment. Mental operations emerge from external operations that become mental through the process of internalization. As noted previously, the concept of internalization is also important in AT. The major distinction between AT and the constructivist approach lies in the fact that constructivism does not consider the role of cultural and historical elements in mental development. Piaget's theory describes human development as a "dyadic relationship" between the natural world and mental world. In contrast, in AT human beings act on natural objects indirectly through the use of culturally developed tools that serve a mediating function. The triadic relationship outlined in AT includes the subject as social entity and the process of a subject's acquisition of knowledge through tools and signs, which are products of historical development. Moreover, in SSAT learning is considered as a complicated, goal-directed self-regulative process.

AT emphasizes that learning must include goal formation, consciousness, and speech. Learning is conceptualized as having a specific structure: training task, student goals and motives, possible actions of students and strategies of performance, feedback, body of information to be imparted, and the process of interaction between the student and instructor. Special attention is paid to development of the ability of students to independently formulate and solve different problems. Mental development is considered not as just absorbing knowledge but the ability to acquire it independently, apply and transfer it to a different situation, and generate new knowledge based on existing knowledge. For this purpose different methods can be used. However, more often, the developmental method of teaching includes a combination of different methods. For example, "discovery" learning can be combined with the algo-heuristic theory of learning developed by Landa (1976). In discovery learning, the student's problem-solving task involves a discrepancy between task requirements, student's knowledge, and task conditions. In contrast, the algo-heuristic method concentrates its effort in developing generalized algorithms and heuristics for solving a particular class of tasks. The self-regulative concept of learning concentrates its effort in developing a number of useful strategies of task performance (Bedny and Karwowski, 2006). According to AT, not only memory but also thought processes and emotionally motivational components of activity are important.

AT takes a distinct perspective on how learning occurs, different from that of cognitive psychology. At the same time, AT does not reject the cognitive approach. Rather, AT and cognitive psychology can be integrated. Therefore, cognition, behavior, and motivation are considered in unity, whereas cognitive psychology focuses, first, on one element of activity, cognition. This approach treats skills and knowledge acquisition in terms of a linear sequence of steps or stages of information processing. Special attention is devoted to memory functions. In addition to this, AT focuses on goal formation and motivation, determining the content of cognitive and behavior actions, strategies of performance, etc. (see Bedny and Karwowski, in this volume).

11.6 Action Theory

Action theory is an important psychological approach that shares some similar features with AT. The concept of action was introduced by sociologist Weber (1947). In action theory, the term *action* resembles the concept of activity in AT. Action refers to purposeful behavior that is directed toward achieving a goal. Action theory integrates the cognitive and motivational components into the study of human behavior (Hechausen, 1991; Gollwitzer, 1996). Action is described as a linear sequence of four stages: predecisional phase (motivational), preactional phase (volitional), actional phase (volitional), and postactional phase (motivational). The model describes a linear sequence of stages that starts with the awakening of a person's wishes, before the goal setting, planning, goal striving, and evaluation of whether the goal has been obtained or further goal pursuit is necessary.

Although the original action theory was claimed to be an attempt at integrating cognition and behavior, in practice it focuses on understanding human motivation. Little attention is given to external behavior itself. Consequently, the action theory is not well adapted to the study of human work and learning.

A newer variant of action theory was developed in Germany, in which goal-oriented behavior is the focus of study. In this second version of action theory, action is organized by the goal, information integration, plans, and feedback. Action can be regulated consciously or via routines (Frese and Zapf, 1994, p. 271). This version of action theory has been greatly influenced by AT, especially the works of Rubinshtein (1957), Leont'ev (1978), Vygotsky (1962), Oshanin (1977), Gal'perin (1969), and Thomashevski (1978).

Because action theory takes many of its ideas from AT, the theories are similar in some respects. For example, a central tenet of action theory is that "people change the world and thereby change themselves." As in AT, personality shapes, and is in turn shaped by, activity (Rubinshtein, 1957). Similarly, concepts such as "operative image," that are considered central to action theory were originally developed and elaborated within AT (Oshanin, 1977; Zavalova, Lomov, and Ponomarenko, 1986). The concept of "action style," recently adopted in action theory, was originally introduced in AT as the "individual style of activity" by Merlin (1973) and Klimov (1969).

However, there are also important differences between the action theory and AT approaches. Unlike AT, action theory does not distinguish between purposeful behavior (action/activity) as the object of study and units of analysis into which the action is decomposed for the purpose of analysis. Furthermore, unlike AT, action theory lacks a precise description of the required units of analysis. In AT and action theory, cognitive and behavioral actions are the major building blocks of activity. At the same time, AT has a set of other precisely defined units of analysis, such as operations, functional micro and macro blocks, and members of human algorithm, classified according to a set of criteria (Bedny and Karwowski, 2006).

Similarly, the central concepts in AT and action theory are defined in different ways. The notion of "goal" in action theory has motivational and cognitive components, "and the action is 'pulled' by the goal" (Frese and Zapf, 1994, p. 274). In AT the goal is a purely cognitive component that is connected with a motive. The motive–goal creates a vector that gives an activity a goal-directed character. It is the motive that pushes the subject to reach a goal. Goals cannot exist without motives. However, the energetic (motivational) components of an activity are distinguished from the cognitive (informational) components. Additionally, AT, unlike action theory, distinguishes between the final goal of a task and intermediate goals of actions.

Similar to the notion of goal, the concept of self-regulation in action theory is fundamentally different from that in AT. For example, in action theory, self-regulation is described in a single-loop cycle with six "actions steps." Action proceeds from goal development to orientation, then to plan generation, decision, execution, and then feedback to the goal again. In AT, self-regulation of activity cannot be presented as a single-loop but is always described as a multiloop model. The units of analysis in action theory are action steps. These units of analysis are not precisely defined. In AT the units of analysis have a precise description as functional mechanisms and function blocks. In action theory feedback is presented as a separate mechanism of action regulation that occurs at the final stage of the model. In AT, self-regulation cannot have feedback only as a final step, but rather as a fundamental property of the entire self-regulation system.

In some ways, the self-regulation model in action theory is more similar to cognitive psychology than to AT. It describes a linear progression from perception to decision making to action execution as these are described in the information-processing system in cognitive psychology. In the Frese and Zapf model, the final step of action/activity always includes motor execution. However, sometimes mental actions can be performed without external behavior.

In addition to German action theory, there is an "approximate theory of action" developed by Norman (1986). Here, action is considered to be purposeful behavior. As in action theory, action is considered a single loop comprising stages of action performance: (1) establishing the goal, (2) formation of the intention, (3) specifying the action sequence, (4) executing the action, (5) perceiving the system state, (6) interpreting the state, and (7) evaluating the system. However, these stages are not precisely described and are extracted without a theoretical foundation. There is no clear-cut separation between stages. The approximate theory of action is different from cognitive information-processing systems because it includes notions of goal and intention. However, the meanings of these ideas are not precisely described.

Finally, we need to briefly consider Shuchman's (1987) concept of situated action. As in all variants of action theory, the term *action* refers to purposeful behavior and includes external behavior and cognition. Shuchman emphasizes the dependence of action on the situation. Action is situated and cannot be planned in advance. In AT, activity includes situated components because it develops on the basis of the principles of self-regulation and cannot be totally predicted because it develops as a process (Rubinshtein, 1973; Bernshtein, 1966; Bedny & Karwowski, 2006). In AT activity is a combination of preplanned and situated components.

11.7   Conclusion

Presently, the most developed directions in psychology in the West are cognitive psychology and its derivative, ecological psychology. Cognitive psychology focuses on internal cognitive processes and views the human mind as an information-processing system. Ecological psychology was derived from cognitive psychology and influenced by the work of Russian psychologists such as Bernshtein on the regulation of human behavior. Unlike cognitive psychology, ecological psychology views human behavior not as simply triggered by internal commands of the mind. Rather, behavior is studied as part of a coordinated "human–external environment" system.

The approaches of cognitive and ecological psychology have yielded many important insights regarding the human mind and human behavior. However, these approaches have some limitations when applied to the study of human work. These limitations stem in part from the fact that cognitive and ecological psychologies do not account for the complex systemic features of cognition and activity. Although many different techniques for the study of human work have been suggested in these approaches, neither approach has an integrated theoretical foundation for the study of human work. As a result, there is presently a need for a psychological theory that can be applied to study human performance.

AT is a direction of psychology that has been successfully applied to the study of human work. General AT is a distinct theoretical approach in psychology that was developed in the former Soviet Union. Unlike cognitive and ecological psychology, AT views human activity as a structured system. This theoretical direction was applied to the study of human work and developed into SSAT (Bedny and Karwowski, 2006). SSAT has a set of precisely defined units of analysis and principles that render it a powerful tool for the study of human performance. In particular, SSAT is adapted to the analysis of complex sociotechnical systems.

We suggest that cognitive and ecological psychology together with general activity theory and SSAT can be combined into a unified direction that can be fruitfully applied to the study of human work. In this chapter we attempt to introduce the reader to AT and thereby facilitate its integration with other approaches in psychology.

References

Al'bukhanova, K. A. (1973). *About Subject of Psychological Activity.* Moscow, Science Publishers.

Andeson, J. R. (1987). *Cognitive Skills and Their Acquision.* Hilsdale, NJ: Lawrence Erlbaum.

Anokhin, P. K. (1962). *The Theory of Functional Systems as a Prerequisite for the Construction of Physiological Cybernetics.* Moscow: Academy of Science of the USSR.

Anokhin, P. K. (1969). Cybernetic and the integrative activity of the brain. In Cole, M. and Maltzman, I. (Eds.) *A Handbook of Contemporary Soviet Psychology.* pp. 830–857.

Bedny, G. Z. (1987). *The Psychological Foundations of Analyzing and Designing Work Processes.* Kiev: Higher Education Publishers.

Bedny, G. and Meister, D. (1997). *The Russian Theory of Activity: Current Application to Design and Learning.* Lawrence Erlbaum Associates. Mahwah, New Jersey.

Bedny, G. Z. and Harris, S. (2005). The systemic-structural theory of activity: Application to the study of human work. *Mind, Culture, and Activity,* 12, 2, 128–147.

Bedny, G. Z. and Karwowski, W. (2004). Meaning and sense in activity theory and their role in study of human performance. *Ergonomia,* V. 26, # 2, pp. 121–140.

Bedny, G., Seglin, M., and Meister, D. (2000). Activity theory. History, research and application. *Theoretical Issues in Ergonomics Science,* # 2, pp. 165–206.

Bedny, G. Z. and Karwowski, W. (2006). *A Systemic-Structural Theory of Activity. Application to Human Performance and Work Design.* Taylor and Francis, Boca Raton, London, New York.

Bernshtein, N. A. (1966). *The Physiology of Movement and Activity.* Moscow: Medical Publishers.

Bernshtein, N. A. (1996). On dexterity and its development. In Latash, M. L. and Turvey, M. T. (Eds.) *Dexterity and Its Development* (pp. 1–244) Lawrence Erlbaum Associates. Mahwah, New Jersey.

Brushlinsky, A. V. (2001). Activity approach and psychological science. *Question of Philosophy,* # 2, pp. 88–94.

Brushlinsky, A. V. (1999). Subject–oriented concept of activity and theory of functional system. *Questions of Psychology,* # 5, pp. 110–121.

Chebykin, A. Ya. (1999). *Theory and Practice of Emotional Regulation of Learning Activity.* Ukraine. Astroprint.

Engestrom, Y., (2000). Activity theory as a framework for analyzing and redesigning work. *Ergonomics,* 7, pp. 960–974.

Engestrom, Y., (1999). Activity theory and individual and social transformation. In Engestrom, Y., Miettinen, R., and Punamaki, R-L, (Eds.), *Perspectives on Activity Theory.* Cambridge: Cambridge University Press, pp. 19–38.

Frese, M. and Zapf, D. (1994). Action as a core of work psychology: A German approach. In Triadis, H. C., Dunnette, M. D., and Hough, L. M. (Eds.). *Handbook of Industrial and Organizational Psychology.* Palo Alto, California: Consulting Psychologists Press. pp: 271–340.

Galactionov, A. I. (1978). *The Fundamentals of Engineering — Psychological Design of Automatic Technological Systems.* Moscow: Energy Publishers.

Gal'perin, P. Y. (1969). Stages in the development of mental acts. In Cole, M. and Maltzman, I (Eds.), *A Handbook of Contemporary Soviet Psychology* (pp. 249–273). New York: Basic Books.

Gal'sev, A. D. (1973). *Time Study and Scientific Management of Work in Manufacturing.* Moscow: Manufacturing Publishers.

Gollwitzer, P. M. (1996). The volitional benefits of planning. In Gallwitzer, P. and Bargph, J. P. (Eds.), *The Psychology of Action.* New York: The Gilford Press, pp. 287–312.

Gordeeva, N. D., and Zinchenko, V. P. (1982). *Functional Structure of Action.* Moscow: Moscow University Publishers.

Heckhausen, H. (1991). *Motivation and Action.* Berlin, NY: Spring-Verlag.

Kaptelinin, V. and Nardi, B. A. (2006). *Acting with Technology,* The MIT Press, Cambridge, Massachusetts; London, England.

Klimov, E. A. (1969). *Individual Style of Activity.* Kazan: Kazahnsky State University Press.

Konopkin, O. A. (1980). *Psychological Mechanisms of Regulation of Activity.* Moscow: Science Publishers.

Kotik, M. A. (1974). *Self-Regulation and Reliability of Operator.* Tallin: Valgus.

Kochurova, E. I., Visyagina, A. I., Gordeeva, N. D., and Zinchenko, V. P. (1981). Criteria for evaluating executive activity. In Wertsch, J. V. (Ed.), *The Concept of Activity in Soviet Psychology,* M. E. Sharpe, New York, pp. 383–433.

Kuuitti, K. (1997). Activity theory as a potential framework for human-computer interaction research. In B. A. Nardi (Ed.), *Context and Consciousness: Activity Theory and Human-Computer Interaction,* The MIT Press. Cambridge, MA, 17–44.

Landa, L. M. (1976). *Instructional Regulation and Control: Cybernetics, Algorithmization and Heuristic in Education.* Englewood Cliffs, NJ: Educational Technology Publication. (English translation)

Lee, T. W., Locke, E. A., and Latham, G. P. (1989). Goal setting, theory and job performance. In Pervin, A. (Ed.), *Goal Concepts in Personality and Social Psychology* (pp. 291–326). Hillsdale, NJ: Lawrence Erlbaum Associates.

Leont'ev, A. N. (1978). *Activity, Consciousness and Personality.* Englewood Cliffs: Prentice Hall.

Leont'ev, D. (1993). Sense–system nature and functions of the motive, *Vestnik MGU, Serija 14. Psychology,* (2), pp. 73–81.

Lomov, B. F. (1977). Directions on developing theories of engineering psychology based on systemic approach. In B. F. Lomov, V. F., Rubakhin, and V. F. Venda (Eds.). *Engineering Psychology,* (pp 31–55). Moscow: Science Publishers.

Merlin, V. S. (1973). *Outlines of a Theory of Temperament.* Perm, Russia: Perm Pedagogical Institute.

Nardi, A. (Ed.) (1997). *Context and Consciousness: Activity Theory and Human-Computer Interaction,* The MIT Press. Cambridge, Massachussets; London, 17–44.

Norman, D. A. (1986). Cognitive engineering. In Norman, D. and Draper, S. (Eds.), *User Centered System Design,* Erlbaum: Hillsdale, NJ, pp. 31–61

Nosulenko, V. N., Barabanshikov, A., Brushlinsky, and Rabardel, P. (2005). Man–technology interaction: some of the Russian approaches. *Theoretical Issues in Ergonomics Science,* # 5, pp. 359–384.

Oshanin, D. A. (1977). Concept of operative image in engineering and general psychology. In Lomov, B. F., Rubakhin, V. F., and Venda, V. F. (Eds.), *Engineering Psychology.* Moscow: Science Publishers.

Petrovsky, A. V. (Ed.) (1986). *General Psychology.* Moscow: Education Publishers.

Piaget, J. (1952). *The Origins of Intelligence in Children.* New York: International University Press.

Platonov, K. K. (1982). *System of Psychology and Theory of Reflection.* Moscow: Science Publishers.

Preece, J., Rogers, Y., Sharp, H., Benyon, D., Holland, S., and Carey, T. (1994), *Human-Computer Interaction.* Addison-Wesley.

Pushkin, V. V. (1978). Construction of situational concepts in activity structure. In Smirnov, A. A. (Ed.), *Problem of General and Educational Psychology,* (pp. 106–120). Moscow: Pedagogy.

Rubinshtein, S. L. (1935). *Foundation of Psychology,* State Pedagogical Publisher, Moscow, USSR.

Rubinshtein, S. L. (1957). *Existence and Consciousness.* Moscow: Academy of Science.

Rubinshtein, S. L. (1959). *Principles and Directions of Developing Psychology.* Moscow: Academic of Science.

Rubinshtein, S. L. (1934). Problems of psychology in Marx's works, Soviet Psychotechnik, # 1, pp.

Rubinshtein, S. L. (1973). *Problems of General Psychology.* Moscow: Academy of Science.

Rubinshtein, S. L. (1922/1986). The principle of creative activity. *Questions of Psychology,* # 4, pp. 101–107.

Shuchman, L. A. (1987). *Plans and Situated Actions: The Problem of Human-Machine Interaction.* Cambridge: Cambridge University Press.

Tikhomirov, O. K. (1984). *Psychology of Thinking.* Moscow: Moscow University.

Thomasevski, T. (1979). *Activity and Consciousness.* Weinheim, Germany: Beltz.

Vygotsky, L. S. (1960). *Developing Higher Order Psychic Functions.* Moscow: Academia of Pedagogical Science. RSFSR.

Vygotsky, L. S. (1962). *Thought and Language.* Cambridge: MIT Press.

Vygotsky, L. S. (1978). *Mind in Society. The Development of Higher Psychological Processes.* Cambridge, MA: Harvard University Press.

Weber, M. (1947). *Theories of Social and Economic Organizations.* New York: Wiley.

Yaroshevsky, M. G. (1992). L. S. Vygotsky and Marxism in soviet psychology. Response to Brushlinsky. *Psychological Journal.* # 5, pp: 97–98.

Zaporozhets, A. V. and Zinchenko, V. P. (1982). *Perception, Motions and Actions.* Moscow: Pedagogy.

Zarakovsky, G. M., (2004). The concept of theoretical evaluation of operators' performance reliability derived from activity theory. In Bedny, G. (Ed.) Special Issue. *Theoretical Issues in Ergonomics Science.* 4, pp. 313–337.

Zavalova, N. D., Lomov, B. F., and Ponomarenko, V. A. (1986). *Image in Regulation of Activity.* Moscow: Science Publishers.

12

Discourse in Activity

Harry Daniels

CONTENTS

12.1 The Development of Engeström's Interpretation of Activity
Theory...248
    12.1.1 Engeström's Account of the First Generation 249
    12.1.2 Engeström's Account of the Second Generation 249
    12.1.3 Engeström's Account of the Third Generation.........................250
12.2 Bernstein 252
    12.2.1 The Potential for Progress in the Development of
    Bernstein's Work 253
        12.2.1.1 Organizational Practices..253
        12.2.1.2 Classification..253
        12.2.1.3 Framing 254
        12.2.1.4 Culture and Language 255
        12.2.1.5 Social Positioning..258
        12.2.1.6 Contradiction and Change in Transmission
        Practices...259
        12.2.1.7 Discursive Hybridity 263
References 264

In this chapter I wish to explore the extent to which two approaches to the social formation of mind are compatible and may be used to enrich and extend each other. It is an extension of thinking that I have been developing in two academic communities: those concerned with activity theory (AT) as developed by Yrjo Engeström, which is derived from the work of the early Russian psychologists Vygotsky and Leontiev, and the studies of the sociologist Basil Bernstein. The two approaches deal with a common theme from different perspectives. The Vygotskian unit of empirical and conceptual analysis was word meaning. In Engeström's hands, the unit of analysis becomes the activity system in which the individual was located. In the attempt to develop an account of social formation, the gaze falls either on the individual in dialogue or the object-oriented activity system. The notion of the object of activity—the problem space or raw material that was being worked on in an activity—is central to the work of Leontiev. The object is that which is explored or transformed according to the goal of activity. Bernstein developed a theory with a language of description that allowed researchers to move from a gaze that lighted first on the social and cultural and permitted a trace to be drawn through principles of regulation at the social organizational level through principles of regulation in and through discourse to possibilities for individual thought and action. The rules of cultural historical formation rather than the object of activity are the focus. It is, as it were, that they were actually examining activity at different scales. To some extent, both attempt to account for cultural historical formation at ontogenetic and cultural levels.

A focus on the rules that shape the social formation of pedagogic discourse and its practices (Bernstein, 2000) will be brought to bear on those aspects of psychology which argue that object-oriented activity is a fundamental constituent of human thought and action (Cole, 1996). The institutional level of analysis was all but absent in much of the early Vygotskian research in the West (see Daniels, 2001). There was no recourse to a language of description that permitted the analysis of object-oriented activity in terms of the rules that regulate the microcultures of institutions. This absence, in some sense, reflects the tension between Vygotsky's work and that of the Russian tradition of AT. It also reflects the difficulty that both traditions have experienced in analyzing the production of discourse in terms of the context of its production. Recent developments in post-Vygotskian theory and AT have witnessed considerable advances in the understanding of the ways in which human action shapes and is shaped by the contexts in which it takes place (Daniels, 2001). They have given rise to a significant amount of empirical research within and across a wide range of fields in which social science methodologies and methods are applied in the development of research-based knowledge in policy making and practice in academic, commercial, and industrial settings (e.g., Agre, 1997; Cole, Engeström, and Vasquez, 1997; Engeström and Middleton, 1996; Daniels, 2001; Lea and Nicoll, 2002; Wenger, McDermott, and Snyder, 2002). This chapter comprises three sections. In the first I will trace the development of Russian social theory through the current manifestation of AT. In the second, I will provide a very brief outline of Bernstein's work on the sociology of pedagogy. I will subsequently argue that these two theoretical developments may be brought into productive relation with one another.

12.1   The Development of Engeström's Interpretation of Activity Theory

Vygotsky provided a rich and tantalizing set of suggestions that have been taken up and transformed by social theorists as they attempt to construct

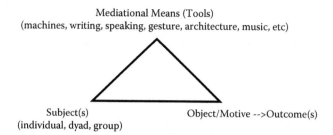

Mediational Means (Tools)
(machines, writing, speaking, gesture, architecture, music, etc)

Subject(s)   Object/Motive -->Outcome(s)
(individual, dyad, group)

FIGURE 12.1
First-generation activity theory model.

accounts of the formation of mind that to varying degrees acknowledge social, cultural, and historical influences. His is not a legacy of determinism and denial of agency; on the contrary, he provides a theoretical framework that rests on the concept of mediation. The means of mediation that have tended to dominate recent discussions are cultural artifacts such as speech or activity. These semiotic and activity-based accounts may be seen as referring to different levels of emphasis within a single process. Wertsch (1998) advances the case for the use of mediated action as a unit of analysis in sociocultural research. Engeström (1993) points out the danger of the relative undertheorizing of context: "Individual experience is described and analyzed as if consisting relatively of discrete and situated actions while the system or objectively given context of which those actions are a part is either treated as an immutable given or barely described at all" (Engeström, 1993, 66). Engeström's concept of activity has developed in response to the challenge embedded in this statement.

12.1.1 Engeström's Account of the First Generation

This first approach drew heavily from Vygotsky's concept of mediation (see Figure 12.1). This triangle represents the way in which Vygotsky brought together cultural artifacts with human actions to dispense with the individual/social dualism. During this period studies tended to focus on individuals.

12.1.2 Engeström's Account of the Second Generation

Here Engeström advocates the study of artifacts "as integral and inseparable components of human functioning," but he argues that the focus of the study of mediation should be on its relationship with the other components of an activity system (Engeström, 1999, 29).

To advance the development of activity theory, Engeström expanded the original triangular representation of activity to enable an examination of systems of activity at the macro level of the collective and the community in preference to a micro level concentration on the individual actor or agent

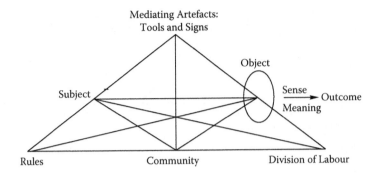

FIGURE 12.2
Second-generation activity theory model.

operating with tools. Although Leontiev was influential on this move, the distinction between action and activity was hidden, and, thus, the concept of goal was more difficult to bring into operational analysis.

Vygotsky was more of a semiotician than an analyst of activity. Despite appeals to a broader conception of social formation, much of Vygotsky's work remains an analysis of semiotic mediation in small-scale settings. This expansion of the basic Vygotskian triangle aims at representing the social/ collective elements in an activity system, through the addition of the elements of community, rules, and division of labor, while emphasizing the importance of analyzing their interactions with each other. In Figure 12.2, the object is depicted with the help of an oval, indicating that object-oriented actions are always, explicitly or implicitly, characterized by ambiguity, surprise, interpretation, sense making, and potential for change. (Engeström, 1999). At the same time Engeström drew on Ilyenkov (1977) to emphasize the importance of contradictions within activity systems as the driving force of change and, thus, development.

12.1.3   Engeström's Account of the Third Generation

Engeström (1999) sees joint activity or practice as the unit of analysis for activity theory, not individual activity. He is interested in the process of social transformation and includes the structure of the social world in analysis, taking into account the conflictual nature of social practice. He sees instability (internal tensions) and contradiction as the "motive force of change and development" (Engeström, 1999, 9) and the transitions and reorganizations within and between activity systems as part of evolution; it is not only the subject but the environment that is modified through mediated activity. He views the "reflective appropriation of advanced models and tools" as "ways out of internal contradictions" that result in new activity systems (Cole and Engeström, 1993, 40). This is a central tenet of Engeström's contribution to his development of AT.

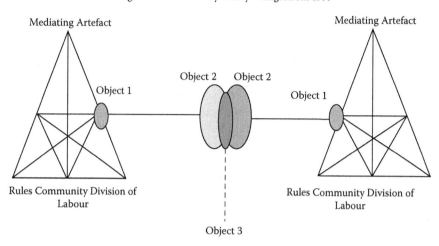

Two interacting activity systems as minimal model for
third generation of activity theory -- Engestrom 1999

FIGURE 12.3
Third-generation activity theory model.

The intent underlying the third generation of AT as proposed by Engeström is to develop conceptual tools to understand dialogues, multiple perspectives, and networks of interacting activity systems. He draws on ideas on dialogicality and multivoicedness to expand the framework of the second generation. The idea of networks of activity within which contradictions and struggles take place in the definition of the motives and object of the activity calls for an analysis of power and control within developing activity systems. The minimal representation that Figure 12.3 provides shows but two of what may be a myriad of systems exhibiting patterns of contradiction and tension. The problem this presents to empirical research is with respect to how the operational model of activity systems or networks is delimited. The delicacy of the model should reflect the complexity inherent in the research question. This is not always easy to discern at the outset of research. Hence, this approach must involve successive moves between theory and data.

> Engeström's approach provides a means of studying learning understood as the expansion through change and development of the objects of activity. This is undertaken through a critical consideration of contradictions within and between activity systems. It lacks a sophisticated account of the regulation of subject–subject relations (and thus, social positioning) and of the situated production, and to some extent the structure and function of the cultural artifacts (such as discourse) that mediate subject–object relations. As Engeström and Miettinen have noted: "The integration of discourse into the theory of activity has only begun." (Engeström and Miettinen, 1999, 7)

12.2 Bernstein

Bernstein's work on analysis and description focuses on two levels: a structural level and an interactional level. The structural level is analyzed in terms of the social division of labor it creates and the interactional level with the form of social relation it creates. The social division of labor is analyzed in terms of strength of the boundary of its divisions, that is, with respect to the degree of specialization. The interactional level emerges as the regulation of the transmission/acquisition relation between teacher and the taught: that is, the interactional level comes to refer to the pedagogic context and the social relations of the classroom or its equivalent. The interactional level then gives the principle of the learning context through which the social division of labor, in Bernstein's terms, speaks. Here, learning context is not necessarily taken as a classroom, but any context in which pedagogic relations obtain.

It becomes possible to see how a given distribution of and principles of control are made substantive in agencies of cultural reproduction, e.g., families and schools. The form of the code (its modality) contains principles for distinguishing between contexts (recognition rules) *and* for the creation and production of specialized communication within contexts (realization rules). The analysis can be applied to different levels of school organization and various units within a level. This allows the analysis of power and control and the rules regulating what counts as legitimate pedagogic competence to proceed at a level of delicacy appropriate to a particular research question.

Bernstein (1990, 13) used the concept of social positioning to refer to the establishing of a specific relation to other subjects and to the creating of specific relationships within subjects. As Hasan notes, social positioning through meanings is inseparable from power relations. Here, the linkage is forged between social positioning and psychological attributes. This is the process Bernstein speaks of when discussing shaping the possibilities for consciousness. This form of analysis allows for the refinement of Engeström's (1999) suggestion that the division of labor in an activity creates different positions for the participants and that the participants carry their own diverse histories with them into the activity. Whereas Vygotsky speaks of mediation and Leontiev of activity systems with a division of labor that sets up different subject positions in relation to an object, Bernstein provides a more detailed analysis of subject positioning, which is ultimately grounded in an analysis of power and control.

The dialectical relation between discourse and subject makes it possible to think of pedagogic discourse as a semiotic means that regulates or traces the generation of subjects' positions in discourse. Within the Bernsteinian thesis there exists an ineluctable relation between one's social positioning, one's mental dispositions, and one's relation to the distribution of labor in society (Hasan 2004).

Here, the emphasis on discourse is theorized not only in terms of the shaping of cognitive functions but also, as it were, invisibly, in its influence on dispositions, identities, and practices (Bernstein, 1990, 3).

Bernstein postulates (1990, 16ff) pertinent concepts to show how this comes about. Socially positioned subjects through their experience of and participation in code-regulated dominant and dominated communication develop rules for recognizing what social activity a circumstance is the context for, and how the requisite activity should be carried out.

In this way Bernstein's work provides a means of translating principles of power and control into principles of communication and through an account of social positioning within pedagogic discourse theorizes the distribution of forms of pedagogic consciousness. As it stands, this approach is not well placed to study the emergent objects of human activity through time. This is something that AT has, in its more recent incarnations, sought to establish.

12.2.1   The Potential for Progress in the Development of Bernstein's Work

12.2.1.1   Organizational Practices

The organizational dimensions of social practice are provisionally sketched in Engeström's version of AT but lack a sophisticated account of the way in which a dominating distribution of power and principles of control generate, distribute, reproduce, and legitimize dominating and dominated principles of communication such as that to be found in Bernstein (2000).

Engeström talks of the division of labor in terms of the horizontal division of tasks between the members of the community and of the vertical division of power and status. Engeström's notion of rules refers to the explicit and implicit regulations, norms, and conventions that constrain actions and interactions within the activity system.

Bernstein uses the concept of classification to determine the underlying principle of a social division of labor and the concept of framing to determine the principle of its social relations and in this way to integrate structural and interactional levels of analysis in such a way that, up to a point, both levels may vary independently of each other.

12.2.1.2   Classification

Classification is defined at the most general level as the relation between categories. The relation between categories is given by their degree of insulation. Thus, where there is strong insulation between categories, each category is sharply distinguished and explicitly bounded, having its own distinctive specialization. When there is weak insulation, then the categories are less specialized, and therefore their distinctiveness is reduced. In the former case, Bernstein speaks of strong classification and, in the latter case, of weak classification. Classification may also be discussed in vertical and horizontal dimensions. For example, the strength of the boundary/distinction between subjects in the curriculum may be described in terms of a horizontal dimension (how different they are) or a vertical dimension (how important they are).

12.2.1.3   *Framing*

The social relations, generally, in the analyses are those between parents and children, teachers and pupils, doctors and patients, and social workers and clients, but the analysis can be extended to include the social relations of the work contexts of industry or commerce. Each category may have its own specificities, but they share the same pedagogic elements. Bernstein considers that from his point of view all these relations can be regarded as pedagogic.

> Framing refers to the control on communicative practices (selection, sequencing, pacing and criteria) in pedagogic relations, be they relations of parents and children or teacher/pupils. Where framing is strong the transmitter explicitly regulates the distinguishing features of the interactional and locational principle which constitute the communicative context …. Where framing is weak, the acquirer is accorded more control over the regulation. Framing regulates what counts as legitimate communication in the pedagogic relation and thus what counts as legitimate practices. (Bernstein, 1981, 345)

Bernstein also provides an account of external framing that refers to the control over communication outside the context of concern. Here is the parallel with the AT notion of community as elaborated in Engeström's work (rather than that to be found in the "communities of practice" approach as expounded by Lave and Wenger [1991]). Crucially, Bernstein allows us to move beyond questions concerning who is a member of the community to questions of relations of control with that community. Above all this form of analysis permits the move between organizational structure and the structure of the discourse.

> Classification refers to *what*, framing is concerned with *how* meanings are to be put together, the forms by which they are to be made public, and the nature of the social relationships that go with it. (Bernstein, 2000, 12)

In that the model is concerned with principles of regulation of educational transmission at any specified level, it is possible to investigate experimentally the relation between principles of regulation and the practices of pupils. Relations of power create and maintain boundaries between categories and are described in terms of classification. Relations of control are revealed in values of framing condition communicative practices. It becomes possible to see how a given distribution of power through its classificatory principle and principles of control through its framing are made substantive in agencies of cultural reproduction, e.g., families or schools. The form of the code (its modality) contains principles for distinguishing between contexts (recognition rules) *and* for the creation and production of specialized communication within contexts (realization rules).

Through defining educational codes in terms of the relationship between classification and framing, these two components are built into the analysis at *all levels*. It then becomes possible in one framework to derive a typology of educational codes, to show the inter-relationships between organizational and knowledge properties to move from macro- to micro-levels of analysis, to delete the patterns internal to educational institutions to the external social antecedents of such patterns, and to consider questions of maintenance and change. (Bernstein, 1977, 112)

The analysis of classification and framing can be applied to different levels of school organization and various units within a level. This is not to argue that schools and families assume the same cultural modalities but rather that Bernstein provides a language of description with which it is possible to describe the cultural particularities that obtain in any specific site of cultural transmission.

In AT the production of the cultural artifact, the discourse, is not analyzed in terms of the context of its production. One cannot "read" the context in the product. The specificities of the artifact are not witnessed in a description that is grounded in the social relations of the context in which it was produced. They are, in Bernstein's work. The language that Bernstein has developed, uniquely, allows researchers to take measures of institutional modality, that is, to describe and position the discursive, organizational, and interactional practice of the institution. Through the concepts of classification and framing, Bernstein provides the language of description for moving from those issues that AT handles as rules, community, and division of labor to the discursive tools or artifacts that are produced and deployed within an activity.

12.2.1.4   Culture and Language

Hasan (1992a, 1992b, 1995) and Wertsch (1985, 1991) note the irony that although Vygotsky developed a theory of semiotic mediation in which the mediational means of language was privileged, he provides very little, if anything, by way of a theory of language use. In an account of the social formation of mind, there is a requirement for theory that relates meanings to interpersonal relations. The notions of meaning and sense discussed by Vygotsky (1987) in *Thinking and Speech* draw attention to the issue but do little to guide an analysis of the regulation of the processes envisaged.

The complexity of mastering scientific concepts is brought home by the distinction between the "sense" (smyl) and the "meaning" (znachenie) of a word.

A word's sense is the aggregate of all the psychological facts that arise in our consciousness as a result of the word. Sense is a dynamic fluid, and complex formation which has several zones that vary in their stability. Meaning is only one of these zones of sense that the word acquires in the context of speech. It is the most stable, unified and precise of these zones.

> In different contexts, a word's sense changes. In contrast, meaning is a
> comparatively fixed and stable point, one that remains constant with all
> the changes of the word's sense that are associated with its use in vari-
> ous contexts. (Vygotsky, 1987, 275–6)

In Chapter 7 of *Thinking and Speech*, Vygotsky discusses the complexities
of the relationships between sense and meaning on the one hand and oral
and inner speech on the other. In this rather beautiful and poetic chapter,
Vygotsky provides what could be taken as the background for the preceding chapters on concept development. The ongoing dynamic between the
use of social speech and relatively stable social meanings in the creation of
particular forms and patterns of personal sense is construed as the motor of
development. The notion of the scientific concept can be seen as a particular
historical cultural form of relatively stable meaning that is brought into productive interchange with the sense of the world that is acquired in specific
everyday circumstances.

However, the absence of an account of the ways in which language both
serves to regulate interpersonal relations and its specificity is in turn
produced through specific patterns of interpersonal relation, and thus social
regulation constitutes a serious weakness. This absence constitutes a significant part of Bernstein's project. He seeks to link semiotic tools with the structure of material activity. In Engeström's (1996) work within activity theory,
the production of the outcome is discussed, but there is less emphasis on the
production and structure of discourse itself.

The challenge is to theorize the Vygotskian tool, or cultural artifact as a
social and historical construction.

Bernstein (1996) refined this discussion through making a distinction
between instructional and regulative discourse. The former refers to the
transmission of skills and their relation to each other, and the latter, to the
principles of social order, relation, and identity. Whereas the principles and
distinctive features of instructional discourse and its practice are relatively
clear (the what and how of the specific skills or competences to be acquired
and their relation to each other), the principles and distinctive features of
the transmission of the regulative are less clear as this discourse is transmitted through various media and may indeed be characterized as a diffuse
transmission. Regulative discourse communicates the institution's public
moral practice, values, beliefs, and attitudes, principles of conduct, character, and manner. It also transmits features of the institution's local history,
local tradition, and community relations. Pedagogic discourse is modeled
as one discourse created by the embedding of instructional and regulative
discourse. This was envisaged in Vygotsky's suggestion that every idea contains some remnant of the individual's affective relationship to that aspect of
reality which it represents (Vygotsky, 1987, 50). However, he did not provide
research with an account beyond the notion of word to the structure of pedagogic *discourse*.

Bernstein (1999) provides a further refinement in that he distinguishes between vertical and horizontal discourse. Horizontal discourse arises out of everyday activity and is usually oral, local, context dependent and specific, tacit, multilayered, and contradictory across but not within contexts. Its structure reflects the way a particular culture is segmented and its activities are specialized. Horizontal discourse is thus segmentally organized. In contrast, vertical discourse has a coherent, explicit, and systematically principled structure that is hierarchically organized or takes the form of a series of specialized languages with specialized criteria for the production and circulation of texts (Bernstein, 1999, 159). Bernstein suggests that Bourdieu's notion of discursive forms, which give rise to symbolic and practical mastery, respectively, and Habermas' reference to the discursive construction of life worlds of individuals and instrumental rationality both relate to parts of a complex field of parameters, which in turn refer to both individual and social experience, and bear upon the model of horizontal and vertical discourse that he seeks to develop. He offers an initial set of contrasts and indicates that many more exist. His lament is for the lack of a language of description of these forms that can serve to generate and relate the possibilities for difference.

Bernstein's (1999) paper serves as an important reminder that the theoretical derivation of "scientific and everyday" in the original writing was somewhat provisional. For example, the association of the scientific with the school does not help to distinguish aspects of formal instruction such as that which adds to everyday understanding without fostering the development of scientific concepts. The association also suggests that the development of scientific concepts must take place in the school and not outside it. Bernstein's analysis is suggestive of a more powerful means of conceptualizing the forms that Vygotsky announced.

It may be as a consequence of the dualist perspective, which remains so powerful, that the emphasis on the interdependence between the development of scientific and everyday concepts is also not always appreciated. Valsiner (1997) distinguishes dualisms from dualities, arguing that the denial of dualism (inner, outer) in appropriation models leads to a denial of the dualities that are the constituent elements in dialectical or dialogical theory. This echoes the Marxist notion of internal relationship in which two elements are mutually constitutive (see Table 12.1). Vygotsky argued that the systematic, organized, and hierarchical thinking that he associated with scientific concepts becomes gradually embedded in everyday referents and thus achieves a general sense in the contextual richness of everyday thought. Vygotsky thus presented an interconnected model of the relationship between scientific and everyday or spontaneous concepts. Similarly he argued that everyday thought is given structure and order in the context of systematic scientific thought. Vygotsky was keen to point out the relative strengths of the two as they both contribute to each other.

TABLE 12.1

Internal Relationship of Elements

	Horizontal Discourse	Vertical Discourse
Evaluative	Spontaneous	Contrived
Epistemological	Subjective	Objective
Cognitive	Operations	Principles
Social	Intimacy	Distance
Contextual	Inside	Outside
Voice	Dominated	Dominant
Mode	Linear	Nonlinear
Institutional	Gemeinschaft	Gessellschaft

Source: From Bernstein (1999), *British Journal of Sociology of Education*, 20, 2, p. 158.

12.2.1.5 *Social Positioning*

Leontiev used the term *activity*, in which subject, object, actions, and operations are mutually present:

> Activity is the minimal meaningful context for understanding individual actions ... In all its varied forms, the activity of the human individual is a system set within a system of social relations ... The activity of individual people thus *depends on their social position*, the conditions that fall to their lot, and an accumulation of idiosyncratic, individual factors. Human activity is not a relation between a person and a society that confronts him ... In a society a person does not simply find external conditions to which he must adapt his activity, but, rather, these very social conditions bear within themselves the motives and goals of his activity, its means and modes. (Leontiev, 1978, 10)

Although Engeström acknowledges that the division of labor in activity results in the creation of possibilities for social position, the implications of Leontiev's account are not fully developed in modern AT. Bernstein's account of social positioning within the discursive practice that arises in activity systems, taken together with his analysis of the ways in which principles of power and control translate into principles of communication, allows us to investigate how principles of communication differentially position subjects acting within particular activity settings. As Hasan (2004) notes, there exists a clear relation between one's social positioning, one's mental dispositions, and one's relation to the distribution of labor in society. The "same" context could elicit different practices from persons differently positioned. Holzkamp elaborates this line of argument by investigating how the normative aspects of social interaction affect subjective interpretations of what actions can even be thought of as possible by individuals in given societal contexts.*

---

* Personal communication, S. Harris.

Socially positioned subjects, through their experience of and participation in pedagogic practice mediated by pedagogic discourse, develop rules for recognizing what social activity a circumstance is the context for, and how the requisite activity should be carried out (after Hasan, 2004). Subject–subject and within-subject relations are undertheorized in AT. The challenge is to develop Bernstein's account of social positioning to allow for the analysis and description of complex activity formations as they change through time.

12.2.1.6 Contradiction and Change in Transmission Practices

The focus of CHAT is on instability (internal tensions) and contradiction as the "motive force of change and development" (Engeström, 1999, 9) and the transitions and reorganizations within and between activity systems as part of evolution. It is not only the subject but also the environment that is modified through mediated activity. Rules, community, and division of labor are analyzed in terms of the contradictions and dilemmas that arise within the activity system specifically with respect to the production of the object. Activity systems do not exist in isolation; they are embedded in networks, which experience constant fluctuation and change. AT needs to develop tools for analyzing and transforming networks of culturally heterogeneous activities through dialogue and debate (Engeström and Miettinen, 1999, 7). Bernstein's work has not placed particular emphasis on the study of change (see Bernstein, 2000). The introduction of the third generation of AT initiated the development of conceptual tools to understand dialogues and multiple perspectives on change within networks of interacting activity systems, all of which are underdeveloped in Bernstein. The idea of networks of activity within which contradictions and struggles take place in the definition of the motives and object of the activity calls for an analysis of power and control within and between developing activity systems. The latter is the point at which Bernstein's emphasis on different layers and dimensions of power and control becomes the key to development of the theory. The minimal representation that Figure 12.1 provides shows but two of what may be a network of systems exhibiting patterns of contradiction and tension.

Lemke (1997) suggests that it is not only the context of the situation that is relevant but also the context of culture when an analysis of meaning is undertaken. He suggests that "we interpret a text, or a situation in part by connecting it to other texts and situations which our community, or our individual history, has made us see as relevant to the meaning of the present one" (Lemke, 1997, 50). This use of notions of intertextuality and of networked activities or network of connections provides Lemke with tools for the creation of an account of "ecosocial" systems, which transcend immediate contexts. Engeström and Miettinen recognize the strengths and limitations of this position. They imply a need for an analysis of the way in which networks of activities are structured—ultimately, an analysis of power and control:

> Various microsociologies have produced eye-opening works that uncover the local, idiosyncratic, and contingent nature of action, interaction, and knowledge. Empirical studies of concrete, situated practices can uncover the local pattern of activity and the cultural specificity of thought, speech and discourse. Yet these microstudies tend to have little connection to macrotheories of social institutions and the structure of society. Various approaches to analysis of social networks may be seen as attempts to bridge the gap. However, a single network, though interconnected with a number of other networks, typically still in no way represents any general or lawful development in society. (Engeström and Miettinen, 1999, 8)

Leontiev (1981) explored this issue from the perspective of development through time. He suggested that in the study of human ontogeny, one must take account of the ordering of categories of activity that corresponds to broad stages of mental development. According to Leontiev,

> In studying the development of the child's psyche, we must therefore start by analyzing the child's activity, as this activity is built up in the concrete conditions of its life … Life or activity as a whole is not built up mechanically, however, from separate types of activity. Some types of activity are the leading ones at a given stage and are of greatest significance for the individual's subsequent development, and others are less important. We can say, accordingly, that each stage of psychic development is characterized by a definite relation of the child to reality that is the leading one at that stage and by a definite, leading type of activity. (Leontiev, 1981, 395)

This analysis of development in terms of stages characterized in terms of particular dominant activities is often associated with the work of Elkonin. In the terms of contemporary CHAT, this account is one of progressive transformation of the object through time. This could be termed a horizontal analysis:

> When we speak of the dominant activity and its significance for a child's development in this or that period, this is by no means meant to imply that the child might not be simultaneously developing in other directions as well. In each period, a child's life is many-sided, the activities of which his life is composed are varied. New sorts of activity appear; the child forms new relations with his surroundings. When a new activity becomes dominant, it does not cancel all previously existing activities: it merely alters their status within the overall system of relations between the child and his surroundings, which thereby become increasingly richer. (Elkonin, 1972, 247)

This concern for the analysis of a "leading" activity is also to be found in the work of Norman Fairclough (1992, 2000).

> Social practices networked in a particular way constitute a social order … The discourse/semiotic aspect of a social order is what we can call an

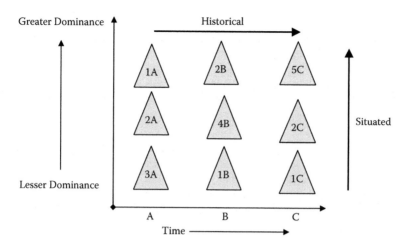

FIGURE 12.4
Dominance in networks of activity systems through time.

order of discourse. It is the way in which diverse genres and discourses and styles are networked together. An order of discourse is a social structuring of semiotic difference—a particular social ordering of relationships amongst different ways of making meaning, i.e., different discourse and genres and styles. One aspect of this ordering is dominance: some ways of making meaning are dominant or mainstream in a particular order of discourse, others are marginal, or oppositional, or 'alternative' ... The political concept of 'hegemony' can usefully be used in analyzing orders of discourse (Fairclough, 1992, Laclau and Mouffe, 1985)—a particular social structuring of semiotic difference may become hegemonic, become part of the legitimizing common sense which sustains relations of domination, but hegemony will always be contested to a greater or lesser extent, in hegemonic struggle. An order of discourse is not a closed or rigid system, but rather an open system, which is put at risk by what happens in actual interactions. (Fairclough, 2004, 2)

Griffin and Cole (1984) noted that in the course of a single session of an after-school activity designed for 7- to 11-year-olds, there could be fluctuations in what activity seemed to be "leading." This could be termed a situated analysis. In Figure 12.4, an analysis of a particular moment in time (A, B, or C) would consider the network of activity systems in which subjects were located and seek to discern the shifts in dominance that take place in short periods of real time in a particular context. For example, at time A activity 1A assumes dominance, whereas at time B activity 2B is represented as dominant or leading. This analysis could be pursued through the application of Bernstein's model to several activities and systems (rather than the one to which it is usually referenced) and also seek to apply his analysis of power and control to the emergence of dominance (1A versus 2A versus 3A). This situated analysis would combine the strengths of AT with its emphasis on networks of activity and the formation of objects of activity with the analytical power and descriptive elegance of Bernstein's work. The implications of different social positions would have to be taken into account as would the recognition that activity systems may be invoked in the absence of the physical presence of all the actors involved (Vygotsky, 1987).

With the following quotation Vygotsky announced the possibility of virtual collaboration without the physical presence of the adult or teacher:

> When the school child solves a problem at home on the basis of a model that he has been shown in class, he continues to act in collaboration, though at the moment the teacher is not standing near him. From a psychological perspective, the solution of the second problem is similar to this solution of a problem at home. It is a solution accomplished with the teacher's help. This help—this aspect of collaboration—is invisibly present. It is contained in what looks from the outside like the child's independent solution of the problem. (Vygotsky, 1987, 216)

The analysis is thus one in which the relational interdependence of individual and social agencies is recognized. The historical analysis would focus on the transformation of dominance through time.

The historical background of the Finnish school of AT posits "networks" of activity systems in which dominance arises at particular moments in both long- and short-term periods. I have placed inverted commas around the word network because I wish to signify a resistance to the notion of network as a connected system within which component parts share some function. Here, I am concerned with the existence of multiple activity systems that may supplant each other and may be mutually transformed. By way of illustration I offer this rather crude example: Suppose a person is both a caregiver (to his or her own child) as well as a professional teacher. If that teacher has a need to collect the child from a nursery at the end of the school day, then the response to class disruption close to the final bell of the day may be very different from what it might be earlier in the day. Here, two activity systems assume a different relationship to one another at particular times of the day. These pulsations in dominance are rarely subjected to rigorous empirical scrutiny. Much of the empirical work that is conducted under the "label" of AT seems to constrain its analysis to one activity system, let alone a network of activity systems, and rarely strays into the analysis of shifts in dominance. Taken together, the implications of the work of Griffin and Cole, Fairclough, Lemke, Leontiev, and Elkonin suggests that such an inquiry should be deployed at both levels, long-term ontogenesis as well as short-term analysis, or even micro analysis. Makitalo and Saljo (2002, 75) argue that it is through the analysis of categories that "people draw on the past to make their talk relevant to the accomplishment of interaction within specific traditions of argumentation." They note, along with Sacks (1992), that categories are activity bound and that their use is inextricably bound up with a particular interactional and moral order (Jayyusi, 1984, 2). Such analyses would share the concern of exploring the way in which subjects are shaped by fluctuating patterns of dominance from the perspective of those actors. However, the emergence of categories is not explored in relation to the principles of regulation of the social setting in which they emerge. The strength of the language of description within the sociology of pedagogy developed by Bernstein is that it explores possibilities for the individual from the direction of an analysis of the rules regulating social circumstances. I want to suggest that there may be some benefit in pursuing the Bernsteinian perspective in the context of the analysis of fluctuating patterns of dominance within networks of activity systems within this framework but from the point of view of the pathway of the object through networks of activity.

12.2.1.7 Discursive Hybridity

To refine an understanding of organizational, discursive, and transmissive practices in such situations, new theories of concept formation that emphasize the complex nature of concepts will need to be deployed. There is a need to develop current work on the predictive relationships between macro structures and micro processes. An important part of the challenge is to show how written and spoken hybrid discourse arises and investigate the consequences of its deployment. In response to this challenge, an understanding of discursive hybridity (Sarangi and Roberts, 1999) may provide an important opening for the development of an understanding of changes in discursive practice as different activity systems are brought into different forms of relation with each other. Research in this field requires a unified theory that can give rise to a coherent and internally consistent methodology rather than a collection of compartmentalized accounts of activity, discourse, and social positioning, which have disparate and often contradictory assumptions.

Let us consider a situation in which several professionals meet to discuss a supposedly common issue: A psychologist, a teacher, a social worker, and a mental-health worker meet to discuss the education and care of a young person who has been expelled from school. We can analyze the historical formation of the professional identities of each actor. They will all have been children, pupils, students, and trainees before moving through professional structures. A historical analysis of the transformations that take place in activity systems could be brought to bear on the formation of dispositions and identities in each of their career trajectories. A situated analysis would need to pursue the ways in which each subject moved (and was able to move) through a short negotiation of possibilities for action as they attempted to understand and work within or impose professional codes. It is here that Bernstein's work on horizontal and vertical structures and pedagogic discourse as an embedded discourse is important. He provides the language of description and the theoretical basis from which to analyze the emergence of "leading" or most powerful activities. Engeström has rightly pointed to the need to analyze contradiction within and between activity systems, and Bernstein provides the theoretical tools to empirically investigate such phenomena in that he connects the analysis of the organizational, discursive, and psychological in a coherent language of description.

To make progress in such empirical work, there is a need for theoretical and methodological development that allows us to identify and investigate:

- The ways in which objects of activity are transformed within the networks of activity systems in which subjects participate
- The circumstances in which particular discourses are produced
- The modalities of such forms of cultural production
- The implications of the availability of specific forms of such production for the positioning of subjects in social space

References

Agre, P. E. (1997) *Computation and Human Experiences*. Cambridge University Press.

Bernstein, B. (1999) Vertical and Horizontal discourse: An essay, *British Journal of Sociology of Education*, 20, 2: 157–173.

Bernstein, B. (2000) *Pedagogy, Symbolic Control and Identity: Theory Research Critique*. Revised edition. Oxford: Rowman and Littlefield.

Bernstein, B. (1996*) Pedagogy, Symbolic Control and Identity: Theory, Research and Critique*, London: Taylor and Francis.

Bernstein, B. (1990) *Class, Codes and Control. The Structuring of Pedagogic Discourse*. Vol. IV: London Routledge.

Bernstein, B. (1981) Codes, modalities and the process of cultural reproduction: A model. *Language in Society*, 10 pp. 327–363.

Bernstein, B. (1977) *Class, Codes and Control Vol. 3: Towards a Theory of Educational Transmissions: 2nd revised edition*. London: Routledge and Kegan Paul.

Cole, M. (1996) *Cultural Psychology: A Once and Future Discipline*, Cambridge, MA: Harvard University Press.

Cole, M. and Engeström, Y. (1993) A cultural-historical approach to distributed cognition, in G. Salomon (Ed.), *Distributed Cognitions: Psychological and Educational Considerations*, New York: Cambridge University Press.

Coulson, S. (2001) *Semantic Leaps: Frame-Shifting and Conceptual Blending in Meaning Construction*. Cambridge: Cambridge University Press.

Daniels, H. (2001) *Vygotsky and Pedagogy*. London: Routledge.

Elkonin, D. B. (1972) Toward the Problem of Stages in the Mental Development of the Child, *Soviet Psychology*, No. 4, pp. 6–20.

Engeström, Y. (1993) Developmental studies on work as a test bench of activity theory, in Chaikin, S. and Lave, J. (Eds.) *Understanding Practice: Perspectives on Activity and Context*, Cambridge: Cambridge University Press.

Engeström, Y. (1996) Perspectives on Activity Theory, Cambridge University Press.

Engeström, Y. (1999) Innovative learning in work teams: Analysing cycles of Knowledge creation in practice, in Engeström, Y., Miettinen, R., and Punamaki, R. L. (Eds.) *Perspectives on Activity Theory*, Cambridge University Press.

Engeström, Y. and Miettinen, R. (1999) Introduction. In Engstrom, Miettinen and Punamaki, R-L. (Eds.), *Perspectives on Activity Theory*. Cambridge: Cambridge University Press, pp. 1–18.

Engeström, Y. and Middleton, D. (Eds.) (1996) *Cognition and Communication at Work*. Cambridge University Press.

Fairclough, N. (1992) *Discourse and Social Change*. Polity Press, Cambridge.

Fairclough, N. (2000) Discourse, social theory and social research: the discourse of welfare reform, *Journal of Sociolinguistics*, 4, pp. 163–195.

Fairclough, N. (2004) The dialectics of discourse, http://www.geogr.ku.dk/courses/phd/glob-loc/papers/phdfairclough2.pdf accessed 30 June 2004.

Fauconnier, G. and Turner, M. (2002) *The Way We Think: Conceptual Blending and Mind's Hidden Complexities*. New York: Basic Books.

Griffin, P. and Cole. M. (1984) Current activity for the future: The zo-ped. In Rogoff, B. and Wertsch, J. V. (Eds.), *Children's Learning in the Zone of Proximal Development*. San Francisco: Jossey-Bass.

Gutiérrez, K., Baquedano-López, P., and Tejeda, C. (1999) Re-thinking diversity: Hybridity and hybrid language practices in the Third Space. *Mind, Culture and Activity*, 6, 286–303.

Hasan, R. (2004) Semiotic Mediation, Language and Society: Three exotripic theories—Vygotsky, Halliday and Bernstein.

Hasan, R. (1995) On social conditions for semiotic mediation: the genesis of mind in society. *Knowledge and Pedagogy: The Sociology of Basil Bernstein*, Alan R. Saadovnik (Ed.), Norwood, NJ: Ablex.

Hasan, R., Fries, P., and Gregory, M. (Eds.) (1995) The Conception of Context in Text, *Discourse in Society: Systemic Functional Perspectives* (Meaning and Choice in Language: Studies for Michael Halliday). (ADPS50). Norwood, NJ: Ablex. pp. 183–283.

Hasan, R. (1992a) Speech genre, semiotic mediation and the development of higher mental functions. *Language Science* 14(4): 489–528.

Hasan, R. (1992b) Meaning in Sociolinguistic theory. In *Sociolinguistics Today: International Perspectives*, Kingsley Bolton and Helen Kwok (Eds.). London: Routledge.

Il'enkov, E. V. (1977) *Dialectical Logic. Essays on its History and Theory*, Moscow: Progress.

Jayyusi, L. (1984) *Categorization and the Moral Order*. London: Routledge & Kegan Paul.

Lea, M. and Nicoll, K. (Eds.) (2002) *Distributed Learning*. Routledge Falmer.

Lave, J. and Wenger, E. (1991) *Situated Learning: Legitimate Peripheral Participation*, Cambridge: Cambridge University Press.

Lemke, J. (1997) Cognition, context, and learning: a social semiotic perspective. In Kirshner, D. (Ed.) *Situated Cognition Theory: Social, Neurological, and Semiotic Perspectives*, New York: Erlbaum.

Leontiev, A. N. (1981) The concept of activity in psychology. In Wertsch, J. V. (Ed.), *The Concept of Activity in Soviet Psychology*, Armonk, NY: M.E. Sharpe.

Leontiev, A. N. (1978) *Activity, Consciousness, and Personality*. Englewood Cliffs: Prentice-Hall.

Mäkitalo, Å., and Säljö, R. (2002) Talk in institutional context and institutional context in talk: categories as situated practices. *Text*, 22(1), 57–82.

Sacks, H. (1992) *Lectures on Conversation*, Vol 1, Oxford, UK: Blackwell.

Sarangi, S. and Roberts, C. (1999) Introduction: discursive hybridity in medical work. In Sarangi, S. and Roberts, C. (Eds.) *Talk, Work and Institutional Order: Discourse in Medical, Mediation and Management Settings*. Berlin: Mouton de Gruyter.

Valsiner, J. (1997) *Culture and the Development of Children's Action: A Theory of Human Development,* 2nd ed., New York: John Wiley & Sons.

Vygotsky, L. S. (1978) *Mind in Society: The Development of Higher Psychological Processes,* Cole, M., John-Steiner, V., Scribner, S. and Souberman, E., (Eds. and Trans.), Harvard University Press.

Vygotsky, L. S. (1987) Thinking and speech. In L. S. Vygotsky, *Collected Works,* Vol. 1, pp. 39–285, R. Rieber and A. Carton, Eds.; N. Minick, trans., New York: Plenum.

Wenger, E., McDermott, R., and Snyder, W. (2002) *Cultivating Communities of Practice.* Harvard Business School Press.

Wertsch, J. V. (1985) *Vygotsky and the Social Formation of Mind.* Cambridge, MA: Harvard University Press.

Wertsch, J. V. (1991) *Voices of the Mind: A Socio-Cultural Approach to Mediated Action.* Cambridge, MA: Harvard University Press.

Wertsch, J. V. (1998) *Mind as Action,* Oxford: Oxford University Press.

13

Movements of the Cane Prior to Locomotion Judgments: The Informer Fallacy and the Training Fallacy versus the Role of Exploration

Gregory Burton and Jennifer Cyr

CONTENTS

13.1 Introduction 268
13.2 Limitations of Prior Research on Nonvisual Locomotion 270
13.3 Exploration with a Cane 275
    13.3.1 Experiment 1 277
        13.3.1.1 Method 278
        13.3.1.2 Data Analysis 279
        13.3.1.3 Results 280
        13.3.1.4 Discussion 286
    13.3.2 Experiment 2 287
        13.3.2.1 Method 287
        13.3.2.2 Data Analysis 288
        13.3.2.3 Results 288
        13.3.2.4 Discussion 289
    13.3.3 Experiment 3 291
        13.3.3.1 Method 292
        13.3.3.2 Results 293
        13.3.3.3 Discussion 294
Acknowledgments 297
References 297

13.1   Introduction

Despite the high numbers of visually impaired individuals who need sensory substitutes for locomotion, science has not had great success heretofore with improving the tools and techniques that were developed in an earlier era. This is noteworthy in that there are few human needs for which the state of the art in 1900 has not been improved. Humans do not travel long distances the same way they did a century ago, nor do they communicate, medicate, or even entertain themselves in the same way. On the other hand, some form of cane for locomotion dates back for millennia, according to a historical analysis (Blasch and Suckey, 1995). The same historical survey dates the use of a long cane rather than a cane equivalent to a walking stick to 1950 after an unsuccessful proposal in the 1870s. But the cane and guide dog are still the prevalent sensory substitutes for active blind persons, despite dozens of travel innovations designed to replace them and hundreds of scholarly papers on the subject. High-tech travel aids are available, but they are largely ignored by the market demographic of active blind and visually impaired persons. The adoption of the long cane is arguably the last popular innovation in this travel tool. A telling manifestation of this trend is given by a survey of the history of education for the blind and visually impaired in Ukraine (Groza, 1985). The substantial improvements in education that were reported since medieval times and in the last century or so were almost exclusively related to improvements in attitude, such as changing perspectives on whether the blind students should be trained in niche skills or to interact with the larger community. No progress was reported in the tools available to the blind person; an analogous survey for any other nation with a good record of education for the blind would show a similar pattern.

I contend herein that this state of affairs is related to a dearth of basic research on cane-aided locomotion, and this lack of research stems in its turn partly from two attitudes that I will describe: the informer assumption and the training assumption. The informer assumption is the attitude that a scientific mystery is to be solved by finding someone who has already solved it (usually by dint of first-hand experience) and then carefully questioning that expert. Those who commit the informer fallacy see no purpose to traditional laboratory research on nonvisual locomotion and, in fact, may see the scientist's indifference to the introspections of the successful blind person as vaguely disrespectful. A related phenomenon that I observe is the scope of what I term the "exhortative literature," a large body of scholarly writing for which the basis of endorsement of a particular technique or belief is impressionistic and personal rather than scientific and quantitative.

The training assumption is that nonacceptance of an offering (similar to a commercial product or medical innovation) on the part of the expected demographic (the buying public or the group with the disorder for which the innovation is intended) can be assumed to reflect inadequate understanding or resistance on the part of the intended consumer, to be potentially remedied by training. The public must be educated to use or to want the product or service (assumes the training fallacy). I would not want to deny that the public and constituencies within the public often do lack adequate understanding of an innovation (e.g., vaccines or fluoride in the drinking water), so I do not use the fallacy label for the notion that training is sometimes needed. However, the habit can be seen in some quarters of promoting training to the exclusion of all else, such as, most importantly, reflection on possible flaws in the product. We obstinately refuse to abandon our paper books for e-books; we decline to invest in viewphones despite their cyclical invention since at least the 1960s in increasingly sophisticated formats; we stick to our old-fashioned college courses with their face-to-face format despite the availability of radio, television, and the Internet to spare us this indignity. The observer committing the training fallacy assumes that the neglected offering would be embraced if the public could be adequately trained to appreciate it. In the case of nonvisual locomotion, this translates into a "we built it, they can learn it" attitude, and a common presumption that people in need of locomotory aids lack the drive to learn the electronic devices (or, with greater political correctness, have drives that are thwarted by rigid habits of their educators and therapists). The fact that the same demographic is highly attracted to other innovations, including electronic ones, is rarely considered.

Both of these attitudes link with a presumption that nonvisual locomotion is centrally and cognitively controlled. If locomotion with a cane were an intellectual activity, it would make sense that its details would be accessible intellectually. Likewise, an intellectual impediment could be reasonably blamed for the nonadoption of tools that seem logically feasible. However, central control as an assumption for action has been criticized for decades. Major influences include the Russian physiologist Nikolai Bernstein, who emphasized the importance of sensation on controlling movement, and the American psychologist James Gibson, who emphasized the physiological and evolutionary illogic of perception being an output of intellectual activity. Bernstein (1996) pointed out the great automaticity of expert action, and how the degrees of freedom necessary to control even a typical movement of the hand would be difficult to control. One of his examples was execution of a dexterous movement in the cold—the difficulty with cold fingers is not loss of strength but loss of sensation. Gibson (1986) emphasized that skillful perception to avoid dangers and exploit opportunities is very different from the sort of abstract perception easily manipulated in psychophysical experiments. There are animals with inferior cognitive abilities or even for whom the existence of cognitive abilities is arguable. However, if there were any animals who could not perceive, they would starve or be eaten. If perception is mandatory but cognition is optional, then perception must not (always) depend on cognition.

I consider my research in locomotion with a cane to be influenced both by the Bernsteinian and Gibsonian perspectives, and I will show how my findings contradict some casual assumptions about cane use that were inspired by the informer assumption and the exhortative strategy. At least for simple, localized locomotory tasks such as judging the crossability of a gap or positioning the feet to pass through an opening, extensive experience with a cane is not necessary for consistent and effective performance, and the evidence suggests that neither classic touch sensations nor the sound of tapping play a critical role in cane use. Rather, a variety of sensations can potentially serve a signal role that contact has been made with a critical point in the environment, such as the bank of a gap, and the perceived posture of the arm constrains the perceived location of that contact and, thus, the affordance of the situation.

Inspired by the eye movement research of Yarbus (1967), I report new experiments examining how the gap is explored, looking for evidence that actors adapt their mode of exploration to counteract the qualities of the cane. Observers in the three studies use canes that differ in length, in mass distribution along the length of the probe, and in mass distribution perpendicular to the length of the probe, that latter manipulation being one that consistently alters how the perceiver holding such a tool feels his arm to be aligned. The boundary between gaps that were crossed in one step or by stepping in was never affected by the probe variable. Manner of exploration was very diverse but the center of exploration was always affected by the probe variable; a model was proposed to resolve the different ways this variable was influenced. The complexity of the pattern underscores the lesser role that mechanical qualities of the cane alone play in this perception relative to qualities of the perceiver's arm holding the cane, as well as the shoulder that elevates the arm, and so forth. This is reminiscent of another emphasis shared by Bernstein and Gibson—the insistence that perception is the result of a system rather than isolated components.

13.2   Limitations of Prior Research on Nonvisual Locomotion

Ten percent of blind and visually impaired Americans (Staff, 1994) use a long cane to assist in locomotion. Research on the perceptual basis for this perception is surprisingly sparse. I have read papers on how canes compare with other sensory substitutes (Strelow, 1985; Skellenger, 1999), and that discuss issues of conspicuity (Blasch and Suckey, 1995), how canes and other aids affect the posture (Gitlin and Mount, 1997), and how the cane should be wielded for maximum efficiency (Uslan, 1978). There is also voluminous research on how humans can perceive the qualities of unseen probes and can perceive qualities of surfaces with unseen probes (Solomon and Turvey, 1988; Carello, Fitzpatrick, and Turvey, 1992; Barac-Cikoja and Turvey, 1999). However, concerning the actual sensations and perceptual processes that allow a human being to augment impaired vision or substitute for lost vision with a hand-held probe, there is quite little. Before I began my own humble researches in 1989, I could find in the literature less than 10 papers on the sensory basis of cane use for locomotion (Potash, 1962; Jansson and Schenk-man, 1977; Schenkman, 1986; Schenkman and Jansson, 1986; Clark-Carter, Heyes, and Howarth, 1986a, 1986b). I have not crossed paths with any subsequent papers that examined the psychophysics or sensory basis of cane use, although research on other aspects of canes and on other sensory questions about perceiving probes continues apace.

Part of the reason for this lacuna is that the question is very difficult. A cleverer and better-resourced researcher may have accomplished more, but my own research after 15 years has mostly established negatives: cane use is not based only on the sound of tapping (Burton, 2000), using a cane for at least simple tasks does not require great experience (Burton and Cyr, 2004), and cane users are little affected by the length of probes (Burton, 1992; 1994). The developing picture of my research is that the posture of the limbs wielding a cane has more to contribute to the sensory basis of cane use than any particular mechanical sensations transmitted by the cane (Burton and McGowan, 1997). I will have more to say about this research later, when I introduce some new findings in this area. First, however, I would like to discuss some attitudes that I perceive as impeding more appropriate progress in this important area.

The problem of locomotion in the blind has garnered much attention from the fields of psychology, rehabilitation, physiology, engineering, and others; this attention has just been focused on other aspects of the problem or has failed to achieve progress in the sensory question that to me seems fundamental. Much of the work done heretofore has been influenced, in my opinion, by a pair of complementary attitudes that I will call the informer fallacy and the training assumption.

To reiterate the working definition given in the preceding text, the informer fallacy is that understanding of a specialized skill is to be obtained by directly questioning one who possesses this skill. This tactic is based on introspection, a psychological method that is obviously long discredited, and the general public is familiar with various counterindications of its effectiveness. For example, the public recognizes that one cannot state in words how to ride a bicycle. On some level, however, the bicyclist *knows* how he rides but is just frustrated in trying to express it. Psychological progress must start with acknowledging that there are also skills for which there is no direct knowledge at all, regardless of format. Humans localize sound by comparing the arrival time and amplitude of compatible sound waves at the two ears but cannot experience this comparison consciously, even if they have read about sound localization in books. This general cognitive impenetrability of important perceptual processes means that understanding of a skill requires objective methods and the cooperation but, for the most part, not the insights of the actor. The skilled actor has insights and experiences that may suggest fruitful directions to the scientist, but does not possess definitive information.

Bernstein (1996) saw cognitive inaccessibility as a crucial strategy for reducing the degrees of freedom for a complex action system, and most people could generate examples such as sound localization without much effort. Nonetheless, I perceive a public attitude that supports consulting the experienced doer in preference to the impartial scientist and, in fact, scientific attitudes can be seen as disrespectful. Scientists invited to talk shows can expect to be excoriated if they cast doubt on the personal experience of another informant on the basis of lack of controls, self-selection, or other familiar data pitfalls. When I was new to recruiting blind and visually impaired informants, I attended a meeting of an organization dedicated to advocacy for the blind. A colleague of mine who belonged to this society had agreed to introduce me weeks before, but at the meeting he finished his introduction by expressing disapproval of using "blind people as guinea pigs." At the meeting itself, the society members suggested that my research would be more productive if it centered on family problems that blind people might experience (an area for which I lack any professional qualification) instead of sensory questions. A published statement of this attitude is provided by Altman (1996): "If the knowledge and experience of the people who live with nonvisual travel is not respected and accepted enough to be placed at the core of this process, then it is little more than an intellectual exercise that will probably do more to preserve the status quo than to bring meaningful advances into people's lives" (293).

An extreme example of this attitude was found in the third edition of a popular perception textbook (Schiffman, 1990). The text gave the example of a blind equestrienne who is nationally ranked. Her method is reported in the textbook in these words: "Her ability to navigate the corners and winding turns on the course during competition are based on the reflective echoes produced by the sounds of her ride" (100). That her talent is incredible is undisputed (I cannot even ride a horse in a line with vision), but the lack of hedging terms in this report of her *technique* would seem to imply that this report is based on rigorous testing. One would hope it was, because considerable empirical weight would be required to overcome (a) the negative argument that the steeplechase environment is probably too loud for effective echolocation and (b) the alternate theory that the horse itself is motivated and skilled at avoiding obstacles. There is a citation in the paragraph that is presented to the reader as essentially equivalent in status to the rigorous studies of Kellogg in 1962 and Cotzin and Dallenbach in 1950 which bracket the equestrian example. However, in fact, the source is to the Rutgers alumni magazine—presumably not a primary source for scientific tests of the rider echolocation theory. This example was withdrawn from later editions of the same textbook.

The attitude that respect for one's subject requires unskeptical acceptance of impressions is also manifest in a general trend in scholarly publications that I have observed. In psychology, at least (and I suspect in other social sciences and related fields), students are trained to emulate the empirical literature. Undergraduate psychological curricula teach concepts of statistical analysis, experimental control, skepticism toward confounds, and so on. Yet, a considerable minority of the publications in some areas is part of a different literature, and assert measures and suggestions without an interest in research of this nature. The reader is urged to adopt or support some step (to try the presented exercise in his or her own class, to include such exercises in all curricula, and to make the exercise a mandatory part of professional training) based on the eloquence of the author and the reported positive impressions of participants. Frequently, data is presented in these articles to document participant satisfaction, rather than a change in some attitude or behavior. I do not recognize this as the empirical literature with which I "grew up," so I will label this body of work the "exhortative literature."

I encountered the exhortative literature recently in trying to find articles on vicarious experiences, the exercises sometimes used as a means of increasing understanding of another time or group, empathy with members of a psychological or social demographic, or so on. It is felt that participants can better understand the challenges facing the blind or elderly, the dangers of crime, the stress of teenage parenthood, or the perils of conformity, by engaging in small-scale simulations of these experiences. Recognizably empirical work on vicarious experiences lends at best mixed support to the efficacy of these techniques. For example, a meta-analysis of Scared Straight–type programs in which students encounter incarcerated criminals indicated that they actually backfire (Petrosino, Turpin-Petrosino and Finckenauer, 2000). Isbell and Taylor (2003) found that students exhibited a familiar ingroup/ outgroup effect regardless of whether they had watched a popular video that is meant to sensitize viewers to outgrouping.

Studies similar to these are greatly outnumbered by articles promoting the use of vicarious experiences for which the supporting data is a positive experience by the administrators or participants in the exercise. Goldstein (1997) recommends an exercise in which participants modify their conversation based on stereotype labels affixed to the foreheads of their conversants; the evidence for the recommendation is that the participants agreed that the exercise taught them something about prejudice and should be adopted in subsequent courses. Hoffman, Brand, Beatty, and Hamill (1985) developed a role-playing board game called GERIATRIX to address the lack of empathy professionals may have for the elderly; the evidence in question is that the participants agreed that they learned some things about their own attitudes. Conill (1998) published an editorial in *Journal of the American Medical Association* describing a program that she believes should be adopted in medical schools nationwide. In this program, medical students spent 24 hours with an artificial disability that required an aid, such as a walker. The evidence in this case was that the students gave positive open-ended feedback and were enthusiastic about the program. None of the three examples given in the preceding text used any sort of inferential statistic to evaluate the change observed, although there are similar articles that do include inferential statistics. The overall number of generally similar articles, which present a suggested therapeutic or educational technique with no more than a satisfaction survey as the empirical basis, is quite large.

The second attitudinal obstruction I see is the training assumption, the predilection to explain the apathy of the intended market for an innovation by pointing to the ignorance or resistance of the audience rather than limitations of the product. Different versions of this attitude show up in all walks of life and are not unrelated to the comfortable bias that leads us to believe that today's teenagers would like music popular when we were teenagers if they knew anything about music. But an excellent psychological example is the unhappy fate of early reading machines for the blind.

Machines that read text and recite it as human speech are now readily available. However, the acceptance that human speech sounds would be necessary is actually a bit of a concession. A much older approach was to render the text as arbitrary pitches because the technology for delivering arbitrary sounds that rose and fell in correlation with the patterns of light and dark in text was much more easily achieved. It was assumed that, as infants, we had all learned one arbitrary code. Learning a second as adults with more motivation and one under our belts should be simple.

However, humans never comprehended the machine speech with arbitrary pitches at rates even remotely equivalent to their comprehension of human speech, even when different arbitrary coding systems were tested (Liberman, Cooper, Shankweiler, and Studdert-Kennedy, 1967). It finally dawned on some of the speech scientists to question the assumption that the first language learned was arbitrary. Liberman began a lifelong campaign to show that speech is a very special form of perceivable and not just a particularly useful pattern of sound. But this required him to question the technology and to overcome the usual impulse to question the training and motivation of the market audience.

Electronic devices for locomotion have been tested for more than a century (Warren and Strelow, 1985—see useful reviews by Brabyn, 1985 and Jansson, 1991). Even 20 years ago an overview could list alternatives such as the Pathsounder, the Mowat Sensor, the Nottingham Obstacle Detector, the Sonic Torch, and the Laser Cane (Warren and Strelow, 1985). I have lurked on a ListServ on which proponents of various approaches would routinely send e-mails pointing out flaws in the others. One inventor was confident enough to state in print that "any other method is depriving blind persons of spatial experience" (Kay, 1985, 138). That blind people generally do not use them has been noted by Jansson (1991), Warren and Strelow (1985), and others and mentioned to me informally by cane users.

It is worth emphasizing the strange discrepancy between the laborious effort invested so far in inventing and testing new electronic approaches and the desultory progress in basic research in how canes are useful. Successive advances in electronic travel aid (ETA) development seem to focus on making the tool easier to use, which cannot hurt, but which presumes that the general approach is correct but just "too much." However, unless I have missed an extensive body of basic research, that general approach is not informed by a thoroughgoing understanding of the phenomenon, because the necessary research has not been done. That device designs are not motivated by specific knowledge about locomotion in the blind was noted long ago (Warren and Strelow, 1985), so the trainer's assumption (build it and they will come) existed even then. As noted eloquently by Brabyn (1985): "In some cases, user performance with the aid was judged inferior to unaided performance, and in most cases the results were inconclusive. Certainly, the production and marketing of the various aids went ahead regardless" (18–19).

I have long wondered if there might well be an analogy between ETAs and the arbitrary artificial speech systems mentioned earlier because most ETAs also convert the information needed by the walker about the distance of obstacles and other aspects of the layout into an arbitrary language. A sonar device, for instance, might deliver a pitch corresponding to the distance of the nearest obstacle. It makes sense but so did the idea of converting text into a series of pitches coding the transitions from light to dark. More research into the natural phonemes of locomotion would be another approach. Among other possibilities, it might be worth reconsidering delivering the results by sound. Although it has been popular to play up the sounds produced by the tapping of a cane as supporting echolocation, there are certainly cane tasks that can be performed well by experienced and inexperienced cane users with sound drastically reduced (Burton, 2000), and cane experts have often noted the high noise level of crucial cane environments. Some ETAs deliver tactile information, but so far none deliver information in haptic terms (i.e., the movement of the entire rod available to the fist) although these are dominant sensations when wielding any hand-held tool.

13.3 Exploration with a Cane

I am asserting that there is a dearth of basic research about the sensory basis of cane usage and, in the absence of this research, it is surprising that scientists have felt so confident about designing electronic aids. Admittedly, research in this area is fraught with frustrations. There seems to be little one can do to a probe to render it useless for a cane, although it probably needs to be rigid and my visually impaired informants frequently describe substituting some found object, such as a broomstick, for a missing cane. At first blush, this casual attitude toward mechanical qualities of the cane is surprising; if the cane is an intermediary between the perceiver and his environment, it might be expected that the qualities of the cane, far from being perceptually "invisible," would influence what is perceived. Yet, the inclination of cane instructors and users to treat the cane as "transparent" is supported by a series of experiments on nonvisual performance of one simple component of locomotion. In judging whether a gap in a path could be crossed, neither naïve sighted persons (Burton, 1992; Burton and Cyr, 2004) nor experienced, visually impaired cane users (Burton and Cyr, 2004) were influenced by the length of the probes. When actors actually cross gaps, cane length does not affect whether or not a gap is crossed or the parameters of the steps used to cross (Burton, 1994). For gap crossing, it appears that the posture of the limbs when a gap is explored with a cane is more critical than the nature of the cane itself; perceivers can judge gaps consistently when they wield only a flashlight that triggers light sensors when it reaches the banks of a gap, but they could not judge consistently when they wielded a cane that was arranged so that the same range of arm movements was necessary for larger and smaller gaps, and their judgments of crossability were increased when the rod was manipulated with weights in a fashion that makes the arm feel as though it is higher in space (Burton and McGowan, 1997). Preliminary experiments (Burton, 2001) suggest that at least one other environmental variable, the location of the center of an aperture, is also perceived in a constant basis with a cane and is similarly uninfluenced by the simple mechanical qualities of the probe.

What is being described here is a perceptual constancy, analogous to familiar optical constancies such as size constancy and shape constancy. Ecological psychology stresses perceptual exploration as the basis for constancies. Gibson (1979) used the phrase "information pickup" to emphasize the active and continuous nature of the act of perception. It is natural, he pointed out, for a perceiver confronting an unclear or unexpected situation to seek additional information from a changed perspective. Support for the unity of exploration and perception ranges from the joint auditory and visual projections to the superior colliculus (Wickelgren, 1971) that allow the eyes to be oriented according to localized sound, to ecological research showing that adaptation to height-altering shoe attachments is facilitated by opportunities to explore (Mark, Balliett, Craver, Douglas, and Fox, 1990). Relatedly, Gibson asserted that perceptual illusions require artificial inhibition of this natural tendency (e.g., the Ames room: Gibson, 1979) although there are contradictory arguments and evidence (DeLucia and Hochberg, 1991). Some static illusions do not survive being regarded from an angle off the normal (Kennedy, Green, Nicholls, and Liu, 1992).

The current experiments are designed to examine the patterns of exploration with canes, assuming that these movements reveal invariants of gaps over different postures and probe qualities. There is a rich tradition of using patterns of exploration to reveal which aspects of a sensed situation play the largest role, such as the famous studies of Yarbus (1967), in which eye movements were shown to be distributed, not uniformly or randomly but in accordance with the meaning of different components of the scene. Subsequent views of a repeatedly shown scene tended to concentrate even more so on a few salient points therein. The influence of these studies extends beyond perception theory; Gandelman (1986) urged art historians to have greater familiarity with Yarbus and related studies.

The following experiments investigated the manner in which observers explore with a cane before crossing unseen gaps. This investigation has a threefold purpose. First, we hope these data on cane movements will indicate the most critical areas of "cane space," following the example of the eye movement paradigm. Second, these experiments require actors to get over a gap but allow them to decide whether to step over or step in; as this decision is different from those required in previous gap-crossing experiments, it allows another test for the lack of influence of mechanical variables that has been the case heretofore. Assuming that the mechanical variables also have no effect on the step-in/step-over decision, our third purpose is to investigate whether these variables do influence how observers wield the cane to make their judgments. The resultant action may be consistent, but this constancy may turn out to be effected by adaptive changes in exploration.

13.3.1 Experiment 1

Experiments on relatively abstract qualities of surfaces have consistently revealed influences of probe qualities. The moment of inertia of the rod, a measure of its resistance to being rotated, influences the perceived separation of two solid objects (Barac-Cikoja and Turvey, 1991), the perceived depth of a surface (Chan and Turvey, 1991), and the perceived distance of a surface struck on an angle (Carello, Fitzpatrick, and Turvey, 1992). In gap-crossing experiments, however, varying the moment of inertia did not result in systematic change in gap-crossing boundaries (Burton, 1992), nor in the steps used to cross gaps (Burton, 1994). Altering the length of the cane, which likewise changes its resistance to rotation, did not influence judgment of crossability for sighted (Burton, 1992; Burton and Cyr, 2004) nor visually impaired individuals (Burton and Cyr, 2004).

In the first experiment, blindfolded actors stood before a gap in a pathway and crossed it, deciding whether to step in or over the gap to make their crossing. Based on previous findings, we predict that the cutoff between gaps that are crossed by stepping over and those crossed by stepping in will be unaffected by rod length. However, observers may have to wield the rods differently when they differ in length. For example, they may explore more, and they may concentrate on different places when using different canes. Little is known of which aspects of an area to be traversed are most salient for nonvisual locomotion. Recommended cane travel techniques do not even agree on whether the cane should be kept in constant contact with the ground, and at least one attempt to accommodate the blind traveler, the strategy of cutting horizontal lines in curb ramps at intersections, met with limited success because most cane users tapped over the lines (Bentzen and Barlow 1995). The practice of "shorelining", or tapping against a predictable boundary like the side of a building, suggests at least one critical area for a different component of locomotion.

There have also been investigations of cane coverage during certain travel techniques, which have revealed, for example, that one popular technique does not really cause the cane to test the area of the ground on which the foot will soon step, as supposed (LaGrow, Blasch, and Del'Aune, 1997). Experiment 1 employs a different strategy for suggesting critical areas for gap crossing, and whether the location of those areas is affected by rod length.

13.3.1.1 Method

13.3.1.1.1 Participants

Eight Seton Hall University students participated for extra credit. Six were female, and two were male. Three height measurements were taken of each participant: standing height, shoulder height, and sitting height. Participants wore shoes for the measurements and in the actual experiment. Standing heights ranged from 165.2 to 180.2 cm. Measures were also taken of the upper arm (point of shoulder to elbow), lower arm (elbow to wrist), lower leg (knee to floor), hand span (distance across palm), and grip. All participants reported being right handed.

13.3.1.1.2 Apparatus and Equipment

The experiment was conducted in a room approximately 6.75 m × 2.5 m. The room contained a walkway composed of four 0.91 m² wooden platforms, of a depth of 0.10 m, raised an additional 0.04 m from the floor for a total height of 0.14 m. The platforms could roll freely along a framework consisting of a pair of metal channels 6.10 m in length and connected firmly by metal crosspieces 0.92 m in width. Tubular safety rails 1.10 m high were attached on both sides of the entire length of the platform. Each platform could roll freely along the framework and be firmly clamped in its place along the channels. Three platforms were securely clamped and the fourth platform remained mobile for adjusting the size of the gap. The upper surface of each platform had grooves cut into it at regular intervals so that if the two platforms were placed next to each other, the location of the junction would not be easily discernible.

In all conditions, participants were asked to wear a blindfold, which was a blackened pair of safety goggles. Two hollow aluminum rods 1.22 m and 1.52 m were used by the participants. A Panasonic video camera was placed perpendicular to the long axis of the walkway at a distance of 4 m from the middle of the walkway. The camera was raised 0.75 m off the ground. The videotape collected 60 frames per second and was positioned so that the gap, which included the edge of the fixed platform and the edge of the mobile platform, was visible.

13.3.1.1.3 Procedure

Throughout the experiment, three platforms (fixed) remained firmly in place along the framework and the fourth (mobile) was moved away from the fixed platform to produce a gap ranging from 0.15 to 0.90 m at increments of 0.15 m. For each trial, the participant stood a few centimeters before the edge of the third platform. The participant was asked not to hold on to the rails at the start of a trial but could use them during the trial if necessary. At the start of each trial, the experimenter set the gap size and the participant was handed a probe. The participants' task for each trial was to cross the gap in each condition. Crossing was defined as starting with both feet on the fixed platform and ending with both feet on the mobile platform. The participants had to decide on each trial whether to cross the gap by stepping over the gap or by taking a step or two on the floor before stepping up onto the mobile platform. This process was videotaped and later analyzed (see data analysis). After the trial, the experimenter took the probe and the participant walked around the walkway back to the edge of the third platform.

Each probe length was paired with each gap length twice for a total of 24 trials per participant, which were presented in a different random order for each subject.

No practice trials were given, and the participant was not allowed to perform any exploratory leg extensions while not blindfolded. However, at the beginning of the experiment, participants practiced walking around the walkway while blindfolded. The participants were informed that different rods would be used but did not know how the rods differed or how many were used. Similarly, participants knew that some gaps would be so big that they could not be crossed at all, and some gaps were so small that they could be crossed easily but did not know the intervals at which the gaps were set.

13.3.1.2 Data Analysis

The video was analyzed to determine (1) the total number of distinct taps during exploration on each trial, (2) the number of times the participant tapped the baseboard in determining whether to step over or step in, and (3) the number of times the participant changed the direction of the cane when trying to make this decision. The baseboard taps could occur on any part of the platform on which the participant was standing (the fixed platform). A "change of direction" was counted when the participant was moving the cane in one direction, made contact with the ground or either of the two platforms, and then moved the cane in the opposite direction. The first motion of the cane after the participant was originally handed the probe was not considered to be a change in direction. If no contact was made, a change of direction was not counted (even though the participant may have swung the stick back and forth several times).

Three coders took part in analyzing the films. The first coder (the first author) recorded the location of each tap from the videotape and then derived the other variables from this record. Up to the point of participant's crossing of the gap, participants varied considerably in their styles of tapping, and in some cases, whether one or more taps had taken place was ambiguous. Thus, a second coder recorded the changes of direction, as well as the number of taps, directly from the videotape. A correlation of .75 was found between these two coders. Owing to this low correlation, a third coder (the second author) was brought in; the first and third coders conferred on some of the trials for which their counts were discrepant. The third coder also recorded the changes in direction and the number of baseboard taps made by each participant, directly from the videotape. A correlation of .90 was found between the first and third coders for changes of direction, .93 for baseboard taps, and .89 for the total taps. The records made by the third coder were utilized for subsequent analysis.

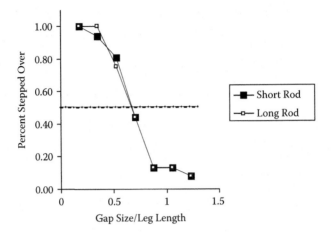

FIGURE 13.1

For Experiment 1, the percentage of trials on which observers crossed a gap by stepping over it rather than stepping in the gap. The percentage is given as a function of the ratio of gap size to the mean leg length of the eight actors.

13.3.1.3 Results

Figure 13.1 shows, for the long and the short probe, the percentage of trials for which the actor crossed over the gap rather than stepping in. The gap lengths have been divided by the mean leg length for the actors. The ogival aspect to these functions for actual crossing is very reminiscent of the equivalent functions for judgment of crossing (Burton, 1992, Figure 6, lower left). In that earlier experiment, participants were asked if they could cross the gap at all and if they could cross with a "natural" step. The instructions in the current experiment were to cross in the preferred manner; we attempted not to bias the participants into attempting to cross over on every trial, so the decision to be made is similar to the second question that was asked of subjects in the previous research. With that in mind, the similar aspect of this function suggests that those earlier participants who judged rather than crossed gaps made realistic judgments; the crossing over of the functions in Figure 13.1 was just below 0.60 m in absolute units and is also close to the crossovers in the judgment functions.

Figure 13.1 also shows essentially identical performance with the short and long canes, as seen before for judgment of gaps for blindfolded-sighted persons (Burton, 1992), visually impaired individuals (Burton and Cyr, 2004), and blindfolded persons crossing a gap of an unknown distance (Burton, 1994). A t-test on the boundaries between gaps crossed over and gaps stepped in corroborates the equivalence apparent in Figure 13.1 because the mean boundaries when actors used the long probe and the short probe were both 0.60 m, $t(7) = 0$. These decisions to step in or over were also made with great consistency; across all eight participants; in only one trial did an actor cross over a gap longer than one he had stepped in on a different trial.

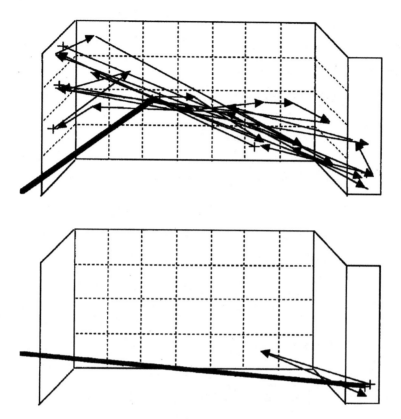

FIGURE 13.2
Schematics for two presentations of the same gap and probe to a participant in Experiment 1 who showed an extreme variation from the first presentation (top) to the second (bottom). The walkway was explored with the short probe. The thick bar represents the first tap, with arrows depicting the movement of the probe thereafter. Crosses signify distinct taps in a fixed location.

These results confirm the constancy found in previous studies; the filmed records were analyzed with the aim of finding aspects of the exploration that were affected by rod length. The participants displayed very different approaches to exploration; Figure 13.2 shows a record of the movement of the probe's tip for two repetitions of the same 1.05 m gap for a single observer.

Interestingly, the trial in which this participant explored more was the second presentation of that gap size–probe length combination.

Schematics of cane movements for both rods are shown in Figure 13.3.

Each part depicts the gap divided into 0.15 m strips lengthwise (parallel to the path of movement) and four 0.25 m strips widthwise. The number of taps in each cell thus defined is symbolized by the darkness of shading in each cell; the first and last columns represent taps on the platforms themselves, and the vertical bank of each platform is also depicted as the column of cells just within the first and last columns, and bordered by dashed lines.

It is clear and logical that the vertical banks of the platforms were the most popular targets for tapping, whereas a secondary cluster of taps is found

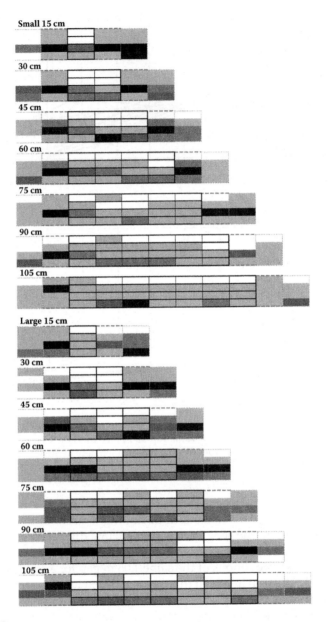

FIGURE 13.3

Schematics for the location of taps for volunteers in Experiment 1 using the short probe (top) or the long probe (bottom) for the seven gap sizes. Each cell bounded by a solid border represents a 0.15 m long by 0.25 m wide space on the walkway; taps on the vertical edge of the starting and ending platform are symbolized in the cells bounded by dashed lines flanking the solid-bordered cells, and the first and last column, surrounded by dotted borders, represents the horizontal surfaces of the start and end platform. The darkness within each cell corresponds to the relative number of taps within the cell accumulated over the eight actors; only cells in which there were literally no taps whatsoever are left unfilled.

between 0.3 and 0.45 m from the edge of the first platform. The concentration of taps on the far bank and platform seems to spread out as the gap increases, from distinct concentrations in gaps of .30 m and less to a general darkening of the figure in the vicinity of the far bank for later gaps. This pattern suggests that for a gap too large to be crossed, it is not necessary for actors to ascertain just how uncrossable it is to perceive its uncrossability. All observers wielded the probes in the right hand and a very clear right-handed bias is evident; gaps of all sizes for both rod lengths have extensive blank areas on the left (top two rows of each schematic), but there are no blank cells whatsoever on the right (bottom two rows of each schematic). Note that a blank cell in Figure 13.3 implies zero taps, for eight observers and two presentations each. Thus, no observers in Experiment 1 felt much of a need to thoroughly explore an entire space before making a judgment.

As a means of quantifying the information in Figure 13.3, a "center of tapping" was calculated for each trial, representing the horizontal location of the average tap for that trial. A tap on the far platform was assigned a 0, and a tap on the footboard of the far platform was assigned a –1. One point was subtracted for each additional 15-cm band, and taps on the footboard of the near platform or on its surface were assigned one or two points less, respectively, than the lowest score for a tap in the gap. For an example, a –4 would be the score for a tap 40 cm before the far platform, because such a tap would fall in the third band extending from –45 to –30 cm; the average for a participant who tapped twice 40 cm from the far platform and once 65 cm from the platform would be $(2*-4 + 1*-6)/3 = -4.8$. Although these scores are partially arbitrary, the physical location can be reconstructed. For example, the mean for participants using long probes to explore 45 cm gaps was –2.867. This signifies that the average tap location was 1.867 bands, or 28 cm before the footboard of the far platform. I analyzed these values for center of tapping rather than the actual physical values because the physical distance would not incorporate the difference between tapping the platforms and tapping the gap.

Figure 13.4 shows the centers of tapping for the six distances and two rods; the 0 point is the far platform, the actors' destination.

There is a clear tendency for the center of tapping to move further away from the far platform as the gap grows wider (the main effect of gap size was significant; $F(5,35) = 30.9, p < .0001$). One contributor to this effect may have been the fact that lower scores were possible for wider gaps (e.g., for the 15 cm gap, –4 was the lowest possible score). The ANOVA confirmed two other trends apparent in Figure 13.4, the effect of rod length ($F(1,7) = 8.0, p < .05$) and the noninteraction of rod length and gap size ($F(5,35) = 2.1, p < .10$). Figure 13.4 shows that the center of tapping was closer to the far platform (farther from the observer) when the rod was longer. The center of tapping could be the result of one of two tendencies: (a) the tendency of perceivers to need information from a particular *place* or (b) the tendency of perceivers to perform a preferred *movement*. If the place is more salient, the rod effect may represent that the contacted place merely felt closer with the longer

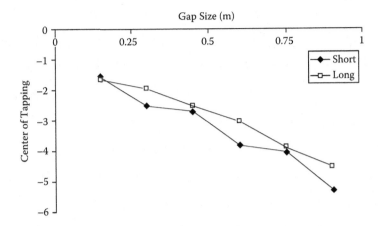

FIGURE 13.4
Center of tapping for the two probes and the first six distances for Experiment 1. The x-axis represents the far platform, to which actors were required to cross.

rod; if the movement is salient, the rod effect may simply be the geometrical result of making the same movement with a longer arm-rod system. The first speculation suggests that longer gaps seemed crossable with shorter rods, a possibility that is at least inconsistent with Figure 13.1 and previous results (Burton, 1992, 1994; Burton and Cyr, 2004). The actual size of the significant difference is 0.38 bands, or 5.7 cm.

The total number of exploratory taps was a function of rod length ($F(1,7)$ = 11.8, $p < .01$), with the gaps explored with the short rod tapped 6.9 times, on average, compared to 8.0 taps for the long rod. Gap size was not significant ($F(5,35) = 0.7$), nor was the interaction ($F(5,35) = 0.8$). Although the coders of the films had an impression that participants were tapping less and less as the experiment proceeded, Figure 13.5 shows no obvious pattern.

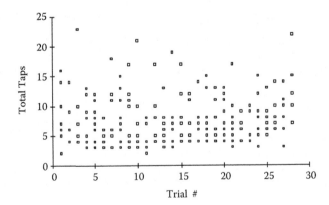

FIGURE 13.5
The total number of taps for the eight observers as a function of trial number; note that the particular gap/probe condition was randomized differently for each observer.

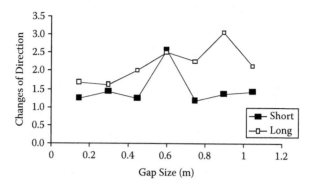

FIGURE 13.6
Mean changes of direction as a function of rod length and gap size for the observers in Experiment 1.

Figure 13.6 shows that trials may differ not only in the number of exploratory taps but also in the number of times a perceiver changed the direction of exploration.

This variable may be taken to represent uncertainty; there were significantly more of these for the long probe (mean of 2.4) compared to the short one (1.7); (F(1,7) = 7.3, $p < .05$). Note that a person could score a zero on this variable if the probe was moved only to a single location before locomotion. As with the total number of taps, there was no effect of gap size (F(5,35) = 1.7) nor was the interaction significant (F(5,35) = 1.0). These variables imply that participants explored more overall and took more "perspectives" when the probe was longer. As an attempt to pin down what participants were looking for, we analyzed the number of times they struck the footboard (i.e., the portion of the platform on which they were standing). Our interest in this variable was based on several factors. Informally, some participants wielding oversized probes in previous studies have reported, even when their overall performance was unaffected, that they felt more uncertain of their own location than the location of the far side of the gap. Relatedly, one strategy that a perceiver could adopt to deal with an inconveniently long probe, or even a probe with an uneven mass distribution, might be to "scale" the perceived distance of the far side of the gap to the distance to their feet as it feels using the same probe. A trial on which participants were more uncertain might feature more of these orienting taps. This variable, however, was not affected by probe length (F(1,7) = .12) or gap size (F(5,35) = 1.7), nor was the interaction significant.

A somewhat complementary variable was the number of taps that took place in the gap, as opposed to either platform or the vertical banks of the platforms (note that total taps ≠ taps in the gap + footboard taps). Rod length had a marginally significant effect on this variable (F(1,7) = 4.8, $p < .07$), with 3.1 taps on average for short probes and 3.6 for long probes. There was a significant effect of gap size (F(5,35) = 9.2, $p < .001$). Because the longer gaps had more room in which to tap, this may not be a very surprising finding, but it confirms that the gap itself was a feature to be explored. Presumably, observers could have adopted a strategy of reaching to the far platform and making a decision based only on the attitude of their limbs when contact was made, but the variables analyzed here indicate that this was not the case.

13.3.1.4 Discussion

Although the analyses reviewed in the previous text did reveal some consistencies, we were struck by the great variability in exploratory styles. Actors differed greatly from one another, and their approach to one presentation of a gap–probe combination was often very different from the next presentation (see Figure 13.2). The major consistencies were as follows. For longer gaps, exploration was concentrated closer to the end of the gap, and more taps fell within the gap rather than on the banks or platforms. Exploration with longer rods was also closer to the end of the gap, and involved more taps and more changes of the direction of cane movement, as well as marginally more taps within the gap itself.

The influence of probe length on exploratory variables despite its lack of influence on the ultimate mode of crossing is reminiscent of anecdotal reports of cane wielders using diverse substitutes for their familiar canes, without noticing a loss of information.

Did probe length affect exploration because it affected arm posture, or because the perceived location of the taps was influenced by length? In previous research with the gap-crossing paradigm (reviewed in the previous text; e.g., Burton, 1992), actors tended to judge gaps explored with longer probes as equally crossable. In the current experiment, if the longer probe caused actors to feel they were reaching farther, it is hard to understand why they would not decline to cross gaps that felt farther away; i.e., Figure 13.1 should have showed an effect of probe length on whether gaps were stepped across or crossed by stepping in. These considerations support that the first possibility, that the influence of rod length on arm posture and movement, was paramount.

These effects may be strictly mechanical (e.g., geometry demands that a longer arm–rod system rotating from an unchanged point falls at a longer distance). However, the perception of posture is also critical. A model developed by Burton and McGowan (1997) asserted that two distinct pieces of information are required for an individual wielding an object to make use of the object's contact with a surface in the world. The fact of contact can be conveyed by a variety of perceptual systems, including haptic, postural, and auditory. But the *location* of contact is largely conveyed by postural information, a claim that Burton and McGowan (1997) believed to be consistent with various findings in the gap-crossing paradigm.

Earlier research has shown ways of altering perceived rod length without altering the posture necessary to wield the rod, and of altering perceived posture without changing the length. The former manipulation can be effected by affixing weights to the rods at different distances from the hand (as in Solomon and Turvey, 1988); the latter, by affixing weights to a crossbar that intersects the main shaft of the probe (as in Burton and McGowan, 1997). Based on the logic in the preceding paragraph, manipulating perceived length should have no effect, and manipulating perceived posture may well have an effect. The next two experiments will consider the effects of these manipulations on exploratory variables.

13.3.2   Experiment 2

The perception of the length of an object wielded in space is largely constrained by the object's moment of inertia, or resistance to being rotated. This quantity is affected by the mass distribution of the object, such that a component of uniform mass will increase the resistance more if it is centered at a greater distance from the rotation point. Thus, the same rod wielded at an end is more resistant to movement than when wielded at the center, and a weighted rod feels longer the further from the hand the weight is affixed (Solomon and Turvey, 1988). Although this basic result has been replicated many times and in various permutations (some of which do require the consideration of additional variables), one surprising finding of Burton (1992) is that the moment of inertia of weighted rods does not affect perceived crossability of a gap contacted with that rod. Considering the two-factor model reviewed in the preceding text, we now believe this is because the perceived posture of the arm–rod system, rather than the perceived length of the rod, is the perception actors rely on for gap crossing. Actors in Burton (1992) may well have perceived the rods as longer, but the rod length was immaterial for perceiving the location of the bank of a gap contacted with that rod.

In Experiment 1 of the current project, rod length again did not affect crossing behavior, but it did affect exploratory behavior. Rod length affects both perceived rod length and the posture necessary to wield the rod; for the former implication to be the cause of the effects of Experiment 1 would seem to contradict previous research. However, the possibility can also be more directly tested by manipulating the perceived length of the rods, using the weight-location strategy. Our prediction is that altering the moment of inertia of rods of equal length will not influence the exploratory variables (center of tapping, total taps, and changes of direction) that were influenced in Experiment 1.

13.3.2.1   *Method*

13.3.2.1.1   *Participants*

Eight Seton Hall University students participated for extra credit; five were female and three were male. The same measurements were taken as in Experiment 1. All participants reported being right-handed.

13.3.2.1.2   *Apparatus and Equipment*

The experiment was conducted in the same room and employed the same walkway as in Experiment 1. The blindfold used in this experiment included padding that conformed to the shape of the eyes and was purchased from Money Talk$, Inc., of Tucson, Arizona. A single hollow aluminum rod, 1.22 m in length, was wielded on all trials; a weight of 150 g was attached to the rod 0.3 m from one end. Mechanically, when the rod was wielded at the end farther from the weight, it resisted movement more than when it was wielded at the end closer to the weight. This yielded probe moments of inertia of 0.106 and 0.218 kg × m² for the two wielding conditions, with moment of inertia calculated from the end of the rod. (The actual axis of rotation of the arm–rod system must necessarily change as the explorer moves his shoulder, elbow, waist, wrist, etc., to make contact with different parts of the pathway. The quantity we use would be an invariant component of the moment of inertia calculated from these varying centers of rotation.) These two conditions were the same as two used in Experiment 3 of Burton (1992), in which probe weighting did influence perceived crossability but not in the order of the moments of inertia of the probes.

The trials were filmed as in Experiment 1.

13.3.2.1.3   *Procedure*

The procedure was the same as in Experiment 1; again seven gap lengths ranging from 0 to 1.05 m were crossed with the two rod conditions and presented twice in a different random order for each participant. For the first explorer, the gap was set up to 6 cm away from the desired distance on one repetition of each trial. Although these errors are less than the interval between gaps, and as the results in the following text indicate little effect of gap length, we did not discard these trials for most variables.

White noise was played for two of the participants while the gap was being set before each trial. As in Experiment 1, no practice trials were given, but participants practiced walking around the walkway blindfolded prior to the first trial.

13.3.2.2   **Data Analysis**

Both authors independently coded the films for the total number of taps and the changes of direction, with both coders working directly with the videotape rather than first recording the location of each exploratory tap. The correlation was .80 for changes of direction and .97 for total taps. A record of the location of the taps was made by the first author and used for center-of-tapping analysis.

13.3.2.3   **Results**

Across all eight subjects, there were three trials in which the individual stepped over a gap longer than one he or she had stepped into on a different trial, indicating a high consistency of performance. The mean crossover/

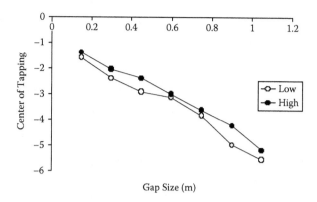

FIGURE 13.7
Center of tapping for the two probes and the seven distances for Experiment 2. The x-axis represents the far platform, to which actors were required to cross.

step-in boundary was 0.58 m when the weight was high on the rod and 0.6 m when the weight was low (i.e., more resistance to rotation). A *t*-test confirmed that this was not a significant difference ($t(7) = -0.6$); both values are quite close to the 0.6 m derived for the unweighted rods in Experiment 1. The corresponding values from Experiment 3 in Burton (1992) were 0.63 m for the high-weight rod and 0.52 m for the low-weight rod. Again, the impression is that moment of inertia does not interfere with judged crossability, with an added indication from the close correspondence in the two experiments that this is a reliable variable.

The weighting condition of the probe was not a significant variable for either of the exploratory variables in the current experiment (total taps: $F(1,7) = 0.1$; changes of direction: $F(1,7) = 0.3$). Gap size was similarly uninfluential (total taps: $F(6,42) = 1.4$; changes of direction: $F(1,7) = 1.0$), nor was either interaction significant. Because the total of exploratory taps was not significant, we did not attempt to decompose it as in Experiment 1.

However, the center of tapping was significantly influenced by rod condition ($F(1,7) = 23.0$, $p < .005$). Gap size was also significant ($F(6,42) = 50.0$, $p < .001$), and the interaction was not significant ($F(6,42) < 1.0$). As seen in Figure 13.7, the center of tapping at every distance for the low-weighted probes was closer to the starting position (farther from the far bank of the gap).

Although these probes should feel longer than the high-weighted probes, the picture is that actors wielding the longer-feeling probes tended to concentrate their explorations at a closer distance. This also contradicts our prediction that the effects on exploratory variables in Experiment 1 reflected purely mechanical aspects of the actors' interaction with the gap.

13.3.2.4 Discussion

In Experiment 1, actors using rods of different lengths moved them differently and concentrated their taps in different places, but did not change their perception of the crossability of gaps. At least three possible explanations were discussed: that gaps explored with longer rods felt shorter, that longer rods affected the mechanical posture needed to move the rod, and that longer rods affected the *perceived* posture that may be the primary datum in perceived crossability. The first and third possibilities seem contraindicated by the unchanging boundary in Experiment 1 between gaps that were crossed by stepping over and those that were crossed by stepping in. The second possibility implied that a probe that felt longer but would not influence exploratory behavior.

The results of Experiment 2 do not fully support this perspective. Performance was very consistent in Experiment 2, and the derived boundaries were in accord with those derived in the previous experiment and for judged crossability in Burton (1992). The probe weighting (and thus, apparent length) did not change crossing behavior, nor did it change the amount of tapping or the frequency of changes of direction of exploration. However, the effect of probe weighting on the center of tapping was not only significant but far from marginal. The exploratory variables of center of tapping versus total taps and changes of direction obviously do not come as a suite. Can the differential influence of rod weighting on the *manner* of exploration and the *place* of exploration be explained?

The failure of center of tapping to "go along" with the other exploratory variables suggests that it is not a variable that influences how large the gap feels to the actor, at least not permanently. The center of tapping may be an exploratory variable that reflects early impressions that presumably are refined (because the action variable does not change) by the sensory information obtained by the taps. Yet, there is a seeming contradiction between the outcome for center of tapping for the two experiments. The longer rods in Experiment 1, the rods that physically intersect with the ground at longer distances, engendered taps at larger distances. The low-weighted probes in Experiment 2, that should feel longer, engendered taps at shorter distances. We will postpone an attempt to resolve this contradiction until the general discussion.

The other distinction between experiments 1 and 2 was expected; the exploratory variables of total taps and changes of direction were affected by rod length but not rod weighting. The discussion after Experiment 1 suggested that this discrepancy could be caused by strictly mechanical factors (the rods in Experiment 1 were of physically different lengths and imposed different geometric interactions among explorer, probe, and surface), perceived length of the probe, or perceived posture of the arm–hand system. Perceived length was an unlikely candidate, given precedents such as Burton and McGowan (1997), and indeed, the probes in the current experiment that were the same length but presumably felt different engendered the same general levels of total taps and changes of direction. A condition in which the arm *feels* as if it is in different postures but actually is not would distinguish the two proposed theories for the results of Experiment 1 (the actual geometrical length of the arm–probe system or its perceived posture). Such a condition was created in Experiment 3.

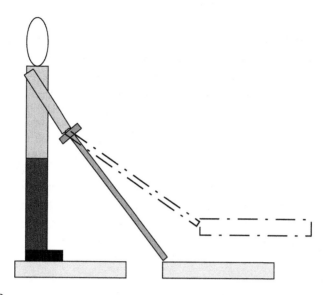

FIGURE 13.8

An individual using the T-shaped rod (as in Burton and McGowan, 1997) with the weight above the hand is likely to feel as if the arm and the probe held in the hand are higher in space than they physically are. The contacted platform should also feel higher and farther away—the greater distance should cause the platform to feel less accessible.

13.3.3 Experiment 3

Previous experiments (Burton and McGowan, 1997) suggest that the perceived posture of the arm wielding the probe has more of an influence on perceived crossability than the perceived qualities of the probe itself. These experiments drew from the discovery of Pagano and Turvey (1995) that the perceived posture of the arm is influenced by the location of weights attached away from the axis running through the arm. Specifically, Pagano and Turvey attached weights to one branch of a T-shaped rod held with the shaft of the T following the long axis of the arm. Using a similar apparatus, Burton and McGowan (1997) skewed the perceived crossability of a gap in the predicted direction—this has been the only manipulation so far that changes the macro variable of gap crossability. The same manipulation changes the perceived center of an aperture tapped with a probe (Burton, 2001).

Experiment 3 employed a procedure analogous to Burton and McGowan (1997) in that actors grasped the intersection of a T-shaped rod with the long portion continuing the long axis of the arm and the short portion perpendicular to it but in the same plane. Weights were attached to the branch above the fist or below the fist. The above-hand weights should pull the perceived long axis of the arm *upward* in space (see Figure 13.8), resulting in the perception of a longer gap. Likewise, the below-hand weight should draw the perceived long axis of the arm *downward*, so that contact with the far surface of the gap is perceived to be closer.

The boundary between gaps crossable by stepping over and those that are stepped in should fall at a smaller gap size, then, for the above-hand condition. Burton and McGowan (1997) provide the most direct precedent for this prediction—the fact that participants in the current research carry out actions rather than merely judge the affordance would lend a powerful corroboration to the previous results. Whether the exploratory variables will be affected is harder to predict; changes of direction and total taps were affected by rod length, but it may have been due to the length itself (in which case there should be no effect in the current experiment) or the perceived posture of the arm holding the probe (in which case there should be an effect). Center of tapping is probably affected, as it always had an effect in experiments 1 and 2, but the effects were apparently contradictory in the first two experiments and thus unpredictable in Experiment 3.

13.3.3.1 Method

13.3.3.1.1 Participants

Six female and two male Seton Hall undergraduates participated for extra credit in an Introductory Psychology course. All were right-handed, although this was not a criterion for participation.

13.3.3.1.2 Apparatus and Equipment

The third experiment was conducted in a different laboratory than the others, and was approximately 9 × 3.5 m in area. Equipment unrelated to the current experiment was also present. The rod that was used was the T-shaped birch rod also used in Experiment 4 of Burton and McGowan (1997). It was constructed with a stem of 0.91 m and a crossbar of 0.30 m. Four annular weights totaling 300 g in mass were attached near the end of one branch, centered at a point approximately 3 cm from the end.

13.3.3.1.3 Procedure

The procedure was similar to that of the first two experiments, with some exceptions. As an attempt to increase the sensitivity of the method, additional gap sizes were included in the middle of the range used in the first two experiments. There were 11 gap sizes: 0.15, 0.30, 0.45, 0.50, 0.55, 0.60, 0.65, 0.70, 0.75, 0.90, and 1.05 m. These gaps were explored with the T-rod held in the fist with the crossbar perpendicular to the long axis of the arm but in the same plane. Each gap was explored once with the weighted arm of the crossbar above the hand and once with the weighted arm below the hand, as in Burton and McGowan (1997). The high-weight condition is expected to lead to the perception of a gap that is farther away and, thus, less crossable than the low-weight condition. The 22 combinations of rod posture and gap size were presented in a completely randomized order, different for each observer.

Some trials were skipped because of procedural error or time constraints on the volunteer.

13.3.3.2   Results

As in the previous experiments, participants were very consistent in their decisions to step over or step across the gaps; across all eight subjects, there were two trials in which the individual stepped over a gap longer than one he or she had stepped into on a different trial. Surprisingly, the mean cross-over/step-in boundaries were not significantly different for the two rod conditions, ($t(7) = 0.2$), rounding to 0.59 m in both conditions. This value is quite consistent with values for Experiment 2 and other precedents, a surprising inconsistency with the results in Burton and McGowan (1997), in which the two rods led to a 0.11 m difference in the mean gap crossed, which would span two gap sizes in the current situation. The numbers reported in that previous research are means over two questions asked of observers, which is whether they could cross the gap at all and whether they could cross without changing their usual step. The means in the current project are much closer to the boundaries obtained for the second question, suggesting that participants are not attempting to cross the maximum possible gap (and, indeed, they were not asked to). Although this comparison was not published in Burton and McGowan (1997), a check of the original analysis shows that the means for the second question were significantly different for the low-weight (0.65 m) and high-weight (0.53 m) rods. So, the discrepancy remains; we do not think it merely suggests that actions (as in this experiment) are more accurate than judgments (as in Burton and McGowan, 1997), because previous experiments in which observers actually locomoted with the help of a cane (Burton, 1994) corroborated results of cane manipulations originally established in a judgment task (Burton, 1992).

The centers of tapping were calculated as in Experiment 1, with the exception that the "short" bands of 5 cm were scored as 1/3 of the regular 15 cm bands (for example, a tap in the 0.70 to 0.75 m band of a 0.9 m gap was scored as −2.33, i.e., one for the footboard, one for the 15 cm between 0.75 and 0.9, and 1/3 for the 0.70 to 0.75 m band). Figure 13.9 shows that the centers of tapping fell consistently further from the far platform as gap size increased, and they fell consistently further away when the weight was above the hand. In this condition, the rod presumably felt as if it was higher in the air, and thus, its end was closer to the far platform; thus, participants may have been reaching to a posturally equal location under the two rod conditions.

For analysis of variance, missing values were set to the opposite condition (i.e., if a participant's 0.45 m condition for the rod weighted below the hand was missing, the value for 0.45 m for the rod weighted above the hand was used). This is presumably a conservative procedure that should, if anything, reduce any difference between rod conditions. Nonetheless, weighting condition was significant ($F(1,7) = 11.2$, $p < .02$). Gap size was also significant ($F(10, 70) = 10.5$, $p < .001$), but the interaction was not ($F(10,70) < 1$), as implied by Figure 13.9.

The dependent variables of total taps and changes of direction were not affected by rod condition or gap size (total taps: weighting condition: $F(1,7)$

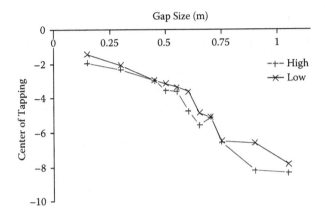

FIGURE 13.9
Center of tapping for the 2 probes and the 11 distances for Experiment 3. The x-axis represents
the far platform, to which actors were required to cross.

= 1.1; gap size: F(10,70) < 1; interaction: F(10,70) = 1.4; changes of direction:
weighting condition: F(1,7) < 1; gap size: F(10,70) = 1.4; interaction: F(10,70) <
1). The absence of these effects in this experiment relevant to their presence
when the rod lengths were varied (Experiment 1) suggests that the effects in
the first experiment were more a matter of the convenience of using the rods
rather than the perception of the surfaces in the situation.

13.3.3.3   Discussion

Diverse perceptual manipulations suggest that judging an environmental
quality and acting on that quality are different. This distinction has been
empirically tested in the ecological school by Heft (1993), for example, finding that judgments of the reachability of an object were more accurate when
the reaching was a component of a functional task than when the judgments
were a task unto themselves. As a philosophy, this distinction has been discussed by authorities ranging from Brunswik to Piaget to William James (see
review by Heft, 1993), and we should add Bernstein (1996) with his emphasis
on the illogic of attributing all of complex actions to the most central control. More recent corroborations come from Goodale and colleagues—for
example, Haffenden and Goodale (1998) found that prehension of central
circles in an Ebbinghaus illusion display was accurate, though abstract estimation of the size of the central circle showed the familiar illusory effects.
These results have been verified, contradicted, and discussed at length; see
Michaels (2000) for discussion from an ecological standpoint.

Visually impaired people are thought to be extremely tolerant of aberrations in their canes; given that it is fairly easy to influence the abstract perception of probes, such as their length (Solomon and Turvey, 1988), the case
of canes for locomotion would seem to be a very practical example of the
judge/act distinction. However, we had two reasons for believing that judg-

TABLE 13.1

Summary of Significant Effects of Probe Manipulations on Four
Exploratory Variables

Probe Distinction	Exploratory Variable	Effect
1. Length	In/Over boundary	None
	Center of tapping	$< .05$; farther from O with long rod
	Total taps	$< .01$; more for long rod
	Changes of direct	$< .05$; more for long rod
2. Moment of Inertia	In/over boundary	None
	Center of tapping	$< .005$; farther from O for high-weighted
	Total taps	None
	Changes of direct	None
3. Perceived arm	In/over boundary	None
	Center of tapping	$< .02$; farther from O for weight on low arm
	Total taps	None
	Changes of direct.	None

ments and actions would act similarly in gap judgment research. First, actors in Burton (1994) reacted much like judges in Burton (1992); neither the length of probes nor their mass distribution influenced judgments of crossability or the dimensions of steps taken to actually cross the gaps. Second, a precedent has recently emerged for actors to be influenced in their actions by the dimensions of probes; actors using T-shaped probes analogous to those in Experiment 3 to position their feet to walk through a passageway moved further to the right when the weighted branch of the T was to the right, and vice versa (Burton, 2001). True, the full passage was not undertaken in that experiment; a new trial was started once actors positioned their feet. However, positioning for movement is far from an abstract or intellectual behavior, and, in fact, actors in that experiment did not even speak during the trials.

This same weighted-branch manipulation was used in the third experiment of the current package, but the findings contradicted those of Burton and McGowan (1997) and Burton (2001); the boundary between gaps crossed by stepping over or by stepping in was the same, even though the interval between gap conditions was considerably less than in Burton and McGowan (1997).

Table 13.1 provides a systematic summary of the effects of the probe manipulations in experiments 1 to 3 on the behavioral variable (whether the actor stepped in or stepped over to handle the gap) and the three main exploratory variables (center of tapping, total taps, and changes of direction).

The behavioral variable is grimly consistent; none of the manipulations change the boundary between gaps crossed with one action and with the other. This topic will be taken up later. The effects on total taps and changes of direction, we believe, as stated earlier, reflect a merely mechanical interference on manipulation that only played a role when probes really differed in length but not when they felt like they did. This leaves center of tapping as the mystery variable; it was always affected by the probe variable but seemingly in inconsistent directions.

We believe the manifold influences on center of tapping are all consistent with the following theory. Actors "want to" concentrate their exploration on the furthest point they can reach by altering only the arm joints. We say "want to" guardedly, without assuming that this is a conscious goal; another evocative but loaded way of expressing this is to classify the furthest point reachable by only arm joint adjustments as the attractor for this exploratory task. Geometrically, this maximum is reached when the angle of the arm and that of the probe held in the fist are as equal as possible. The fact that actors made the same ultimate decisions regardless of probe shows that they must make some corrections and refinements during exploration; indeed, the bulk of the exploration might be safely interpreted as the process of making those refinements and gathering new information for feedback. In these experiments, the corrections are often instigated by the impression that the previous exploratory thrust failed to reach the attractor; despite an appropriate arm movement, the actor feels as if he or she has not reached the attractor point, so the actor moves the probe again.

Imagine reaching with closed eyes for a table edge with a pencil in each hand. If the left hand's pencil is shorter than the right hand's pencil, then the "attractor," the maximum point that can be reached without altering any joints beyond the arm, will be farther from the observer for the right hand's longer pencil. This situation and result are both consistent with Experiment 1. If both pencils are the same length but they have been weighted so that the right hand's pencil *feels* longer (Experiment 2), uniform movements will make contact with equivalent spots on the table top, but the right-hand contact point will feel too close, as though the arm–pencil system could have reached further. So, a corrective action will be undertaken to move this pencil further away, to a point farther from the observer. Likewise, if the observer uses a probe that affects the perceived posture (Experiment 3), he may contact the physically correct maximum point but feel as though his reach has fallen short.

These results suggest that center of tapping is a viable and informative variable in studying exploration of spaces without vision, but given the fact that the ultimate over/in decision was unaffected by any of the manipulations in experiments 1 through 3, it is clearly not "the" variable that could be used in any direct way by the observer to determine a crossing strategy. At this point, it seems unlikely that there is any such thing as "the" variable. Posture (Burton and McGowan, 1997; Burton, 2001) has a large role in forming the initial perception, but this initial perception is obviously soon calibrated by numerous other perceivables. In contrast to our original assumption, it seems probable that observers using a T-shaped rod would also begin to correct their locomotion if they had been allowed to continue after their judgments in Burton and McGowan (1997). This variability makes nonvisual locomotion no different from visual; an observer staring up at the famous Pozzo cathedral ceiling may well believe that he could ascend in a balloon right through the roof, but this conclusion would soon be disabused if he had an actual balloon.

Acknowledgments

The authors would like to thank Albert Montano, Dino Juliano, Eileen Porro, Marisa LaMonte, and John McGowan for assistance with the conduct and analysis of these experiments; Anna Apostu for help with the article by Groza; as well as the Center for the Ecological Study of Perception and Action at the University of Connecticut for loan of some of the equipment. Portions of Experiment 3 were presented at the XIth Conference on Perception and Action at Storrs, Connecticut, in June 2001. This research was supported by Grant 1-RO3-MH-54535-01 by the National Institute of Health awarded to the first author.

References

Altman, J. 1996. Fitting the white cane to the real world. *Journal of Visual Impairment and Blindness* 90: 292–293.

Barac-Cikoja, D. and Turvey, M. T. 1991. Perceiving aperture size by striking. *Journal of Experimental Psychology: Human Perception and Performance* 17: 330–46.

Barac-Cikoja, D. and Turvey, M. T. 1999. Anisotropy in the extended haptic perception of longitudinal distance. *Perception and Psychophysics* 61: 1522–36.

Bentzen, B. L. and Barlow, J. M. 1995. Impact of curb ramps on the safety of persons who are blind. *Journal of Visual Impairment and Blindness* 89: 319–28.

Bernstein, N. A. 1996. On motor control. In *Dexterity and Its Development*, Latash, M. L. and Turvey, M. T. (Eds.), 25–44. Mahwah, NJ: Erlbaum.

Blasch, B. B. and K. A. Suckey. 1995. Accessibility and mobility of persons who are visually impaired: A historical analysis. *Journal of Visual Impairment and Blindness* 89: 417–22.

Brabyn, J. 1985. A review of mobility aids and means of assessment. In *Electronic Spatial Sensing for the Blind*, Warren, D. H. and Strelow, E. R. (Eds.), 13–27. Dordrecht: Martinus Nijhoff.

Burton, G. 1992. Nonvisual judgment of the crossability of path gaps. *Journal of Experimental Psychology: Human Perception and Performance* 18: 698–713.

Burton, G. 1994. Crossing without vision of path gaps. *Journal of Motor Behavior* 147–161.

Burton, G. 2000. The role of the sound of tapping for nonvisual judgment of gap crossability. *Journal of Experimental Psychology: Human Perception and Performance*, 26: 900–916.

Burton, G. and Cyr, J. 2004. Gap crossing decisions in the sighted and visually impaired. *Ecological Psychology* 16: 303–18.

Burton, G. and McGowan, J. 1997. Contact and posture in nonvisual judgment of gap crossability. *Ecological Psychology* 9: 323–354.

Burton, G. 2001. Centering as a task for dynamic touch. In *Studies in Perception and Action VI*, Burton, G. and Schmidt, R. C. (Eds.), 33–36. Mahwah, NJ: Erlbaum.

Carello, C., Fitzpatrick, P., and Turvey, M. T. 1992. Haptic probing: Perceiving the length of a probe and the distance of a surface probed. *Perception and Psychophysics* 51: 580–98.

Chan, T.-C. and Turvey, M. T. 1991. Perceiving the vertical distances of surfaces by means of a hand-held probe. *Journal of Experimental Psychology: Human Perception and Performance* 17: 347–58.

Clark-Carter, D. D., Heyes, A. D., and Howarth, C. I. 1986a. The efficiency and walking speed of visually impaired people. *Ergonomics*, 29: 779–89.

Clark-Carter, D. D., Heyes, A. D., and Howarth, C. I. 1986b. The effect of non-visual preview upon the walking speed of visually impaired people. *Ergonomics*, 29: 1575–81.

Conill, A. 1998. Living with disability: A proposal for medicine education. *Journal of the American Medical Association*, 279: 83.

DeLucia, P. R. and Hochberg, J. 1991. Geometrical illusions in solid objects under ordinary viewing conditions. *Perception and Psychophysics* 50: 547–54.

Farmer, L. W. 1980. Mobility devices. In *Foundations of Orientation and Mobility*, Welsh, R. L. and Blasch, B. B. 357–412. New York: American Foundation for the Blind.

Gandelman, C. 1986. The scanning of pictures. *Communication and Cognition* 19: 3–24.

Gibson, J. J. 1979. *The Ecological Approach to Visual Perception*. Boston, MA: Houghton-Mifflin.

Gibson, J. J. 1986. Notes on direct perception and indirect apprehension. In *Reasons for Realism*, Reed, E. and Jones, R. 289–294. Hillsdale, NJ: Erlbaum.

Gitlin, L. N. and Mount, J. 1997. The physical and psychosocial benefits of travel aids for persons who are visually impaired or blind. *Journal of Visual Impairment and Blindness*, 91: 347–59.

Goldstein, S. B. 1997. The power of stereotypes: A labeling exercise. *Teaching of Psychology* 24: 256–58.

Groza, T. A. 1985. History of education of visually impaired children in Ukrainian SSR. *Defektologiya* 2: 69–75 [Russian].

Haffenden, A. M. and Goodale, M. A. 1998. The effect of pictorial illusion on prehension and perception. *Journal of Cognitive Neuroscience*, 10: 122–36.

Heft, H. 1993. A methodological note on overestimates of reaching distance: Distinguishing between perceptual and analytical judgments. *Ecological Psychology*, 5: 255–271.

Hoffman, S. B., Brand, F. R., Beatty, P. G., and Hamill, L. A. 1985. Geriatrix: A role-playing game. *The Gerontologist*, 25: 568–572.

Isbell, L. M. and Tyler, J. M. 2003. Teaching students about in-group favoritism and the minimal groups paradigm. *Teaching of Psychology* 30: 127–130.

Jansson, G. 1991. The control of locomotion when vision is reduced or missing. In *Adaptability of human gait*, Patla, A. E. (Ed.), 333–57. NY: North-Holland: Elsevier.

Jansson, G. and Schenkman, B. 1977. The effect of the range of a laser cane on the detection of obstacles by the blind. Report 211, Department of Psychology, University of Uppsala.

Kay, L. 1985. Sensory aids to spatial perception for blind persons: Their design and evaluation. In *Electronic Spatial Sensing for the Blind,* Warren, D. H. and Strelow, E. R. (Ed.), 125–40. Dordrecht: Martinus Nijhoff.

Kennedy, J. M., Green, C. D., Nicholls, A., and Liu, C. H. 1992. Illusions and knowing what is real. *Ecological Psychology,* 4: 153–72.

LaGrow, S., Blasch, B. B., and Del'Aune, W. R. 1997. Efficacy of the touch technique for surface and foot-placement preview. *Journal of Visual Impairment and Blindness,* 91: 47–52.

Liberman, A. M., Cooper, F., Shankweiler, D., and Studdert-Kennedy, M. 1967. Perception of the speech code. *Psychological Review,* 74: 431–61.

Maravita, A., Husain, M., Clarke, K., and Driver, J. 2001. Reaching with a tool extends visual-tactile interactions into far space: Evidence from cross-modal extinction. *Neuropsychologia,* 39: 580–85.

Mark, L. S., Balliett, J. A., Craver, K. D., Douglas, S. D., and Fox, T. 1990. What an actor must do to perceive the affordance for sitting. *Ecological Psychology,* 2: 325–66.

Michaels, C. 2000. Information, perception, and action: What should ecological psychologists learn from Milner and Goodale (1995)? *Ecological Psychology,* 12: 241–58.

Pagano, C. C. and Turvey, M. T. 1995. The inertia tensor as a basis for the perception of limb orientation. *Journal of Experimental Psychology: Human Perception and Performance,* 21: 1070–87.

Petrosino, A., Turpin-Petrosino, C., and Finckenauer, J. O. 2000. Well-meaning programs can have harmful effects! Lessons from experiments of programs such as Scared Straight. *Crime and Delinquency* 46: 354–79.

Potash, L. 1962. Correlates of the tactual and kinesthetic stimuli in the blind man's cane. *American Foundation for the Blind Research Bulletin,* 1: 117–129.

Schenkman, B. N. 1986. Identification of ground materials with the aid of tapping sounds and vibrations of long canes for the blind. *Ergonomics,* 29: 985–98.

Schenkman, B. N. and Jansson, G. 1986. The detection and localization of objects by the blind with the aid of long-cane tapping sounds. *Human Factors* 28: 607–18.

Schiffman, R. 1990. *Sensation and Perception: An Integrated Approach,* 3rd ed. New York: Wiley.

Solomon, H. Y. and Turvey, M. T. 1988. Haptically perceiving the distances reachable with hand-held objects. *Journal of Experimental Psychology: Human Perception and Performance,* 14: 404–27.

Staff. 1994. Demographics Update. *Journal of Visual Impairment and Blindness,* 88 (1, Part II), 4.

Skellenger, A. C. 1999. Trends in the use of alternative mobility devices. *Journal of Visual Impairment and Blindness,* 93: 516–21.

Strelow, E. R. 1985. What is needed for a theory of mobility: Direct perception and cognitive maps—lessons from the blind. *Psychological Review,* 92: 226–48.

Turvey, M. T., Fitch, H. L., and Tuller, B. 1982. The Bernstein perspective: I. The problems of degrees of freedom and context-conditioned variability. In *Human Motor Behavior,* Kelso, J. A. S. (Ed.), 239–52. Hillsdale, NJ: Erlbaum.

Uslan, M. M. 1978. Cane technique: Modifying the touch technique for full path coverage. *Journal of Visual Impairment and Blindness* 72: 10–14.

Warren, D. H. and Strelow, E. R. 1985. Historical overview. In *Electronic spatial sensing for the blind,* Warren, D. H. and Strelow, E. R, (Ed.), 1–12. Dordrecht: Martinus Nijhoff.

Wickelgren, B. G. 1971. Superior colliculus: Some receptive field properties of bimod-
    ally responsive cells. *Science* 173: 69–72.
Yarbus, A. L. 1967. *Eye movements and vision* Haigh, B. (Trans.), New York: Plenum.

Section V

Emotional Regulation of Activity and Education

14

Emotional Intelligence: A Novel Approach to Operationalizing the Construct

E. L. Nosenko

CONTENTS

Abstract..303
14.1 Introduction...304
14.2 The Principle of the Unity of the External and the Internal in
      Psychic Determination of Behavior...307
14.3 Internal and External Components of Activity and Their
      Manifestation as a Dynamic Unity 312
14.4 Some Empirical Findings Illustrating the Relevance of the
      Suggested Approach to the Operationalization of Emotional
      Intelligence..314
      14.4.1 Method 314
14.5 Results and Discussion 316
14.6 Conclusions and Future Research..321
References 323

Abstract

This paper sets out the theoretical foundation of EI accounting, following the principle of the dynamic unity of the "external" and the "internal" in a multilevel psychic determination of human activity. This principle has been guiding the research of Russian and Ukrainian psychologists since the Soviet era. In this regard, EI is conceptualized as the unity of its *dispositional* component, including ontological (essential) and phenomenological (gained with experience) personality characteristics, with a *perceptually appraised* component, manifested as the characteristics of behavioral acts and the emotions accompanying them, which reflect different types of conscious regulation of behavior and the level of EI attained by the individual. Operationalizing EI can be based on singling out different types of behavioral acts: (a) impulsive, situationally conditioned, and initiated on the sensory-perceptive level (manifesting the domination of the "external" over the "internal" in the psychic determination of behavior); (b) voluntary and mediated by thinking (which reflect domination of the internal over the external); and (c) *suprasituational* or driven by sets, values, convictions (which reflect a harmonic balance of the external and the internal).

Empirical findings are presented that confirm the efficacy of the suggested criteria for assessing EI and prove that EI is reflected in individual consciousness in the form of psychological well-being, self-esteem, and preferred coping strategies. Future directions of EI research are presented.

14.1   Introduction

The construct of emotional intelligence (EI), a systematic theoretical account first proposed by American psychologists (Salovey and Mayer, 1990; Mayer, Di Paolo and Salovey, 1990), evoked the interest of specialists worldwide, including those in Ukraine (Davies, Stankov and Roberts, 1998; Mehrabian, 2000; Newsome, Day and Catano, 2000; Petrides, 2000; Furnham, 2001). Awareness of the significance of the phenomenon grew when research indicated that EI determined success in life to a greater extent than IQ (Goleman, 1995). Studies followed on the role of EI in the workplace (Cooper and Sawaf, 1998; Wessinger, 1998), followed by the publication of practical guides intended to help people master EI and achieve emotional literacy (Hein, 1997).

Ukrainian scholars came to know about the state of the art in the investigations of EI mainly from the publications of the author of this paper, which appeared in conference proceedings published by Dnipropetrovsk and Kharkov national universities. In those publications, EI was traced back to Gardner's (1983) embryonic form and two ramifications of EI—intrapersonal and interpersonal—specified by Gardner in the monograph *Multiple Intelligences* (Gardner, 1993) were analyzed.

Later on we substantiated a trait-oriented approach to operationalizing the dispositional component of EI with reference to the "Big Five" personality traits, each of which has a strong affective core (Nosenko and Kovriga, 2001). Simultaneously, the ideas of the trait emotional intelligence and the results of psychometric investigations with reference to established trait taxonomies were published by Petrides and Furnham (2001). We also suggested singling out, alongside the dispositional (inner) component of EI, its perceptually appraised (outer) component manifested in behavior in the form of intensity, frequency, and modality of the emotions, experienced by people in their everyday life and interaction with other people (Nosenko and Kovriga, 2002).

When translating into the Ukrainian language a number of publications on EI available mainly in English and trying to operationalize the construct in terms of the Russian and Ukrainian methodological traditions, we realized that the level of the theoretical investigation of the nature of EI and, thus, the completeness and adequacy of its operationalization lagged behind the somewhat hasty attempts at developing various kinds of practical guides and training programs for enhancing EI in both work and educational settings.

Other authors express similar dissatisfaction with the level of theoretical conceptualization of the construct, which speaks to the inadequacy of diagnostic techniques designed for assessing EI (Petrides and Furnham 2000, 2001).

The revised model of EI proposed by Mayer and Salovey (1990) consisted solely of the following abilities: to perceive, appraise, comprehend, express, and control emotions. Goleman (1995) presented an account of the concept that encompassed not only abilities, but also personality variables, such as impulsiveness, assertiveness, and optimism. In an attempt to emphasize the importance of the adequate operationalization of EI for its measurement, researchers proposed a differentiation between the trait EI and the ability EI (Petrides and Furnham, 2001). The former encompassed behavioral dispositions measured through self-report, while the latter concerned actual abilities. In the opinion of those authors, actual abilities (earlier referred to as "information-processing EI," an appropriate term unfortunately neglected later on) ought to be measured with maximum-performance tests. These types of tests, though, are very difficult to design, mainly because the criteria for defining correct answers must be based on the deep theoretical substantiation of the approach to the operationalization of the construct, which requires further explorations.

Under the approach, which will be explicated later in this paper, a different subdivision of the components of EI is suggested for its operationalization with internal (dispositional) components of EI singled out from the external (perceptually appraised) components (Nosenko and Kovriga, 2001). This suggested, as will be shown in the following section of this paper, an alternative approach to operationalizing both internal and external components of EI, which differs from that employed in maximum-performance tests and is rightly criticized by the authors who used them in empirical studies (Davies, 1998). Davies et al. concluded that the trait EI dimensions are indistinguishable from the established trait dimensions and are subsumed under the Big Five personality dimensions, having a strong affective core, whereas ability dimensions tend to have low reliabilities and entail considerable problems with defining correct responses objectively. Our own empirical data (Nosenko and Kovriga, 2002) yielded similar results. But, nevertheless, it is important to stress the point that the trait EI component (distinguishable or not from the personality trait dimensions) is definitely associated with the latter. So, reserving the terminological discussion for later, it is worth highlighting here that the necessity of distinguishing dispositions from abilities (or whatever term might be chosen as a descriptor for the construct singled out in EI alongside with the trait EI) is felt by the majority of authors.

It is interesting to note that Petrides and Furnham, explaining why they would want to study EI as a trait rather than a cognitive ability, state that the two constructs mentioned above "are not mutually exclusive and may coexist" (Goleman, 1995, p. 427]. They stress the point that there is no reason why operationalization of the trait EI should preclude that of the latter and vice versa. Empirical data show, indeed, that the trait EI is likely to be implicated in a variety of behaviors and subjective judgments. In order to avoid these semantic inconsistencies between traits and abilities, Petrides and Furnham proposed two alternative labels for these two fundamentally different constructs, i.e., *emotional self-efficacy* (for the dispositional facet) and *cognitive emotional ability* (for the ability facet). This terminological differentiation does not, though, help to resolve a difficult problem in determining objectively correct responses to maximum-performance test items because, as the above mentioned authors themselves admit, "an ability EI scale designed to measure, say, 'emotion perception' would comprise items with correct and incorrect responses, whilst the respective trait EI scale would consist of typical self-report items ... " (Goleman, 1995, p. 428).

The fact that it is particularly difficult to apply truly veridical criteria in scoring EI tasks prompted many researchers to investigate the construct of EI as a constellation of dispositions and self-perceived abilities, rather than as a class of cognitive-emotional abilities.

Summing up a brief review of the theoretical research on the approaches to the operationalization of EI, it is possible to conclude that both trait EI component and its ability component (to use the most frequently occurring term) warrant future investigation.

The investigation can be based on a number of facets of EI, identified by Petrides and Furnham through content analysis of the salient literature of American specialists, including EI tests (Salovey and Mayer, 1997; Goleman, 1995; Bar-On, 1997).

The aim of this paper is to set forth a theoretical foundation for the conceptualization of EI, based on the Russian and Ukrainian methodological traditions, leading to a novel approach to the operationalization of the construct. The suggested conceptualization encompasses the facets of EI singled out with reference to the original American research.

The choice of the topic of this paper was prompted, alongside the novelty of the phenomenon itself and its evident practical significance, by a number of other considerations. Among them, it is worth mentioning, first of all, the assumption, shared by Ukrainian specialists in the field of emotions, that it is the stable features of emotionality in the structure of personality that determine the success of self-actualization, efficacy of coping with stress, and even the professional achievements of people. In this light, EI can easily be ascribed the status of a system-forming factor in the structure of personality.

Besides, Ukrainian psychologists have had experience with investigating some psychic phenomena that are closely related to EI. Empirical data have proven the correlation between high levels of psychological shrewdness and emotional creativity with the level of professional efficacy achieved by people in the sociologic professions (including psychological counseling and teaching).

The present paper also has a "supraobjective," so to speak, which is to acquaint the American and other foreign colleagues with some theoretical issues of contemporary Russian and Ukrainian psychology, which will be referred to in the following section of this paper for the substantiation of the declared novel approach to operationalizing EI. This might appear to be of interest, as the former schools of Soviet psychology, which had been developing until recently in isolation from other world psychology studies, is still associated by our American colleagues predominantly with the names of Pavlov, Vygotsky, and Luria.

14.2 The Principle of the Unity of the External and the Internal in Psychic Determination of Behavior

Among the key theoretical principles that have been directing research efforts of Russian and Ukrainian psychologists since the Soviet era, the principle of dynamic unity of the external and the internal in psychic determination of behavior deserves mentioning. Different levels and types of consciousness determine different types of behavioral acts: an impulsive act, resembling a conditioned response; a behavioral act mediated by reasoning; and a behavioral act driven by the system of values and sets acquired by the personality (a deed). Thus, consciousness as the highest form of the development of the human psyche determines the changes in the inner characteristics of human activity. The latter simultaneously determines changes in the psychological regularities of manifested behavior. Concrete manifestations of dynamic unity (Rubinshtein, 2006) alter in the process of psychic development, which finds reflection in the growth of self-driven activity as a fundamental characteristic of the human psyche. The application of the principle of unity of cognition and behavior for the study of human work is discussed in great detail in G. Bedny, W. Karwowski, M. Bedny (2001) and G. Bedny, W. Karwowski (2006).

As given in a contemporary psychological dictionary (i.e., divinition, considered theoretically substantiated), self-driven activity is characterized by the determination of human behavioral acts by the nature of the subject's state at the moment of enacting, unlike reactivity, which is determined by the preceding situation. Human activity can reach the highest *suprasituational* level of its determination, i.e., can be driven by the sets and values which surpass the frameworks of the concrete aims of the activity. Activity can be considered as a dynamic strategy which adapted to the situation based on mechanisms of self-regulation (Bedny, Karwowski, 2006).

Thus, increase in the overt behavior of the self-driven activity in comparison with the impulsive (reactive) type of behavior is a manifestation of the growth of the role of the internal in the psychic determination of behavior. This fundamental assumption is the essential guiding assumption for singling out different levels of the voluntary control of emotional processes as well. On the basis of this assumption, one can claim the existence of the intellectual (conscious) component in the determination of the overt behavior, driven by emotional processes. Following the logics of the above reasoning, impulsive behavioral acts can be assumed not to be mediated by an intellectual component, i.e. to be situationally conditioned. This type of behavior mirrors, so to speak, the emotional coloring of the situation ("Someone slammed the door on me—and I will slam the door in return"). This type of impulsive behavior can be considered a form of the external manifestation of the lowest level of the inner psychic activity, when the behavioral act is triggered at the sensory-perceptive rather than intellectual level. Behavioral acts, mediated by thinking, unlike the impulsive ones, are characterized by relating the present situation to the past, by anticipating its likely consequences. The presence of an intellectual component in determining the external manifestation of emotions can, thus, be a criterion for diagnosing the level of EI attained.

As evident from the facets of the external manifestation of EI, singled out by American psychologists, the ability of the person to suppress impulsive responses is considered one of the key characteristic features of EI.

Thus, the psychological reality, so to speak, of the phenomenon of EI is proven by the ability of the human being to reflect on the external situations and respond to them in different forms. This sort of reflection is anticipatory in nature.

Different forms of psychic reflection of reality function at the following levels of behavioral activity:

- Sensory-perceptive
- Conceptual thinking
- Psychic reflection (the highest level), which encompasses social experiences shared by individuals

The fact that the same person can behave differently under similar conditions, i.e., can stick to social norms and cultural stereotypes in one situation and resort to the impulsive type of behavior in another similar situation, proves the possibility of exercising a voluntary control over one's emotions and of the modifiability of EI.

This speaks in favor of singling out EI as a specific human faculty, which appears at a comparatively high level of psychic reflection of reality. The possibility to reflect reality at this high (ideal) level of its comprehension liberates the human being from the constraints of the immediate situation. Thus, behavior can be determined not only by the immediate situation, but by past awareness of socially accepted norms of behavior and other components of the inner human world. It is reasonable to conclude that EI can be considered as one of the forms of reflection in behavior of this relative independence of the person in an immediate life situation in which the emotional process takes place. The increasing levels of this independence find manifestation in different forms of behavioral acts, analyzed above: an impulsive one (characteristic of the low level of situational independence); an act mediated by reasoning (characteristic of a higher level of situational independence); and a suprasituational act (characteristic of the highest level of situational independence).

One can figure out different forms of the covert and overt manifestations of the nature of the internal determination of human activity. For instance, in the sphere of interpersonal conflict interaction, the existence of individual inner sets relating to emotional response (associated with the highest level of EI) can find an overt manifestation in a certain incongruence (incompatibility) of the modalities of emotional expression of the interacting partners.

The recognition of this type of specificity of psychic processes at the highest level of the development of human consciousness makes it possible to claim that EI is characterized, first and foremost, by the manifestation of covert activity in the form of behavioral acts mediated by reasoning, no matter what kind of physiological changes might accompany the emotional process under a concrete situation.

There are prominent contemporary psychologists with research based on the principle of dynamic unity in the psychic determination of human activity who claim that the growth of the role of the internal in determining human behavior finds its manifestation both in the ability of the individual to anticipate the consequences of one's activity and in making self-initiated conscious choices of the concrete forms of the activity.

That is why it is doubtful to expect that all emotionally intelligent people should necessarily be optimists. People, whose stable features of emotionality are characterized by domination of negative emotions, namely, sorrow and fear, appear to be very sensitive to the emotions of other people and actively maintain agreeable interaction with others, i.e., appear to be capable of demonstrating a sufficiently high level of interpersonal EI. This makes it possible to hypothesize that people pay a different "price" for their efforts to achieve EI and the domination of positive emotions in behavior is not the major characteristic feature of EI.

Returning to the discussion of the problem of the dynamic unity of the external and the internal in psychic determination of human behavior, it is necessary to point out that the human psyche, which can be considered to be the "internal" in reference to the surrounding world, encompasses its own psychic "internal" and psychic "external." As to the criteria of singling out those two facets of the human psyche, one can ascribe the status of the internal to the intellectual component of EI and that of the external to its emotional component.

EI also seems not only to reflect a certain aspect of the internal human world (or the individual theory of the reality, including self-theory, to use the terminology suggested by American psychologist Epstein (1990), but also to determine concrete forms of consciously regulated behavior.

Because EI can be claimed to be the major aspect of the manifestations of the inner world of the personality, the approach to its operationalization should be elaborated with account of the characteristic features of the human psychic reality as a whole. Let us analyze some of them in more detail to realize the specificity of their application for studying the phenomenon of EI. Similar to the human way of life, which is subject to changes, EI does not remain unchanged, either. The interaction of the external and the internal in the subjective human world is not only a form of their coexistence as a unity, but also a necessary condition of their development. The human being acquires in the course of the life activities the properties that are neither completely predetermined by the external conditions nor inherited. In describing the formation of the human inner world, the researcher singles out its three facets which seem, in our opinion, to be relevant for operationalizing EI as well, namely: an interindividual facet—the realm of interpersonal relations; an intraindividual facet, characterizing the engrossment of the individual into the internal world which enables the human being to get liberated, to a certain extent, from the situational constraints and social obligations; and a metaindividual facet, characterizing the extent to which the individual can influence other people. The researchers, studying the formation of the individual inner world, claim that the mechanism of its formation is manifested through interiorization and exteriorization of the experience gained. The process of interiorization is interpreted as the formation of an inner plan to enact some outer activity, i.e., use an intellectual process. It takes place in the course of the interiorization of the external forms of activity and is again manifested in them. It can be postulated that similar processes make up the mechanism of EI formation: the human being interiorizes knowledge and observations (as well as the results of one's own experience) as to the efficacious forms of emotional responding, which promote better adaptation to the immediate physical and social environment, as well as exteriorizes them in the form of the corresponding behavioral acts. Exteriorization, of course, requires certain will-driven efforts, since the regulation of the internal and the external in enacting the psychic processes makes the foundation of the conscious regulation one of the mechanisms that manifests itself in the changes in meaning of the acts. Maybe in this very direction through changing the meaning of the acts it is reasonable to seek ways of developing EI and singling out its "structural units," the majority of which constitute an emotional process. The characteristic features of this emotional process should, thus, be chosen as the source of the criteria for the operationalization of EI and for assessing the level of its development. The concrete behavior of the human being in different life situations is secondary, mediated by some internal process. Without understanding the mechanism of the latter, it is impossible to develop EI. We stressed the unique role played by the psychic processes in the formation of the human inner world, which he rightly declared (Nosenko and Kovriga, 2001, 2002).

One can agree that psychic processes are structural units of the inner world of personality. Bearing this in mind, emotional processes (their frequency, conditions of triggering, modality) can be postulated to be structural units of EI, which should be studied both as the criteria for assessing the level of EI attained and as the basis for its purposeful development. It is known that psychic processes are sort of mediators between the internal and the external worlds of the human being at any level of activity. This definition of the role of psychic processes seems to open up new vistas for developing an adequate psychological approach to operationalizing EI, based on the analysis of the dynamics of emotional processes.

To crown the picture of interaction between the external and the internal human worlds as a dynamic unity, mediated by psychic processes, it is necessary to dwell on two more psychological categories: *attitude* and *meaning*. We can develop Lazursky's ideas as to the existence of the internal and external facets of the psyche (in his terminology "endopsyche" and "exopsychy"), starting with the point that an integrated individual system of the subjective evaluative attitudes of personality to the surrounding world makes the psychological core of the personality. It encompasses the interiorized experience of relationships with other people in the immediate social environment and that of the process of interaction of the personality with the surrounding world at large.

In this light, EI can be claimed to encompass a certain system of the subjective evaluative attitudes of the individual to the world and others that are manifested in behavior. It is the integrated individual system of evaluative attitudes that enables the human being to perform behavioral acts at the suprasituational level in the form of deeds, rather than impulsive acts. The regulation of life activity based on the system of values, determining the sense of life, enables the human being to "liberate oneself" from the constraints of the immediate situation, to regulate one's behavior, to set and achieve distant aims. It is worth mentioning in this connection an attempt to order twenty-four major psychological concepts in the form of a unifying categorical network. With reference to this network, one can easily comprehend the multilevel determination and the structure of the human psyche. The concept of EI can be encompassed into that categorical network as well. The following hierarchy of concepts represents emotional functioning in the categorical network: *affectivity, emotions, feelings,* and *sense.* Whereas affectivity corresponds to the so-called protopsychological level of emotional functioning, emotions and feelings are related to the two levels of *psychosphere* and the concept of the *sense* corresponds to the highest level of psychic determination of behavior, termed *noosphere.* At the level of the noosphere emotional functioning is based on the *sense.* In a hierarchy of concepts, represented by an *organism,* an *individual,* the *self,* and a *personality,* the emotional functioning, based on the sense (system of individual meanings), is related to the concept of *personality.* It seems reasonable to postulate that EI as an integral component of the individual inner world of the personality corresponds to the highest level of emotional functioning—that based on the system of meanings reflecting the *sense* of life.

According to the model of the multilevel regulation of social behavior, the choice of different forms of behavior is determined by the following factors:

1. Situational sets that determine the operational aspect of behavior, which are not fully controlled by the individual

2. Fixed social sets and norms, determining the manner of behavior, repeated by the individual under similar situations

3. Dispositional sets, which, once formed, determine the dominating line of behavior in this or that sphere of activity

4. A system of personality values, determining the sense of life and the adequate means of its attainment.

The choice of the above mentioned forms of behavior are postulated to be governed by the characteristics of the so-called spatial structure of the choice area. Ukrainian psychologists suggested the description of the "spatial structure of the choice area" in terms of the following dimensions:

- An infinite area of choices (not predetermined by the individual and not fully comprehended)
- An area of choices possible for the concrete individual
- An area of alternative choices (choices that are interrelated with the inner sets of personality and situational conditions)

Thus, *choices* reflect both the internal and the external world of the individual as well.

Concluding the description of the interaction of the external and the internal, it is necessary to point out that researchers single out at least the following three major forms of interaction:

1. Domination of the external over the internal, which finds manifestation in the external locus of control, prevalence of the external motivation of behavior over the internal one; extraversion over introversion; impulsive, situationally determined behavior over the behavior governed by the inner sets of the individual.

2. Domination of the internal over the external, which is manifested in behavioral tendencies, characterized by the bigger role of consciousness in making decisions as to the choice of the forms of activity, its regulation, and control.

3. Harmonic balance of the external and the internal in determining individual behavior.

14.3 Internal and External Components of Activity and Their Manifestation as a Dynamic Unity

In this section, an approach to incorporating the above-presented theoretical ideas into the operationalization of EI is suggested. This is the first attempt to introduce the problem of EI into the content of investigations aimed at singling out those aspects of the dynamic unity of the external and the internal in the psychic determination of behavior, which explains how human beings can consciously control their emotions and, in the long run, achieve the level of self-regulation that determines the success of their life activity.

The major postulates of the approach can be summed up as follows.

1. Because the inner world of the personality reflects all the aspects of its functioning, including interaction with the surrounding world, other people, and awareness of oneself as the subject of life activity, EI can be regarded as an aspect of the manifestation of the subjective (inner) world of the personality. EI reflects the degree of the positive attitude of the personality to the *world* in which one can achieve success; to the others as worthy of the agreeable attitude; to *oneself* as capable of striving to achieve the goals of life and worthy of self-respect.

2. As the core of the personality is represented by an individual's integrated system of the *evaluative attitudes* to reality, EI as a form of the manifestation of the inner world of personality encompasses personal ideas to a certain *order in the world* and as to what is acceptable and what is unacceptable in one's behavior. In the categorical network of the hierarchically ordered concepts (referred to in the previous section of the paper) the ideas of the *order in the world* are termed *logos*.

3. Human activity, including interaction with others, can be performed at different levels: sensory (perceptual, imaginative, and verbal) and logical (Lomov, 2006).

   Activities can also include impulsive acts and voluntary acts. These acts in different proportions can be combined, depending on the level of regulation:

   - Sensory-perceptual level, very often including situational or *impulsive* acts
   - Intellectual and imaginative levels (mediated by reasoning), including, first of all, *voluntary acts* (situationally relevant acts)
   - Personal regulation of activity—the highest of which also may be present. This level of regulation of activity includes the level of personal sets, ideals, and values (suprasituational), resulting in *deeds*.

The functional model of self—the regulation of activity (Bedny, Karwowski, 2006)—includes cognitive and emotionally motivational mechanisms. This model links emotions, motivation, cognition, and behavior in a unitary system. It demonstrates the influence of emotionally motivational mechanisms on the interpretation of an event.

The above-presented description of the inner structure of human activity correlates with different levels of EI, the operational criteria of which can be specified in terms of the *degree of mediation* of the emotional process by the intellectual one. The ability of the person to perform one's activity at the suprasituational level (i.e., beyond situational constraints) in the form of the deeds rather than impulsive acts speaks in favor of the possibility to define EI as a form of manifestation of the human *intellect*. This opens up new vistas for specifying criteria of its assessment and formation. The nature of determination of activity can be regarded as the *inner facet* of the external manifestation of EI.

4. Because psychic processes are claimed to be the means of exercising interaction of the internal and the external in psychic determination of the human behavior, the level of development of EI can be assessed on the basis of characteristics of the *emotional processes*. The possible operational indexes of the level of EI can be: (a) congruency/incongruency of the external characteristics of the emotions, experienced by the person; (b) the emotional coloring of their antecedents; (c) frequency of occurrence of emotional processes, and (d) modality of emotions, their intensity, and the like. Those are the external indexes of EI.

5. Examing the results of the latest investigations on personality psychology, it was seen that over 90% of differences in behavior observed in different life situations (134 situations classified by the functional principle were analyzed) can be attributed to the Big Five dispositional traits, namely, 60% to emotional stability; 25% to extraversion, and 13% to openness to new experience. Personality dispositions can be regarded as a mediating *internal* component of EI. Dispositional characteristics can be grouped into the ontological (essential) and phenomenological acquired by experience. The former encompasses dispositional traits, whereas the latter reflects individual knowledge, ideas, values, and experience.

6. EI can be regarded as reflected in the individual consciousness both integratedly in the form of *psychological well-being*, and differentially, in the form of an adequate *self-esteem* and efficacious *coping strategies* chosen in critical and decision-making life situations.

14.4 Some Empirical Findings Illustrating the Relevance of the Suggested Approach to the Operationalization of Emotional Intelligence

14.4.1 Method

The suggested approach to the operationalization of EI was tested on a group of 150 subjects, students of Dnipropetrovsk Medical Academy (aged 18–22). The details of the organization of the empirical study are published in our monograph (Nosenko, Kovriga 2002), so they will be referred to below only partially.

The major idea of the study was to test the randomly chosen sample of subjects on the assumed multidimensional dependent variable of EI, which was operationalized in terms of the indexes substantiated in the previous section of the paper. It was implied that the subjects might naturally differ on the level of the EI they had achieved, and with the help of the cluster analysis statistics (the algorithm of K-means), it might be possible to divide them into several subgroups (clusters) with maximum differences in the means of the indexes chosen for the operationalization of the multidimensional conceptual variable of EI.

We decided to perform cluster analysis on the basis of the means of those indexes, which were chosen to represent the internal facets of EI (in the meaning of the term "internal" specified in previous sections of the paper). The algorithm of K-means of the cluster-analysis statistics allows us to preliminarily define the number of clusters that the researcher anticipates differing in the means of the operational dependent variable. We chose to have the sample split into three clusters, anticipating that they will correspond to the high, intermediate, and low levels of the *internal* facets of EI. We reasoned that if the differences in those means appeared statistically significant, it would be possible by analyzing the intercluster differences in the means of the external indexes of EI and the scores of the subjects on the EI test taken by them, to evaluate the adequacy of the chosen approach to the operationalization of EI.

The operational variables representing the internal facets of the multidimensional conceptual variable in this empirical study included the following groups of the theoretically substantiated internal facets of EI.

The first group included the internal *ontological* facets of EI, represented by personality traits, which constitute the "Big Five" factor personality model. They were measured with the help of the taxonomic inventory, adapted to the Ukrainian culture by Ukrainian psychologists.

To the second group the internal *phenomenological* facets of EI were included, represented in this empirical study by the ego-control and ego-resiliency scores. The latter were measured with the help of ego control and ego resiliency scales, developed by American psychologists J. Block and J.H. Block (1980). The inventories were translated and adapted to the Ukrainian culture by the author of this paper. Ego control and ego resiliency indexes were considered to reflect phenomenological internal facets of EI, because they are formed in the process of socialization and individual activity of the person and are determined by the conditions of life activity and experience gained by the person. To the same group of the phenomenological internal facets of EI, we also included the ambiguity intolerance index measured with the help of Norton's inventory (1975). Ambiguity intolerance is, to a great extent, determined by the conditions of socialization and is formed with experience, though it might have inborn characteristics as well. In this aspect it correlates with the personality trait of openness to new experience, included in the ontological internal facets of EI.

The third group of EI facets included the indexes of *sensitivity* to emotionally charged situations. This group of facets was postulated to reflect the external aspect of the manifestation of the internal in the determination of EI (specified above in the first two groups of facets). Sensitivity to the emotionally charged situations was operationalized through the scores of the psychological well-being scales (Ryff, 1989) as well as through the self-esteem scores (including the evaluation of one's character strength, wisdom, happiness, emotional resistance, and health). Additionally, the third group of EI facets included the trait and state anxiety scores and the indexes of coping behavior efficacy reflecting the degree of *self-confidence* of the subjects in stressful and responsible decision-making situations. The latter were measured with the help of the technique developed by Endler and Parker (1995). We complemented the procedure of assessing the efficacy of coping behavior by calculating the coping efficacy coefficient, measured as the ratio of the scores characterizing the frequency of the *problem–focused* strategy to the sum of the *emotion-focused* and *avoidance* strategies frequencies.

The choice of the psychological well-being index for the operationalization of EI needs a brief explanation. The analysis of interpretations of the psychological well-being construct shows that the majority of researchers resort to it when trying to specify *positive psychological functioning*, which, in our opinion, can be claimed to be the key *outcome* of the high level of EI. In fact, all the six major constructs of psychological well-being (Ryff, 1989)—namely self-acceptance, maintaining positive relationships with other people, autonomy, ecological mastery, setting goals in life and striving for self-perfection—are closely related to the major facets of EI and, in a way, can be viewed as a sort of *consequences* of EI. The space in this paper does not allow for elaborating on the substantiation of the latter claim, which seems appropriate in light of the claims of EI theorists concerning the role of EI as a better predictor of life success than IQ.

14.5   Results and Discussion

The results of the cluster analysis of the operational variables of the *internal facets* of EI, conceptualized as presented above, yielded three clusters of subjects with maximum differences in the means of the chosen variables.

Thirty-seven subjects (cluster number 1) had the lowest in the sample scores on psychological well-being, ego control, ego resiliency, introversion, agreeableness, conscientiousness, openness to new experience, and the coefficient of coping strategies, efficacy alongside the highest scores on ambiguity intolerance, neuroticism, trait and state anxiety, and frequency of resorting to emotion-focused coping strategy.

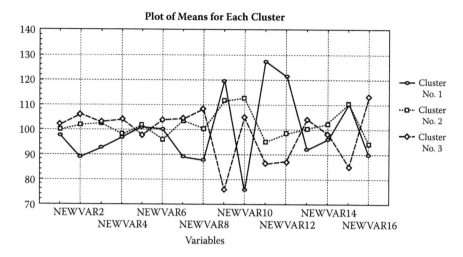

FIGURE 14.1

Results of clusterization (Nosenko). Variables: 1—academic proficiency; 2—psychological well-being; 3—ego control; 4—ego resiliency; 5—ambiguity intolerance; 6—introversion; 7—agreeableness; 8—conscientiousness; 9—neuroticism; 10—openness to new experience; 11—situational anxiety; 12—trait anxiety; 13—problem-focused strategy; 14—avoidance strategy; 16—coefficient of coping efficacy.

The opposite third cluster encompassed 58 subjects who had, on the contrary, the highest scores on those operational variables for which in cluster 1 there were the lowest scores, and vice versa. The intermediate second cluster included 55 subjects. Statistically significant differences of the means were registered between the first and the second clusters on three of the five trait factors: agreeableness, conscientiousness, and openness to new experience (at $p<0.5$ by t-test) and between the second and third clusters on four of the five personality traits: introversion, conscientiousness, neuroticism, and openness to new experience. The differences between the first and third clusters were significant on all five personality traits .

Thus, the data confirmed that clusterization of the sample on the hypothesized mediating (internal) facets allowed for the singling out of statistically different groups.

As for the *phenomenological* internal facets of EI, differences in the intercluster means appeared to be significant on ego control between the first and third clusters and between the second and third clusters. On the basis of these findings, we concluded that the growth of the level of ego control in the subjects of the third cluster corresponds to the highest scores on the implied ontological facets of EI, namely "conscientiousness." Differences in the means of ego-resiliency scores were discovered to be significant between the first and second clusters and between the first and third clusters, which also differed on agreeableness. Similar differences were registered for the ambiguity intolerance (which negatively correlated with the differences on the openness to new experience).

As for *sensitivity* to stressful factors, statistically significant differences were found between the cluster means on the following: psychological well-being, anxiety (both trait and state), self-evaluation of character, health, and emotional stability, as well as on the frequency of choosing problem-focused and emotion-focused coping strategies. The subjects with the lowest level of psychological well-being have the lowest scores on problem-focused coping and the highest on emotion-focused coping.

Summing up the analysis of the above mentioned empirical findings it is possible to claim that the singled-out three clusters of subjects significantly differed on their *resistance to stress* and *psychological adaptability*, as well as on the *awareness* of positive life functioning.

Having singled out the subgroups of subjects, statistically different on the hypothesized internal facets of EI, we aimed to analyze whether there existed significant differences between the subgroups on the *external* manifestation of EI and on the *EI scores* measured with the help of existing EI inventories. To obtain the former, the subjects were requested to monitor their emotional states during fourteen continuous days and to make appropriate notes in the a suggested diary of emotional states, specifying the modalities of emotions, their antecedents, intensity, duration, and the like.

Because the Ukrainian version of the Bar-On Emotional Quotient Inventory, which we would prefer to use for the purpose of this empirical study, was not available, we suggested that our subjects take an Internet variant of the EI inventory (offered in different languages at Queendom.com Serious Entertainment). The inventory comprises of both self-assessment measures of EI and maximum-performance items, i.e., descriptions of different variants of behavior under specially chosen situations. The results of testing with the mentioned inventory were sent to the subjects via Internet with brief commentaries as to the level of their EI attained and the aspects that require perfection. Of course, the maximum-performance section of the tests is strongly culturally-bound and we could do only an approximate assessment of our subjects' EI levels. Few subjects (less than 20% of the sample) were scored as possessing *above average* levels of EI. The majority was assessed as having *average* or *below average* levels. The analysis of the distribution of our subjects scoring above average, average, and below average EI levels (according to the results of the Internet EI test) showed that all the subjects who scored above average and average EI levels belonged either to the third or to the second clusters in our distribution. This was done according to the ontological and phenomenological dispositional characteristics of EI, and was substantiated in this paper. The subjects scoring below-average level of EI appeared to be members of the opposite (first) cluster in our distribution. This can be regarded as partial confirmation as to the adequacy of the suggested alternative approach to assessing EI.

The data drawn from the Diaries of Emotions filled in by our subjects (we analyzed 113 self-observations of the emotional states, registered by the subjects belonging to the first cluster; 211 observations, made by second-cluster members; and 191 observations, made by members of the third (highest level) cluster.

The observation data were operationalized so as to assess: (a) the *adequacy* of *the identification* of one's own emotions; (b) the tendency to regulate one's own negative emotions in interpersonal interaction; (c) the frequency of experiencing positive and negative emotions in the different situations of one's life activity; and (d) the *frequency* of occurrence of the so-called *prosocial* emotions in the experience of the subjects (on the basis of which one can judge awareness of the norms of adequate behavior in the process of interpersonal interaction, about openness to new experience, positive or negative self-perception, etc.).

Analysis of the *frequency* and *intensity* of the overt manifestations of the emotional processes yielded the following findings:

The number of *positive* emotions experienced by the subjects compared with the number of *negative* emotions increases as the level of dispositional facets of EI grows (the differences between the three clusters are significant by ϕ^* Fisher at $p < 0.02$).

As for the intensity of *positive* emotions (self-assessed by the subjects on a ten-point scale), we have not found anything statistically significant between the first and the third clusters, whereas the intensity of *negative* emotions appeared to be the lowest in the third cluster, i.e., in the group of subjects with the highest implied level of dispositional EI.

As the subjects were requested to identify, in their Diaries of Emotions, the *antecedents* of the emotions they experienced, we could assess the percentage of *unidentified* negative and positive emotions (for which the subjects did not identify antecedents).

It was found that the subjects from the first cluster (with the lowest dispositional EI level) failed to identify the antecedents of 30% of their positive and 34% of negative emotions, registered in the diaries. The subjects of the second (intermediate level) cluster could not identify 19.5% of their positive and 14% of negative emotions, whereas the subjects of the third cluster (with the highest level of dispositional EI) failed to identify only 9.8% of positive emotions and identified all the negative emotions they experienced (intercluster differences are significant by ϕ^* Fisher criterion at $p < .05$).

We consider that this operational index of EI adequately reflects the *degree of domination* of the conscious (intellectual) component in the emotional process over the subconscious one. If the subjects experience difficulties in recollecting the antecedents of emotions they experienced during a day, they are likely to have responded to some situational stimuli impulsively, i.e., on the sensory-perceptive level, not mediated by reasoning (otherwise, they would have remembered better what caused their emotional responses).

Thus, the above discussed operational criterion for assessing the external aspect of EI seems to be quite sensitive to and informative of both the ability to identify one's own emotions and the "specific weight" of the impulsive acts in behavior.

We also calculated, differentially, the number of negative emotions caused by: (a) interaction with other people; (b) dissatisfaction with one's own behavior, and (c) the life events outside the immediate personal environments of the subjects.

It was found that the subjects of the first and the third clusters significantly differed on the number of negative emotions experienced in the course of interaction with other people, whereas the subjects of the intermediate level second cluster experienced significantly more negative emotions (than the subjects of the other two clusters) accounted for by dissatisfaction with one's own behavior. The latter finding deserves some commentary. It might be proof that the subjects of the second cluster are more critical about their own behavior because they take more efforts to control it consciously, while the subjects of the first cluster act impulsively, without sufficient awareness of the level of adequacy of their behavior. As to the fact that the subjects of the third cluster experience less negative emotions in connection with all the classes of antecedents, it speaks in favor of the statement that they have already formed their own individual systems of sets and values, in accordance with those that govern their own behavior. This type of behavior we have characterized above is a suprasituational one.

It is necessary to comment on another observation connected with the analysis of the frequency of positive emotions experienced by subjects who belonged to different clusters. It is interesting enough that the percentage of subjects who experienced positive emotions in connection with their own activity turned out to account for 75% of the third cluster, 41% of the second cluster, and 52% of the first cluster. (The differences between the first and third clusters and the second and first clusters are significant by ϕ^* Fisher criterion at $p < .001$). Why do the subjects of the intermediate cluster experience even fewer positive emotions than the subjects of the first cluster? The explanation may be the same as that suggested above for the differences in negative emotions. The process of the formation of EI requires conscious efforts and is not an easy one. That is why the people with the lowest level of EI might be more satisfied with themselves than those who are still in the phase of striving to achieve it. At first, people who have already attained the highest level of EI feel happier, since they have managed to overcome the constraints of the immediate situations and act on the suprasituational level.

The analysis of the frequency and diversity of *prosocial* emotions (diversity was measured with the help of type/taken ratio) allowed for the following observations. Among the emotions reportedly experienced by the subjects of the first cluster, the four *basic* emotions (anger, fear, sorrow, and joy) prevailed; the subjects who belonged to the second and first clusters reported other types of emotions as well.

Besides, these subjects differed from the subjects of the low level cluster in the ability to give clear verbal definitions of the interrelationships of their emotions with antecedents and to specify their meaning (as exemplified by the following definitions: "pride in my friend's achievements," "rage at dishonest behavior," and "serenity while visiting church").

The analysis of the intensity of emotions showed significant differences between the opposite clusters. Differences were assessed not only on the basis of the subjects' self-scaling their emotional intensity (on a ten-point scale), but also by calculating the *ratio* of the emotions, the intensity of which is obliquely reflected in their modality (such as rage, anger, and despair) versus less intensive emotions (such as anxiety, irritation, and surprise).

Concluding this section of the paper, it can be stated that the general hypothesis on the dynamic unity of the external and the internal, in the determination of emotional behavior, has been confirmed.

14.6   Conclusions and Future Research

1. On the basis of theoretical substantiation and some empirical findings presented in the paper, EI is arguably mediated by the subjective inner world of the personality, represented by the *ontological* (essential) and *phenomenological* (gained with experience) characteristics of the latter. The ontological characteristics appear to be subsumed by the following four "Big-Five" personality traits, each of which has an emotional core: neuroticism (emotional stability), conscientiousness, agreeableness and openness to new experience. The phenomenological characteristics are represented by the sets, ideals, values, and knowledge of what is good and what is bad, and of life experience, which find reflection in forms of ego control, ego resiliency, and sensitivity to stressful life situations experienced on the perceptual level and displayed in behavior.

   The above specified ontological and phenomenological characteristics of the subjective inner world of personality can be ascribed the status of a *personality dispositional component* of EI.

2. The dispositional (internal) component of EI is reflected in the individual consciousness in the forms of psychological well-being, adequate self-esteem, and efficacious coping strategies, chosen by the individual in critical and responsible life situations. In a way, the last three forms of reflection (of one's EI in consciousness) can be viewed as a sort of *consequence* caused by EI.

3. The *external* manifestations of EI are characterized by the level structure, with "the internal of the external" being represented by the *nature of choices* of different types of behavioral acts in the real life situations, including: (a) a wide variety of situational possibilities, not differentiated by the individual (entailing the impulsive behavioral acts); (b) individually acceptable possibilities (resulting

in behavioral acts mediated by reasoning); (c) alternatives, differentiated by the individual in accordance with a system of sets, values, and meanings (entailing behavioral acts performed on a suprasituational level).

4. The major "structural unit" of the external manifestation of EI is an *emotional process* characterized by qualitative and quantitative properties, such as intensity and frequency of occurrence; the sign of the emotion that accompanies the emotional process; its modality; congruence/incongruence of the emotional process to its antecedents, and frequency of the so-called unidentified emotions.

5. The major *criteria* for assessing the level of EI attained by the individual should be investigated in the characteristics of the dynamic unity of the internal and the external in the determination of behavior and emotions accompanying it. The more behaviors and emotions are dominated by an intellectual component, the higher the level of EI.

The following description of the hierarchy of levels of EI can be suggested.

- The *lowest* level of EI is characterized by impulsive behavioral acts initiated on the sensory perceptive level by enacting the activity with domination of the external component over the internal, low level of awareness of the possible impact of the activity with low self-control, and high situational dependence of the activity. Self-esteem might be inadequately high and coping strategies are dominated by avoidance.

- The *intermediate* level of EI is characterized by self-driven, external activity performed on the basis of reasoning and voluntary efforts, which might be reflected in the individual consciousness in the form of negative emotions. This level of EI can be interpreted as characterized by the domination of the internal over the external; by a high level of ego control; and by domination in coping behavior between problem-focused strategies and the emotion-coping ones. This level of EI is characterized by adequately high self-esteem.

- The *highest* level of EI is characterized by the highest level of the development of the "inner world" of personality. It is based on the awareness of the individual regarding those alternatives of behavior that correspond to his/her system of sets, values, and ideals. This allows performing behavioral acts on the suprasituational level to feel liberated from the constraints of a concrete situation. The choice of the ways of behavior is made without excessive voluntary efforts because the choice reflects the system of the established social stereotypes, which were formed under the influence of convictions and are fully comprehended. Sufficient level of ego control has an internal locus and the individual displays, in behavior, a moderate degree of sensitivity to the emotion-provoking stimuli. The individuals with a high level of EI experience a high level of psychological well-being and have adequate self-esteem. Their coping behavior is characterized by the harmonic balance of the problem-focused and avoidance strategies with low levels of emotion-focused coping. In terms of the characteristics of the unity of the external and the internal in the determination of behavior, it can be described as a harmonic unity.

6. The internal and the external components of EI appear to be *jointly* and *simultaneously* manifested in overt behavior and emotions, determined by dispositions, the nature of choices of different types of behavioral acts, and the characteristics of emotional processes. Thus, the term *information-processing cognitive component* (mentioned in the introductory part of this chapter) for denoting what we call *the external* component of EI might be a luckier one than the *ability* component.

Future research on EI should be directed at revising its operationalization and, thus, the inventories designed for its assessment. This paper opens up some new directions for future research and presents an approach to operationalization, which will make the practice of assessing EI, at least, free of culturally-bound constraints.

References

Bar-On, R. 1997, *Bar-On Emotional Quotient Inventory*: Technical Manual. Multi-Health: Toronto.

Bedny, G. and Karwowski, W. 2006. *The Systemic–Structural Theory of Activity. Applications to Human Performance and Work Design.* Taylor and Francis. Boca Raton, London, NY.

Bedny, G. Z., Karwowski, W., and Bedny, M. (2001). The principle of unity of cognition and behavior: implication of activity theory for the study of human work, *International Journal of Cognitive Ergonomics*, 5, 4, 401–420.

Block, J. and Block, J. H. 1980, The role of ego-control and ego-resiliency in the organization of behavior. *Development of Cognition, Affect and Social Relations*. 13: 39–101.

Cooper, R. K. and Sawaf, A. 1997, *Executive EQ: Emotional Intelligence in Leadership and Organizations.* Grosset Putnum: New York.

Davies, M., Stankov, L., and Roberts, R. D. 1998, Emotional intelligence: in search of an elusive construct. *Journal of Personality and Social Psychology* 75: 989–1015.

De Raad, B. and Kokkonen, M. 2000, Traits and emotions: a review of their structure and management. *European Journal of Personality*. 14: 477–496.

Endler, N. S. and Parker, J. D. A. 1995. Assessing a patient's ability to cope. In Butcher, J. N. (Ed.), *Practical Considerations in Clinical Personality Assessment*. New York: Oxford University Press. pp. 329–352.

Epstein, S. 1990. Cognitive–experiential self-theory. Pervin, Lawrence, A. (Ed.) *Handbook of Personality: Theory and Research.* New York: The Guilford Press. pp. 165–192.

Gardner, H. 1983. *Frames of Mind: The Theory of Multiple Intelligences.* New York: Basic Books.

Gardner, H. 1993. *Multiple Intelligences: The Theory in Practice.* New York: Basic Books.

Goleman, D. 1995. *Emotional Intelligence.* New York: Bantam Books.

Hein, S. 1997. *EQ for Everybody: A Practical Guide to Emotional Intelligence.* Gainseville, FL: Alligator.

Lomov, B. F. 2006. *Psychological Regulation of Activity. Selective Works.* Academy of Science Institute of Psychology, Moscow, Russia.

Mayer, J. D., DiPaolo, M., and Salovey, P. 1990, Perceiving affective content in ambiguous visual stimuli: a component of emotional intelligence. *Journal of Personality Assessment,* 54: 772–781.

Mayer, J. D. and Salovey, P. 1997. What is emotional intelligence? In *Emotional Development and Emotional Intelligence: Educational Implications,* 2nd ed., Salovey, P. and Sluyter, D. (Eds.). Basic: New York; 3–31.

Mehrabian, A. 2000. Beyond IQ: broad-based measurement of individual success potential or emotional intelligence. *Genetic Social and General Psychology Monographs* 126: 133–239.

Newsome, S., Day. A. L., and Catano, V. M. 2000. Assessing the predictive validity of emotional intelligence. *Personality and Individual Differences,* 29: 1005–1016.

Norton, R. 1975. Measurement of ambiguity tolerance. *Journal of Personality Assessment.* 607–619.

Nosenko, E. and Kovriga, N. 2001. Trait-oriented approach to operationalizing emotional intelligence. *Abstracts of the 7th European Congress of Psychology.* Great Britain, London. July 1–6, p. 211.

Nosenko, E. L. and Kovriga, N. V. 2002. An approach to operationalizing emotional intelligence as a dynamic unity of its internal and external components. In B. Rammstedt, R. Riemann (Eds.) 11th European Conference of Personality. Friedrich-Schiller-Universitat, Jena, 21.7–25.7.2002. Conference Program and Abstracts. Lengerich: Pabst Science Publishers. p. 232.

Petrides, K. V. and Furnham, A. 2000. On the dimensional structure of emotional intelligence. *Personality and Individual Differences,* 29: 313–320.

Petrides, K. V. and Furnham, Adrian. 2001. Trait emotional intelligence: psychometric investigation with reference to established trait taxonomies. *European Journal of Personality,* 15: 425–448.

Rubinshtein, S. L. 2006. *Basis of General Psychology.* Piter Publisher, Russia.

Ryff, C. D. 1989. Happiness is everything or is it—explorations on the meaning of psychological well-being. *Journal of Psychology and Social Psychology.* 57: 1069–1081.

Salovey, P. and Mayer, J. D. 1990. Emotional intelligence. *Imagination, Cognition and Personality,* 9: 185–211.

Steiner, C. 1997. *Achieving Emotional Literacy.* Bloomsbury: London.

Tapia, M. 2001. *Measuring Emotional Intelligence.* Psychological Reports 88: 353–364.

Wessinger, H. 1998. *Emotional Intelligence at Work.* Jossey-Bass: San Francisco.

15

Emotional Regulation of the Learning Process

Olexiy Y. Chebykin and S.D. Maksymenko

CONTENTS

15.1 Introduction 325
15.2 Purposes and Methods of Our Study 328
15.3 Method of Emotion Regulation in the Learning Process 331
15.4 Verification of the Emotion Regulation Method 333
15.5 Conclusions 337
References 337

15.1 Introduction

Emotions and cognitive processes are closely linked. Their interaction constitutes the most important of psychic regulations. Understanding the mechanisms that influence emotions involved in human cognitive activities is one of the major challenges of modern psychology (Chebykin, 1999).

Our mind operates in three ways: cognitive, affective, and motivational. The cognitive includes functions such as memory, reasoning, judgment, and abstract thought. Affective includes emotions, moods, evaluations, and other feeling states. Motivational is the sphere of the personality, including a biological urge or learned goal-directed activity. Cognition and affect together make up emotional intelligence (EQ).

The concept of EQ derives partly from earlier ideas about social intelligence, which was first identified by Thorndike who defines social intelligence as the ability to understand people. EQ was included in Gardner's description of inter- and intrapersonal intelligences in his theory of multiple intelligences in 1983. Gardner (1983) presented seven types of intelligence, namely, verbal, musical, logical, spatial, kinesthetic, interpersonal, and intrapersonal. Afterwards, he added naturalist and existential dimensions. The interpersonal intelligences consist of the ability to understand others. Intrapersonal intelligence is the ability to develop an accurate model of the self and use it effectively to operate throughout life. Moreover, he described these skills as necessary for social interaction and the understanding of one's own emotions and behaviors. However, Salovey and Mayer (1995) re-conceptualized interpersonal and intrapersonal intelligences under a broader label of EQ and proposed a more comprehensive framework of EQ.

Many research disciplines are useful in the study of education, such as ethnographic research, case studies, grounded theory, participative inquiry, clinical research, and phenomenological research (Mertens, 1998). Interpretive approaches such as phenomenology focus on understanding the nature of reality through people's experiences via subjectively constructed processes and meanings. This tends to generate an epistemology where phenomena have defined realities. The phenomenological approach emphasizes the subjective processes of the situation. The aim of this approach is to determine what an experience means for those who have had the experience and are able to provide a comprehensive description of it. This approach is interested in ways in which a phenomenon is experienced, rather than in the nature of the phenomenon itself (Altricher and Somekh, 1993; Mertens, 1998; Pring, 2000; Scott and Usher, 1996).

Saarni (1990) outlines the following eleven components and skills of emotional competence:

1. Awareness of one's emotional state
2. Ability to discern others' emotions
3. Ability to use the vocabulary of emotion and expression
4. Capacity for empathic involvement in others' emotional experiences
5. Ability to realize that inner emotional state need not correspond to outer expression
6. Awareness of cultural display rules
7. Ability to take into account unique personal information about individuals and apply it when inferring their emotional state
8. Ability to understand that one's emotional-expressive behavior may affect another
9. Capacity for coping adaptively with aversive or distressing emotions by using self-regulatory strategies
10. Awareness that the structure or nature of relationships is in part defined by both the emotional immediacy or genuineness of expressive display
11. Capacity for emotional self-efficacy

According to Denham (1998), emotion understanding is knowledge of the causes and consequences of emotions. This understanding has many important implications. For instance, research has found that emotion understanding aids in self-control, as well as in the development of a child's theory of mind (Saarni, 1990). The understanding of emotion allows children to tie situations, subjective emotional states, and expressive signals together into coherent emotional experiences (Denham, 1998). The understanding of emotion involves the following aspects described by Denham (1998):

1. Labeling emotional expressions
2. Identifying emotion-eliciting situations
3. Inferring the causes and consequences of emotion-eliciting situations
4. Using emotion language to describe their own emotional experiences and clarify those of others
5. Recognizing that their own emotional experiences can differ from others' emotional experiences
6. Awareness of emotion regulation strategies
7. Knowledge of emotion display rules
8. Knowledge that more than one emotion can be felt at the same time, even if they conflict
9. Understanding of complex social and self-conscious emotions such as guilt

Studies of the regulation of emotions use the main achievements in psychology for understanding emotions and EQ. They lead to fundamental results in teaching technologies. The main aim in this field is to promote the active participation of a student's emotional sphere in the learning process. Unfortunately, nowadays this is still far from being achieved.

Seval Fer (2004) claims that in contrast to IQ, which is considered relatively stable and unchangeable, research (Ashforth, 2001; Cherniss and Goleman, 1998; Cooper, 1997; Goleman, 1995) has indicated that EQ is acquired and developed through learning and repeated experience at any age. This understanding of EQ and mental development in general is derived from Vygotsky's (1978) idea about the zone of proximal development (ZPD). He first applied this idea in the context of testing and instruction. He asserted that the ZPD is the difference between a child's "actual development as determined by independent problem solving" and the level of "potential development as determined through problem solving under adult guidance." Application of this concept to education in general and to EQ in particular have become known in the West. In this context, EQ skills are becoming more important as society creates new challenges for youth. We can shape our EQ by learning not only well-developed intellectual abilities, but also social and emotional skills. Intellectual ability is essential for being successfully educated and for being a contributing member of society. EQ is also equally essential, and can help people study to their potential and develop healthy interpersonal relationships. Understanding one's own emotional processes can have far-reaching effects for social functioning and the quality of life. However, Richardson (2000) indicates that young people who lack social and emotional competence might end up becoming self-centered and unable to empathize and relate to others. According to Goleman (1995), the most troubling findings were in national surveys in the United States, in which more than 2,000 children were rated by parents and teachers in a longitudinal study. The results indicated that children had become more impulsive, disobedient, angry, lonely, and depressed. Parents had less time to spend with children, children spend more time in front of a TV or computer, and they were not getting the basic needs for the emotional foundation that they needed. The teen crime rates, drug-abuse rates and so on are only now helping us realize that this situation makes the young emotionally more at need. Moreover, longitudinal studies indicated that children with social deficiencies suffered both socially and academically (Chebykin, 1999).

Proceeding from the above statements, we conducted investigations into the emotional state of students in various stages of learning. According to Bedny and Karwowski (2006) cognitive, emotional, and motivational components of activity are integrated into the holistic self-regulative process. Emotions and motivation are considered as functional mechanisms or functional blocks of the self-regulative process. The emotionally evaluative block is involved in the evaluation of the significance of task or situation. The motivational function block determines what aspects of activity will be induced. These two function blocks interact with other functional mechanisms and particularly with such functional mechanisms as "goal" and "assessment of difficulty of situation." For example, if a student evaluates a goal as not being personally significant (emotionally evaluative stage) and very difficult due to his low self-efficacy, this decreases motivation (the inducing component of activity) and the student can avoid the goal. A model of self-regulation of activity elucidated not only the natural manifestation of different emotions but tried to discover ways of regulation. Our study along the lines of the theoretical data presented by Bedny and Karwowski demonstrates that there is a possibility to control the emotions of students during the learning process and to shift emotions in the necessary direction.

Further development of the system of education is linked to the increasing informational flow and consequently to the sharp intensification of the learning process. In these conditions the positive emotional components in the learning process become very significant. However, we claim that traditional learning techniques result in the increase of negative emotions among students and therefore in the decrease of teaching effect.

15.2   Purposes and Methods of Our Study

The purposes of our studies were to demonstrate the influence of emotions on the learning process and to find the effective methods to provide the optimal emotional regulations in teacher–student interaction.

We suggested that the understanding of emotional regulations mechanisms and their correlations with informative and dynamic characteristics of the learning process would allow:

- Development of a model of emotional regulations to structure conditions promoting the learning process
- Determination of a set of positive emotions that would lead to increasing the effectiveness of learning
- Working out of adequate teaching techniques for emotional regulations during the learning process

We studied the experimental groups of students with emotional states in a wide range of parameters. The standard techniques were used: external observation and introspection, questionnaires, and psycho-physiological and psychological methods of investigations.

Psychological components of emotion regulation in the learning theory can be divided into three groups:

Situational–relevant. In this group can be integrated emotionally charged situations in the learning process such as "dramatizing," "overcoming obstacles," "novelty" (Vygotsky L., 1978, 1981, 1987, Leont'ev A. N., 1978). Our study demonstrates that such situations are mainly necessary for the intensification of the learning process.

One-way oriented. A concept that mostly relied on the motivational characteristics of learning processes at all stages, based either on the effect *ional* or cognitive components (Rogers, 1961). In this case, the specially selected didactic material and specific techniques have to be used by the teacher.

Purpose oriented. This deals with the meaning of individual emotional characteristics of the learning process (Rubinshtein, 1959, etc.) or trying to understand the entire set of emotions appearing during this process (V.V. Davydov, 1986).

Our analysis of well known results and the recent data on the interrelations between the emotions and motivation, goal-formation, analysis of cognitive processes, and conscious and unconscious components of activity (Bedny and Karwowski, 2006) allowed us to evolve the structure of emotional regulations of the learning process (see Table 15.1).

The table shows three stages of learning process.

- At the initial, *approximately-motivated stage* (orientation stage) the motives that are determined by the external didactic material dominate.
- At the second, *executive stage* the motives are connected with surrounding things and phenomena.
- At the third, *reflective-evaluated stage* the objective-oriented systems are developed.

TABLE 15.1

Levels of Emotional–Motivational Regulation in the Learning Process

Basic Components of Development Training	Level of Regulation		
		Emotional	
	Motivate	General	Specific
Approximately motivated	Prevalence of effective external motives directed at didactic material	Emotions are like a dynamic characteristic	Initial analysis of learning tasks is emotionally identified and is either consolidated in one's consciousness or not
Executive	Prevalence of an inner motive involving general goals, dealing with facts, understanding the qualities of things and phenomena	Correlation of emotional and cognitive processes	New emotions catalyze gnostical functions of induction and correction of learning operations
Reflective–evaluative	Prevalence of an inner motive; elaborating objective systems of general and concrete phenomena; establishing main principles	Inclusion of emotions into gnostic process of consciousness	Display of heuristic functions of learning practical operations

This scheme determines our strategy of research aimed to classify emotions and to elucidate the regulation methods.

Further we tried to determine situations, in which emotions are caused that provide a successful regulation in the learning. We tested students of high school of different levels, students of pedagogical colleges and the educators. Special procedures to investigate the emotions were developed. As a result of this investigation, the following emotions linked to the learning process were identified: surprise, curiosity, boredom, resentment, perplexity, doubt, inspired, indifference, enthusiasm, interest, inquisitiveness, delight, disappointment, joy, anger, fear, shame. *We claim that these emotions are typical for the education process.*

At the first stage of learning the emotionality is caused by the dynamic characteristics of a person.

At the second stage of the learning process an internal motive is formed linked to the general goals of learning. This provides the intensification of the interaction between the emotional and cognitive processes.

At the final stage the domination of the inner motive takes place. We suggest that the emotions are actively engaged here into the Gnostic functions of the consciousness. It shows the interest, inquisitiveness and enthusiasm. We denoted these emotions as "desirable" (wishful).

At all stages of the learning process we observed so-called "experienced emotions." They include not only positive emotions, but also negative ones as well as boredom, perplexity, doubt, indifference, disappointment, anger and fear. In real practice of the learning process the correlation of emotions has a complex and occasionally an ambivalent nature.

The determination of the optimum role of specific emotions in the learning process demands an accurate emotions diagnosis and further correction of emotions by teacher. We suggested that the teacher's accuracy in recognizing emotions depends on his ability to analyze the particular features of emotion expression.

To verify the hypothesis 71 teachers were invited to take part in the experiment. Two groups of features were identified: expressive and subjective ones. It was found that there are 19 subjective features of individual emotions, and the expressive ones have 12 features. Besides that it was shown that individual emotions have one to three identical expressive features. The main expressive features are perplexity and surprise, surmise, doubt, boredom, indifference, disappointment, curiosity, inquisitiveness and interest, enthusiasm and inspiration, anger and fury, joy and delight.

The characteristic features are following:

Perplexity—widely open eyes and raised brows with partly open lips

Indifference—blank and indifferent look, yawning, fiddling and quiddling

Curiosity—bright and shining eyes, fixed look with partly open lips

Enthusiasm and interest—animation and shining eyes, swiftness, and unresponsiveness to external actions

Anger and fury—knitted brows, screwed up eyes, slightly parted lips

Joy and delight—smile, shining eyes, merry and joyful look

The expressive changes in the eye and mouth area are mostly often indicators of emotions. We widely used photos of students in experiments. As a result it was found that the process of actual recognition of emotions by teachers can be improved during the teaching process.

Each set of similar emotions contains some features (surprise, boredom, interest, doubt, fury, joy, dread, shame, and resentment), which can be recognized by a teacher (see Table 15.2).

15.3 Method of Emotion Regulation in the Learning Process

Recent findings in emotional intelligence support the concept of confluent education, which holds that effective learning develops in the interaction of

TABLE 15.2

Discerned Peculiarities of Emotions

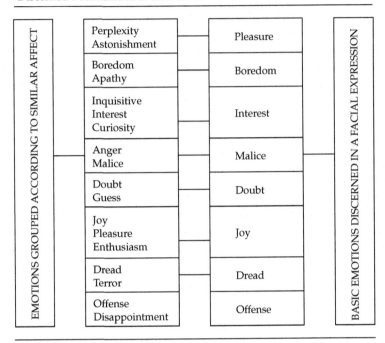

EMOTIONS GROUPED ACCORDING TO SIMILAR AFFECT		BASIC EMOTIONS DISCERNED IN A FACIAL EXPRESSION
Perplexity Astonishment	Pleasure	
Boredom Apathy	Boredom	
Inquisitive Interest Curiosity	Interest	
Anger Malice	Malice	
Doubt Guess	Doubt	
Joy Pleasure Enthusiasm	Joy	
Dread Terror	Dread	
Offense Disappointment	Offense	

cognitive and emotional domains. Therefore, effective educational practice requires attention to the development of many forms of intellect through formal teaching practice as well as informal teaching practice. Current research and practice both firmly demonstrate that the growth of ethical or principle driven behavior—a critical component of emotional intelligence—develops through numerous informal interactions both in and out of class.

The experience of leading teachers in this field were the basis of formulating general methods for students' emotional sphere regulation. The following methods are used:

- Active correction, such as expressive, intonational, contact, confidentiality, encouragement, attention switching, etc;
- Increase of the emotional effect by preparation in advance of purpose-oriented didactic material, vivid emotional signs, improvised topical commentaries, dramatization, musical effects, etc.

Our system of emotional regulation methods includes:

- The self-regulating correction of the emotional states of the teacher (autogenic training)
- The control of emotional sphere of students

It is necessary also to underline two additional interrelated groups of regulation methods:

- The expressive intonation, trope and stylistic methods
- The use of traditional lore, humor, parables, catch words, etc.

15.4   Verification of the Emotion Regulation Method

In another experiment, 54 teachers (with 1100 students) were involved. More than 100 lessons on various subjects were conducted. The experimental lessons were conducted by postgraduate students.

The observed discrepancy between actually experienced and desirable emotions of the students at the final (and partly at the performance) stages of the educative process can be explained by different abilities of students for achieving learning at a fixed stage. It was found in the final stage that the reduction in diversity of manifested emotions with the majority of students was due to the dominance of arbitrary forms of self-regulation at that stage.

The 6- to 7-year-old children showed a higher rate of surprise in comparison with 7- to 8-year-old pupils. The studies show that described methods were helpful to control student's emotions during the teaching process. The creation of situations of suddenness, brightness, and contrast was used in order to excite surprise and other emotions. At the performance stage the theatrics, success-and-encouragement, and comfort techniques were applied. These produced empathy in the pupils.

In another study, we observed 98 teachers, and studied 46 experimental and 52 control lessons. A range of selected emotions (indifference, anger, dread, resentment) was chosen. It was found that negative emotions were revealed less frequently in the experimental studies than in the control groups. A detailed analysis of data has revealed that there are some students in each group who more strongly manifested negative emotions in certain situations than other children. This must be taken into consideration by teachers in their work.

In experimental studies when our methodology is used, negative emotions of all levels and types can be corrected. In the case of a quantitative indication of diagnosed emotions after correction in experimental and control studies, statistically reliable differences were obtained. The effective correction of negative emotions was achieved due to quick detection and employment of adequate control actions.

We found that in the process of realization of our methodology, any activity that is carried out with too high an intensity for a long period inevitably results in deterioration of the psychic conditions of students. This is mainly manifested by children of the lower grades. The age difference of students in the process of realization of our emotion regulation methodology plays a significant role in providing positive results.

We have been led to conclude that a creative teacher must:

- Pay attention to the activity and moods of students under different teaching conditions
- Follow the peculiarities in changes of functional states of students before and after the application of regulation methods

The efficiency of rehabilitation of emotional conditions during the teacher process is determined by the shifts in the moods and activity caused by the regulation method. The greatest shift in the diagnosed parameters was observed at experimental studies in comparison with control studies in mathematics and reading. These data were obtained for children in the beginning grades. For the second grade students these differences were more expressed at studies of natural history. For the third grades students for all diagnosed data (with the exception of the lessons of mathematics) statistically reliable differences in all diagnosed parameters were noted.

We found that the important factor in the rehabilitation of emotional state of students in the learning process is the following regulation techniques (the corresponding effectiveness in percents is shown):

- Switching of attention to another activity (75%)
- Different combinations of motor and cognitive actions (50–68%)
- Combination of cognitive and motor actions with a commentary (72%)

The largest drop in indicators of functional abilities of students in conditions of our methodology using was observed at studies of mathematics and natural history. In these cases they varied from low to medium levels. At the same time it has been revealed that the use of our rehabilitation methods effectively improves the abilities of children. The most effective rehabilitation techniques are different combinations of cognitive and motor actions.

We investigated psycho-physiological and psychological characteristics (60 parameters) for 120 students. The study was aimed to determine three key factors: self-regulation (self-control, confidence, etc.); expressivity (emotionality, psychomotor tonus, etc.); and empathy (introversion, concern, trustfulness, anxiety, etc.). The experimental data are consistent with our long-term observations in teacher colleges. The most important task was to find the elements determining emotional stability of the teacher and the ways to diagnose this stability.

We also developed a procedure for the diagnosis of emotional maturity of teachers. Such procedure must be valid and reliable, providing a simple treatment of the results, and manageable. Out of 800 considered assertions 36 were selected related to self-regulation, expressiveness, and empathy. Later, six more questions were added to test the sincerity of answers. The questions were general, and only a small percentage (1–3%) of answers in different samples were insincere. The validity of the procedure was confirmed by the agreement of testing results for the diagnosed indicators with the preliminary self-assessment by each subject. The reliability of testing was assessed by repetitive experiments with the same group of subjects (140 persons). The stability of results in various cases was moderate at the level of statistically significance.

A special workshop, "Principles of Emotional Regulation in the Educative Process," was organized. The program comprised two main parts.

The first one was devoted to teaching students about behavior under conditions that lead to emotion-provoking situations. Typical emotions of children and the signs to recognize them were analyzed. We showed typical examples of emotion-provoking situations that appear in the teaching of mathematics, physics, and social science disciplines. The second part of the seminar was devoted to the techniques of simultaneous self-control and control of the emotions of students. We have seen that the participants of seminars do not always exhibit the necessary abilities to satisfy these requirements.

In our study we utilize also formative experiments. This method of study, which was suggested by Vygotsky (1981), emphasized that the activity of a subject involved in an experiment can be developed under the guidance of the person conducting it. This experimental method emerged from study of the relationship between development and learning, according to the Vygotsky learning guide on mental development. He introduced the concept of the zone of proximal development, defined as the gap between a child's actual performance and the level achievable with the help of an adult. From this it follows that the task of the teacher is to move ahead of the student's development, leading him or her into a zone of proximal development and increasing the efficiency of the learning process. In our study we paid attention first of all to the emotionally motivational aspects. Analysis during the process of development helps scientists and practitioners better understand the activity of subjects. Only through analysis of the acquisition stage can we discover better methods to employ in the teaching process. In the formative experiment related to the approbation of our methods for emotion regulation, 62 students of the teaching faculty were engaged. In the framework of the experiment the emotional maturity of students were diagnosed in terms of their motivation to learn in the field of their chosen profession, and their abilities to perform the main types of emotional regulation required in the teaching process (predictive, procedural, and final).

The following results were obtained: 31 students showed domination of the procedural type of emotional regulation, 23 students exhibited the domination of the final type of the emotional regulation. Index of emotional stability in all groups changed mainly between 7.08 and 8.38 with the maximum value of 12.0. After the formative experiment the values of the mentioned parameter increased by 4.1 points in comparison with the initial values and this differences was statistically significant.

The results showed that the most effective methods of self-control and control of the emotions of students included self-command (90±4%), switching attention (87±0.4%), and ideomotor training (79±0.5).

Thus, after the initial experiment with students the visible changes in the development of basic types of emotional regulation, emotional stability, and motivation were observed. Among the favorable conditions that lead to the best results were:

- Adequate simulation of emotionally-provocative situations, which were necessary for students to understand different emotion-regulating techniques
- The individual work of a psychologist, a specialist in didactics, and a teacher during lesson preparation and for final assessment of the results.

A general analysis of the results of emotional regulation in the educative process showed that the most important factor is the ability of students to foresee an emotion-provoking situation (i.e., to suggest several alternatives to its possible development and to suggest several methods of control) and the development of the pupils' skill to maintain a cheerful and productive atmosphere during the lesson. Systematic development of emotional stability against a background of clearly expressed empathy indicates that the pupils have acquired the necessary techniques of self-control. Emotional control affects the development of students' expressiveness. We found that the most important emotional training for students is the formation process of emotion regulation. So it can be argued that emotional maturity is an important diagnostic characteristic of students' (entrants included) preparedness to master the teaching profession.

In conclusion it can be mentioned that a multidimensional study of the control of emotions in the educative process led to a useful insight in the problem of linking emotional intelligent to the educational process. We determined the nature of psychological mechanisms and conditions for the emotional regulation of teaching and learning processes. We defined specific emotions of their changes at typical stages of the educative process. The principles of the emotional regulation method have been identified. The signs of emotional expression have been described. We underlined the necessity to combine the control of emotions of students and self-control of the teacher.

Our studies have revealed that emotional regulation in the learning process can not be performed separately from the motivational and cognitive components of activity. The cognitive component comprises the informative and dynamic characteristics of the educative process. The motivation component is the dominating director of the activity; and the emotional component, the interaction of the emotional with the evaluative aspects of the activity.

15.5 Conclusions

As a result of the analysis of obtained results it can be concluded:

1. The system of application of the methods of psychic regulation of emotional state of students in the learning process is developed.

2. Methods of emotional regulation include, first, blocking of mechanisms of interaction of the motives with typical emotional processes and, secondly, blocking of conditions that cause emotion-provoking situations.

3. The unrealized emotional potential of students in the learning process is due to the discrepancy between actually experienced and desirable emotions at different stages of the process.

4. An analysis of subjective and expressive signs of specific emotions of students makes it possible to identify the groups of equivalent emotions. The existence of such groups of emotions easily recognized at the level of expressive signs.

5. The conditions have been determined that provide emotional regulation in the learning process with using didactic material that leads to the increased emotional charge, correction of situational emotions, and changing their functional role.

6. Increased emotional charge of the didactic material at the motivational stage of the learning process was achieved by insertion conditions into the process such as suddenness, brightness, and contrast. At the performing stage other situations were more useful, such as the use of theatrics, success and encouragement, and comfort techniques. At the reflection and assessment stage, the emotion-provoking situations were complemented by the extremality, deficit of assessment criteria, and disinformation.

7. Control of the functional states of the pupils in order to rehabilitate their capacity to work in learning conditions is achieved through psychomotor techniques (motor reactions combined with speech and music, etc.) conducive to comforting emotions.

8. Emotional regulation in high school can be successfully performed by teachers who review the special psychological training that we have outlined.

References

Altricher, H., Posh, P., and Somekh, B. (1993). *Teachers Investigate Their Work: An Introduction to the Methods of Action Research*. London: Routledge.

Ashforth, E. (2001). The handbook of emotional intelligence. [Book Review]. *Personnel Psychology, 54*(3), 721–725.

Bedny, G. Z. and Karwowski, W. (2006). *A Systemic-Structural Activity Theory. Application to Human Performance and Work Design,* Taylor and Francis Boca Raton, London, NY.

Chebykin, A. (1999). *Theory and Practice of Emotional Regulation of Learning Activity.* Ukraine: Astroprint.

Cherniss, C. and Goleman, D. (1998) *Bringing Emotional Intelligence to the Workplace.* http://www.eiconsortium.org/research/technical_report.pdf

Cosaro, W. and Eder, D. (1990). Children's peer cultures. *Annual Review of Sociology, 16,* 197–220.

Danville, I. L. The Interstate Printers and Publishers, Inc. Salovey, P. and Sluyter, D. J. (1997). *Emotional Development and Emotional Intelligence.* New York: BasicBooks.

Davydov V. V. (1986). *Problems of Developing Education,* Moscow, in Russian.

Denham, S. (1998). *Emotional Development in Young Children.* New York: Guilford Press.

Díaz, R. M., Neal, C. J., and Amaya-Williams, M. (1990). The social origins of self-regulation. In Moll, L. C. (Ed.), *Vygotsky and Education: Instructional Implications and Applications of Sociohistorical Psychology.* Cambridge: Cambridge University Press.

Fer, S. (2004). *The Qualitative Report,* Vol. 9, Number 4, pp. 562–588. http://www.nova.edue/ssss/QR/QR9-4/fer.pdf

Gardner, H. (1983). *Frames of Mind: The Theory of Multiple Intelligences.* New York: Basic Books.

Goleman, D. (1995). *Emotional Intelligence: Why It Can Matter More than IQ.* New York: Bantam Books.

Izard, C. E., Trentacosta, C. J., King, K. A., and Mostow, A. J. (2004). An emotion-based prevention program for Head Start children. *Early Education and Development, 15,* 407–422.

Krejcie, R. V. and Morgan, D. W. (1970). Determining sample size for research activities. *Educational and Psychological Measurement, 30,* 607–610.

Leont'ev, A. N. (1978) *Activity, Consciousness and Personality,* Englewood Cliffs: Prentice Hall.

Mertens, D. M. (1998). *Research Methods in Education and Psychology: Integrating Diversity with Quantitative and Qualitative Approaches.* London: Sage.

Pring, R. (2000). *Philosophy of Educational Research.* London: Continuum.

Richardson, R. C. (2000). Teaching social and emotional competence. *Children and Schools, 22* (4) 246–252.

Rogers, C. (1961) *On Becoming a Person.* Boston: Houghton–Mifflin.

Rubinshtein, S. L. (1959). *Principles and Directions of Developing of Psychology,* Moscow: Academic Science.

Saarni, C. (1990). Emotional competence: How emotions and relationships become integrated. In Thompson, R. A. (Ed.), *Nebraska Symposium on Motivation: Vol. 36. Socioemotional Development* (pp. 115–161). Lincoln: University of Nebraska Press.

Salovey, P. and Mayer, J. D. (1990). Emotional intelligence. *Imagination, Cognition, and Personality, 9,* 185–211.

Scott, D. and Usher, R. (1996) *Understanding Educational Research.* London: Routledge.

Vygotsky, L. S. (1978). *Mind in Society: The Development of Higher Psychological Processes.* Cole, M., John-Steiner, V., Scribner, S., and Souberman, E. (Eds.). Cambridge, MA: Harvard University Press.

Vygotsky, L. S. (1981). The genesis of higher mental functions. Wersch, J. V. (Ed. and Trans.). *The Concept of Activity in Soviet Psychology.* Armonk: M. E. Sharpe.

Vygotsky, L. S. (1987). *Thinking and Speech.* In Reiber, R. W. and Carton, A. S. (Eds.), Minick N. (Trans.), *The Collected Works of Vygotsky, L. S. Vol. 1: Problems of General Psychology.* New York: Plenum.

Wood, D., Bruner, J. S., and Ross, G. (1976). The role of tutoring in problem solving. *Journal of Child Psychology and Psychiatry, 17,* 89–100.

16

Emotional Resources of the Professional Trainer

G.V. Lozhkin

CONTENTS

Abstract..341
16.1 Formulation of the General Problem 341
16.2 Methods and Procedures...348
16.3 Results 349
16.4 Conclusions..353
References 354

Abstract

In this chapter a gnosiological analysis of the category "psychological potential" of the professional trainer is carried out. With the help of a narrative method, the activity of the trainer is reproduced and reconstructed on a psychological level. Also, the trainer's professionalism is identified. It is established that among the leading professionally important qualities of the trainer, the level of the person's emotional resistance plays an important role, stabilizing the emotional factors. The symptoms of "emotional burnout" of trainers during their professional development are determined. Besides, a comparative analysis of the phases of "emotional exhaustion" of experts of different sports is carried out. Personal, organizational, and role factors that influence the occurrence of the "burnout" syndrome of sports trainers are determined.

16.1 Formulation of the General Problem

Training of high-level sportsmen depends on many factors: economic maintenance, organization of training and competitive processes, moral and material stimulation, planning, selection, restoration, medical and scientific maintenance, and the traditions and popularity of the particular sport. These factors are, of course, of varying importance, but there is yet another factor that is significant—the level of development of the sport, as well as the achievement of the sportsman, is appreciably determined by the work of the trainer. It is quite clear that the trainer performs the function of personality formation of the pupil. Hence the need to understand the effects of the professional trainer's activity— a challenge faced by psychology, which has already been co-opted in finding solutions for the important problems that afflict sports.

It is not enough to analyze just the traditional displays of behavior, the standard characteristics, of the trainer. This conclusion follows from a generalization of the practice of sports and is based on certain theoretical points. Therefore, we think that a paradigm of psychological potential enables us to find out and explain changes of psyche, as a sportsman as well as a trainer. The inclusion of potential is rather important; it is determined, first, by the logic of development of scientific psychology, which demands generalizations concerning variations of mentality and reveals the general tendencies of behavior of the person in dynamic conditions present in the sphere of sports.

During the accumulation of potential, the complete mental organization of the person is fixed. Besides, numerous interrelations between the general displays of psyche and its individual, specific forms in concrete conditions (for example, training and competitive activity) are integrated simultaneously.

The current status of the idea of potential in sports reflects the obvious prevalence of a physiological component over two others: social and actually psychological. Lack of conceptual development has resulted in neglect of the concept of "psychological potential." The resources of regulation of behavior can be considered a certain potential that provides a steady level of performance of a certain work. The essence of the concept of "psychological potential" is defined thus—the ability of the trainer to carry out an activity that spans a long period in the career of a sportsperson (Lozhkin, 2003).

So, the realities of modern professional sports dictate the study of the trainer as the subject of professional activity, as the complete personality who chooses his or her own ways in sports (Voljanjuk, 2004). The key moment in the creation of psychological potential in the professional trainer is the point at which the person comprehends his or her own abilities and achievements. It is very important to future generations of sports trainers that they are able to use previous trainers' experience.

There are different ways to transmit experience. Some of the trainers publicize it in interviews, articles, and reports (I. Turchin, A. Gomel'sky, S. Kolchinsky, etc.), and others in the form of dissertations (D. Tyshler on fencing, S. Vojtsehovsky on swimming, V. Savchenko on boxing, A. Tarasov on hockey, V. Drjukov on modern pentathlon, etc.). It is also disseminated in the form of memoirs, which can be considered both a model of activity and an ensemble of those personal qualities that predetermine success. Internalizing these experiences, the trainer forms his or her own experience. A portrayal of previous experience, one's own judgment, and personal history is called a nar-

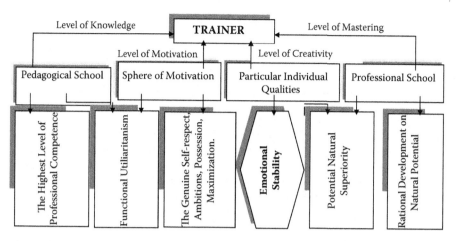

FIGURE 16.1
Acmeological invariant of the professionalism of the trainer.

rative. The narrative, as a method based on psychology, was promoted in the 1980s. It employs a flexible model that makes it possible to understand reality because it is reality itself, and it may be considered at diagnostic, symbolic, and associative levels (Chepeleva, 2004). The original sources of the narrative data are personal documents (memoirs, notes, diaries, etc.), as well as materials on interviews and conversations. Using the narrative method in sports helps focus attention on unique aspects of the trainer's life and the person's approach to the description of his or her career.

The content analysis of 76 memoirs of outstanding trainers in different kinds of sports has helped identify their most important professional and personal qualities, reconstruct, at a psychological level, their activities, and prove the acmeological invariant of their professionalism (see Figure 16.1).

Although the structure and components of Figure 16.1 are self-evident, some explanations are necessary. It is quite probable that the invariant unites those mental characteristics that are able to organize and coordinate different interactions of the trainer with reality. Four global criteria may be singled out among them: knowledge, motivation, creativity, and mastering.

Each of these global criteria is characterized by its own sets of individuals and may be measured (either empirically or theoretically. The level of mastery depends on motivation, knowledge, and creativity. Creativity depends on knowledge, which comprises an understanding of the methods of training the sportsmen, etc. Professionalism of the trainer assumes availability of schools, pedagogical (educational institution) and professional (the trainer's environment).

Six groups of qualities of the professional trainer, which were identified as a result of the content analysis of memoirs, are given in the bottom part of Figure 16.1. First, it is necessary to single out the following:

1. Potential natural superiority, i.e., the presence of inborn potential, emphasizing its intellectual superiority in many fields.

2. Rational development of natural potential in family, school, sports, and institute.

3. Real ambition. The trainer needs recognition. The increased social requirement to act is necessary for the trainer. Territorial psychology is shown here, and the trainer expands his professional territory.

4. Highest level of professional competence.

5. Functional utilitarianism, which means that the trainer may be a good family man or efficient housewife, a perfect father or loving mother, but family and friends are not a chief concern. The main aim of the professional trainer is exaltation of the mind, manifested as internal satisfaction.

6. Emotional stability, understood as immunity to emotionally saturated factors.

Well-known professional trainers point out a significant individual quality: emotional stability. In the pedagogical and psychological literature emotional stability is treated in different ways: nature of the personality, mental condition, characteristics of the psyche, and integral features of the personality. In this study the emotional stability of the trainer is understood as immunity to emotionally saturated factors.

Because of a great variety of emotionally saturated factors (those that work up emotions directly, as well as those that cause their expressiveness), the trainer's mental state may become either positive or negative, or remain conditionally "neutral"; "triggering mechanisms" of emotional experiences may occur as the result of the activity or accompany it. The phenomenology of positive and negative emotions is characterized by a double-edged influence on the efficiency of professional activity: they may either increase or lower it.

Various combinations of positive and negative emotions during the trainer's activity cause some degree of mental tension, depending on which characteristics of the activity may remain unchanged, may be improved upon, or may be worsened (Nemchin, 1983). In other words, the parameters of activity serve as a peculiar indicator—the productive expression of those mental tensions that occur in difficult conditions.

First-degree tension may be so named only relatively because the signs of tension are not observed clearly or their displays are insignificant. In this condition, "inclusion" into an extreme situation does not occur. The subject does not consider the situation to be difficult, or one that requires the mobilization of effort to overcome it and to achieve the set goal. So, the subject of activity is not under any compulsion to overcome difficulties that are present in the situation objectively. The individual is not interested in the result of activity and is not anxious about the possible consequences.

The most general characteristics of the second degree of mental tension are the mobilization of mental activity and increase of social activity. In moderate tension, not only is motivation to achieve the objective and enthusiasm for vigorous action clearly shown, but satisfaction from the activity is also experienced. A moderate degree of mental tension is characterized, in a practical way, by a general improvement of the quality and efficiency of mental activity and represents a mental state in which the abilities of the subject to achieve the goal are revealed completely.

Considering structural–functional peculiarities of strongly expressed mental tension, it is important to note that disorganization of mental activity and a strongly manifested sensation of general physical and mental discomfort are typical responses. The physical discomfort is combined with negative emotions, anxiety, inner unrest, and expectation of failure and other unpleasant consequences in extreme situations.

Research has proved that under the influence of tension response forms move toward extreme points of the scale, *inhibition* and *excitability* (Nemchin, 1983). Passiveness, restrained mental processes, and even "emotional inertness," displayed as apathy and negative indifference, characterize the inhibitive type of response. Energetic extroversion, bustling, talkativeness, and rapidly changing decisions characterize the excitable type of response. In the state of tension, behavior is characterized by predominance of stereotypical answers.

Nevertheless, an estimation of the influence of signs and the progressive movement of the emotional state to nervous mental tension depends on the peculiarities of the regulatory characteristics of an individual and also of special forms of unconscious mental activity that provide a temporary reduction of conflict.

The phenomenon of the emotional burnout syndrome having been first identified as a state of exhaustion combined with a feeling of uselessness can be defined by the quantitative character of the trainer's professional activity and the exhaustion of mental-energy resources (Lozhkin and Voljanjuk, 2004).

The emotional burnout syndrome is one of the little-known phenomena of personal collapse. It is a response to prolonged stress in interpersonal relationship.

An analysis of the literature (Waneberg and Gold, 2001; Formanjuk, 1994; Maslach, 1978) shows that psychologists got interested in the problem of "burnout" not long ago. Meanwhile, in the English-language literature over 2000 articles devoted to this phenomenon have been published. The term was introduced in 1974 by H.J. Frainderburger, an American psychiatrist, to characterize the psychological state of healthy people who communicated with clients in the emotionally demanding task of providing professional help.

Nowadays, researchers distinguish about 100 symptoms connected with "mental burnout." At the same time there is no agreement among them, and reasons for and mechanisms of burnout are hardly studied.

Researchers discovered certain interrelationships between development of the syndrome and the manner of reacting to stress. The most vulnerable are those who display an aggressive response and try to overcome their rivals by any means, as well as those who consider their professional activity to be of special significance.

Many researchers offer various models for studying this phenomenon. However, to our mind, Maslach and Jackson's three-dimensional construction is the most successful. The scientists distinguish and describe three components of burnout: emotional exhaustion, depersonalization, and reduction of personal achievements (Maslach, 1978). Most specialists agree that these three elements should be taken into consideration to define the existence and degree of burnout. Nevertheless, we are certain that to the definition of *professional burnout* it is necessary to add a fourth factor—predisposition to conflict (Lozhkin, 2003).

Most scientists think that the characteristics of mental burnout (Waneberg and Gold, 2001) include the following:

- Exhaustion, physical as well as mental, which is the loss of interest, energy, and faith
- Negative reaction to other people
- The reduced level of self-estimation, negative "I-conception"
- Despondent mood, cynicism, apathy, sensation of failure, and depression
- Reaction to constant daily stress

Thus, it is possible to conclude, that mental burnout represents personal collapse owing to emotionally complicated or intense relations in the human–human system. In this connection some authors designate the mental burnout syndrome as "professional burnout" (Lozhkin and Vyday, 1999). This process lasts for a certain period and does not arise suddenly. As a rule, the person passes through stages, leading to professional burn out (Waneberg and Gold, 2001). The stages depend on the strong and weak sides of the person, and also on certain requirements of professional work. It is necessary to note that professional trainers are major candidates for burnout. Long-term active emotional reaction of the trainer either to successes or failures of their sportsmen or sportswomen is accompanied by emotional exhaustion. The emotional exhaustion is characterized by three parameters: tension, resistance, and exhaustion, with corresponding symptoms (Lozhkin, 2003; Lozhkin and Voljanjuk, 2004).

Emotional exhaustion leads to inadequate emotional reaction, to the economy of emotions that does not allow resisting psychotraumatic pressures. The development of all the components of emotional exhaustion testifies to a decrease of mental resources. Thus, the trainer cannot control the level of excitation; his or her ability for psychological and then social adaptation to changeable conditions is reduced. A condition of personal indifference appears, and then a reduction of professional work.

In reality the trainer's condition is characterized by persistent changes, the subjective value of which is determined by features of emotional and intellectual dimensions of the mind. The nature of any change, or any phenomenon, is understood by a person with the help of comparison with something already familiar or known. Changes may have an external character and may be accompanied by uncertainty, novelty, and vagueness. On subjective reflection it may be affected by forms of ambiguity and uncertainty of images, ideas, memories, desires, requirements, etc. So, the processes of overcoming the discrepancy between objective reality and subjective reflection determine the emotional activity of the trainer, and the substance of behavior regulation processes.

In behavioral activity it is possible to distinguish three base components: the cognitive, the emotional, and the regulative. In fact, the last component allows coordination of all the components in working toward the end result.

The cognitive component consists of the ability to understand one's own feelings, as well as the skill to verbalize emotional experiences and their causes. Research (Lozhkin and Voljanjuk, 2004) has determined that problems of cognitive deficiency appear as a result of emotional exhaustion. Because of this, trainers lose their ability to stand intense emotional experiences and cannot comprehend or process new knowledge and cope with modern tendencies. Next, they try to escape from their professional responsibilities and direct all efforts to regulating their own state and behavior.

There are problems related to emotional regulation that are also rather complicated. A deficiency of positive emotions and an obsession with negative experiences are typical of situations of failure in a professional trainer's activity. The emotional problems of the subject may be the consequence of cognitive-resource deficiency and due to their exceeding allowable limits.

The regulative component also includes an expressive component, which combines two qualities: an ability to show and express emotional experiences using mimicry, gestures, movements, and intonations, and the skill to restrain a display of emotions. An important characteristic of emotional exhaustion of the personality is the acceptance of responsibility for one's feelings together with an awareness of the freedom to express them.

The objectives of research consisted in revealing and evaluating the emotional burnout syndrome in qualified sports teachers. The following tasks concretize the objectives:

1. To reconstruct the activities of trainers on a psychological level and prove their professionalism
2. To determine the nature of symptoms of emotional burnout among trainers during their professional development
3. To carry out a comparative analysis of the formative phases of emotional exhaustion among trainers due to a particular sport
4. To find out personal, organizational, and role factors that influence the development of burnout syndrome in sports teachers

5. To work out and approve a system of psychoprophylactic actions that will lead to the eradication of the professional burnout syndrome of trainers

16.2   Methods and Procedures

During the work the following complementary and interdependent methods of scientific and pedagogical search were employed: analysis of and generalization from the literature, pedagogical observation, questionnaires, and statistical processing of results. To find out and evaluate the emotional burnout syndrome, the modified method of Boyko (Lozhkin and Povjakel, 2003) was used. This method helps find out what phase of the progression of emotional burnout—tension, resistance, or exhaustion—the subject of professional activity is going through. Also, it helps to determine the expressiveness of the dominating symptoms (see Table 16.1).

The use of content and quantitative indices, counted for different phases of syndrome development, allows us to get a complete volumetric description of a specialist's personality.

In the research 89 trainers of winter Olympic sports took part (64 men and 25 women aged from 21 to 64 years). Also, there were 47 track-and-field athletics trainers and 43 football trainers of different qualification levels (from the third category to the highest one).

The experimental group was divided into three subgroups taking into consideration previously discovered correlation regarding causality (Lozhkin and Vyday, 1999), which confirms that length of service and age may contribute to burnout: (1) 37 people with average service of 7 years and an average age of 30, (2) 40 people with average service of 22 years and an average age of 46, and (3) 12 people with average service of 32 years and an average age of 56.

TABLE 16.1

Dominating Symptoms in Emotional Burnout Phases

Tension	Resistance	Exhaustion
1. Experiencing psychotraumatic conditions	1. Inadequate selective emotional reaction	1. Emotional deficit
2. Self-dissatisfaction	2. Emotional and moral disorientation	2. Emotional "pushing aside"
3. Being at the end of tether	3. Economy of emotions sphere widening	3. Personal "pushing aside" (depersonalization)
4. Anxiety and depression	4. Reduction of professional duties	4. Psychosomatic and psychovegetative breaking

16.3 Results

The results of psychological research studies on acting trainers for different kinds of sport are presented in the tables. These results were obtained in joint research work with Volyanyuk (Lozhkin and Volyanyuk, 2004).

The analysis results show that 15 people, i.e., 17% of the respondents, did not experience emotional changes that could be categorized as burnout symptoms. The average age of these specialists was 40 years, and length of service was 17 years. The total points for the emotional burnout syndrome did not go beyond 36 in this category. The rest had developing or already-formed symptoms of emotional burnout. The results of the data from general analyses were transformed, as shown in Table 16.2.

The analysis results in Table 16.2 show that the formation phase of emotional burnout does not differ much among the trainers of the first, second, and third groups. The second phase of emotional burnout of trainers develops regardless of the age and length of service. It is important to stress that phases of *tension* and *exhaustion* are also manifested; the first phase was found only among sport trainers of the second and the third groups, whereas the third phase was found only among specialists of the first and the second groups.

Research on the nature of symptoms present, depending on the phase of emotional burnout in the explored groups, showed that the first phase, tension, had not formed in 46%, was in the formation process in 41.6%, and had already formed in 12% of the respondents.

In a comparative study of the results obtained (see Tables 16.3 through 16.5), the conditions "being at the end of the tether" and "self-dissatisfaction" were not found in respondents of the second and the third groups among already formed, or dominating, symptoms. This fact may affirm that they can realize judging dissonance, which is present at the emotional, cognitive, and behavioral levels. At the same time, the existence of this ability in qualified trainers is accompanied by a high level of anxiety about their social status—concern about the possibility that the symptoms would strengthen.

TABLE 16.2

Data on Emotional Burnout Phases in Trainers of Winter Sports

Group	Length of Service (Years)	Age (Years)	Tension	Resistance	Exhaustion
$n^1 = 37$	1–15	21–48	35.8 ± 3.2	50.1 ± 2.5	40.3 ± 3.1
	X = 7.75	X = 30.6			
$n^2 = 40$	15–27	39–64	42.1 ± 2.7	53.9 ± 3.2	41.2 ± 3.0
	X = 22.4	X = 46.1			
$n^3 = 12$	27–45	45–64	37.5 ± 3.0	58.1 ± 3.7	35.7 ± 3.1
	X = 32.6	X = 56.8			

Note: n = number of subjects in the groups (37; 40; 12); X in column 2 = average length of service; X in column 3 = average age.

TABLE 16.3

Special Features of Emotional Burnout Symptoms in Trainers of the First Group[a] (in Points)

Extent of Phase Formation	Tension					Resistance					Exhaustion				
	n	1	2	3	4	n	1	2	3	4	n	1	2	3	4
Not formed	20	7.6	4.6	2.1	6.6	8	14.9	8.4	1.9	4.3	14	2.4	7.9	6.6	3.6
Forming	13	12.4	10.5	10.4	14.6	22	**20.3**	12.5	10	8.7	17	15.1	9.6	13.5	7.6
Formed	4	**19.2**	11.5	**16.7**	**21.5**	7	**22.5**	13.6	**16.9**	16	6	14.3	10.8	**27.8**	15

Note: n = total number of subjects in the first group (n = 37)

1, 2, 3 and 4 = a category of the dominating symptoms in the emotional burnout phases according to Table 16.1.

TABLE 16.4

Special Features of Emotional Burnout Symptoms in Trainers of the Second Group[a] (in Points)

Extent of Phase Formation	Tension					Resistance					Exhaustion				
	n	1	2	3	4	n	1	2	3	4	n	1	2	3	4
Not formed	15	5.5	6.0	3.8	8.2	10	16.2	5.8	2.4	4.8	17	6.3	7.3	7.4	4.0
Forming	19	14.2	8.4	7.5	15.7	12	**21.5**	4.6	9.3	9.9	15	11.1	5.9	**18.2**	7.5
Formed	6	**21.1**	13.6	13.3	**19.0**	18	**23.3**	15.8	**17.8**	16.3	8	15.6	12.8	**26.5**	15.3

Note: n = total number of subjects in the second group (n = 40)

1, 2, 3 and 4 = a category of the dominating symptoms in the emotional burnout phases according to Table 16.1.

TABLE 16.5

Special Features of Emotional Burnout Symptoms in Trainers of the Third Group[a] (in Points)

Extent of Phase Formation	Tension					Resistance					Exhaustion				
	n	1	2	3	4	n	1	2	3	4	n	1	2	3	4
Not formed	6	9.1	6.3	2.0	4.8	1	15	17	2.0	0	8	4.3	8.8	9.5	6.9
Forming	5	13	6.8	9.8	**17.6**	5	**20.4**	10.2	10	10.6	3	7.7	10	13	11.3
Formed	1	**25**	13	12	**30**	6	**21.1**	12.1	**16.8**	17.6	1	12	8	**30**	10

Note: n = total number of subjects in the third group (n = 12)

1, 2, 3 and 4 = a category of the dominating symptoms in the emotional burnout phases according to Table 16.1.

That is, we think, the main reason for the domination, in the tension phase, of symptoms such as "experiencing psychotraumatic conditions" and "anxiety and depression."

It must be noted that the given symptoms are characteristic of the trainers in the first group, too; however, the cause of their occurrence is different. In the first place, it is connected with the difficulties that arise at the adaptation and initial professional orientation stages.

Further analysis of the results obtained has shown that the resistance phase is not formed in 21.4% of the interrogated sports teachers, is formed in 34.8%, and is in the process of being formed in 43.8% of the group. It is seen that among the symptoms that are manifested in this phase, the most commonly expressed are "nonadequate selective emotional reaction," and "expansion of spheres of economy of emotions and reduction of professional obligations," so that it is possible to explain the trainer's desire to develop psychological protection. At the same time, it is necessary to note that frequently repeating negative emotional states reduce the trainer's frustrated tolerance, the basis of which is the ability to evaluate a real situation adequately and detect a way out of the problem. The presence of these symptoms may also speak to the economical expenditure of power resources.

Next is the exhaustion phase, the formation of which testifies to the full use of the mental resources by the expert. The subject of the professional activity is not capable of any adaptation or resistance in this phase; the person cannot control the level of excitation or resist psychotraumatic influences. It is established that the given phase is not formed in 43.8% of the interrogated trainers. It is formed in 16.8%, and it is in the process of formation in 39.4%. The analysis of results submitted in Tables 16.3 through 16.5 shows that the symptom "individual separation" is dominant in this phase, and its high parameters are fixed both in beginners and in skilled sports teachers. The reason of such results is seen in the objective loss expressed to a greater or lesser extent by the subject of an opportunity to be ideally submitted in the ability of the pupils or colleagues to live.

Examples of comparative analysis of emotional burnout demonstrate that the levels of tension and resistance are higher in trainers of winter sports and track-and-field athletics compared to football trainers (Figure 16.2).

The analysis of results presented in Figure 16.3 has shown that the formation of emotional burnout phases is not essentially different in trainers of winter sport and track-and-field athletics. It is important to note that the level of emotional exhaustion in football trainers is lower in comparison with that of trainers of track-and-field athletics and winter sports, which means the former preserve their emotional resources for their activity. At the same time it must be noted that the high parameters in the resistance phase, for trainers of winter sports and track-and-field athletics, is connected with a charge of emotional resources due to a possible decrease in the success rate of the trainer's activity.

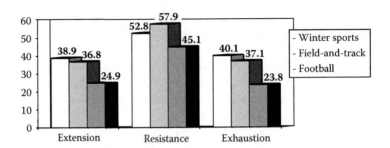

FIGURE 16.2
Formation of phases of emotional exhaustion in trainers of different kinds of sport.

In a number of studies (Waneberg and Gold, 2001), high levels of quantifiable burnout were found more in women trainers compared to male trainers, whereas the present study does not reveal any proven gender difference.

Long-term research has helped establish that in the formation of burnout syndrome in sports teachers a significant role is played by the following factors:

1. Personal factors: reactance, authoritarianism, a level of empathy in combination with a fanatical fidelity to business and reaction to stress, aggression, and apathy from failure to achieve desirable results in the short term, dissatisfaction with professional growth, etc.

2. Role factors: role conflict, role uncertainty, role overload, etc.

3. Organizational factors: destabilizing organization of activity, the importance of a problem, duration and excessive loading in situations of intense interpersonal relations, insufficient material equipment, lack of opportunity for relaxation and emotional unloading, high-grade rest, weak economic stimulation, etc.

Also, it is necessary to note that the factors related to professional activity correlate with burnout more than the individual characteristics of professionals (Volyanjuk, 2003).

Thus, the presence of real mechanisms of negative influences in the professional environment impacting the sports teacher's personality make the task of developing and introducing appropriate psychoprophylactic actions necessary. In essence, these consist of the following:

1. Definition of short-term and long-term objectives, which not only provide feedback but also testify that the trainer is right. This enhances long-term motivation because the success in achieving short-term objectives raises the degree of self-esteem and confidence.

2. For maintenance of mental and physical well-being, time-outs, i.e., rest from work and other loadings, are very important. Reducing the amount of training loadings as well as the intensity of physical exercise promotes mental health.

3. Mastering psychological skills and techniques such as relaxation, ideomotion, and positive internal speech promote the reduction of stress, which is associated with professional burnout.

4. Retaining a positive outlook on life and work. For this purpose an environment that provides social support is necessary, because it will help support a positive point of view.

5. The control of the emotions arising after a planned work.

6. Keeping fit. An increase or excessive decrease of body weight negatively influences the level of self-esteem and promotes the development of burnout syndrome. Physical fitness promotes mental stability.

16.4   Conclusions

Emotional resources are characterized by a certain level of development of the personality, openness to emotional experience, comprehension of one's own feelings, and ability to express emotions adequate to the situation so that one's affairs can be managed flexibly and creatively.

The phenomenon of emotional burnout delimits an interval during which the trainer is capable of active and productive work with pupils. It is very important when psychological characteristics, which are most significant for the regulation of mental activity, have been formed. Thus, they become a condition of stability and determine the duration of the trainer's professional activity.

The formed phases of emotional burnout among the trainers of the first, second, and third group of examinees have no essential distinctions. Regardless of age and experience the second phase of emotional burnout is in a stage of formation among sports teachers.

A study of the development of emotional burnout symptoms among qualified sports teachers gives an opportunity to intervene in time in the progression of this condition. Also, it provides an opportunity to find ways (or even to work out a model) to prevent the syndrome and thus remove its negative influence.

References

Chepeleva, N. V. (Ed.), 2004. *Problems of Psychological Germeneutics.* Kiev: Millennium.

Formanjuk, T. V. 1994. A syndrome of emotional burnout as a parameter of a teacher's professional disadaptation. *Questions of Psychology*, No. 6, pp. 57.

Lozhkin, G. V. 2003. Psychological potential of the qualified athlete. *VII International Scientific Congress Modern Olympic Sports and Sports for Everybody.* Moscow: Russian State University of Physical Education, Sports and Tourism, Vol. 2. pp. 272–273.

Lozhkin, G. V. and Povjakel, N. I. 2003. Practical psychology in systems man-machine: *Teaching Aid.* Kiev: Academy of Public Administration, pp. 244–251.

Lozhkin, G. V. and Voljanjuk, N. J. 2004. Emotional burnout of the coach. Materials of the international scientific conference of psychologists of sports and physical training, Rudikovskiy Reading. Moscow: Russian State University of Physical Education, Sports and Tourism, pp. 50–52.

Lozhkin, G. V. and Vyday, A. 1999. Mental "burnout" of the leader. *Personnel: Scientific Journal.* Kiev: Academy of Public Administration, pp. 44–50.

Maslach, C. 1978. Job burnout—how people cope, *Public Welfare,* Spring, pp. 24–46.

Nemchin, T. A. 1983. *A Condition of a Nervously-Mental Tension.* Leningrad: Leningrad State University, 244 pp.

Voljanjuk, N. J. 2003. Emotional burnout of the sports pedagogue. Herald of Kharkiv National University named after Karazin, V. N. Kharkiv, No. 599. pp. 67–72.

Voljanjuk, N. J. 2004. Use of a biographic method in the psychologo-pedagogical researches. *Practical Psychology and Social Work.* 1 (58) 37–40.

Waneberg, R. S. and Gold, D. 2001. Bases of psychology of sports and physical training. In Lozhkin, G. V. (Ed.), *Collection of Work in Sports and Physical Training.* Kiev: The Olympic Literature, pp. 13–25.

Section VI

Personality

17

Good Judgment: The Intersection of Intelligence and Personality

Robert Hogan, Joyce Hogan, and Paul Barrett

CONTENTS

17.1 The Academic Study of Intelligence 359
17.2 An Evolutionary Model of Intelligence: MetaRepresentation............. 361
17.3 Measuring Intelligence 363
17.4 The Structure of Intelligence: The Ubiquitous Two-Component Model.. 364
17.5 An Adaptive Model of Intelligence: Power, Structure, and Style 366
17.6 Defining Personality.. 368
17.7 Reasoning Style 369
17.8 Conclusion 372
References 373

> No psychologist has ever observed intelligence; many have observed intelligent behavior. This observation should be the starting point of any theory of intelligence. (Chein, 1945, p. 111)

Managers and executives must frequently decide how best to allocate scarce financial and human resources. Each decision they make is the end result of a problem-solving exercise. The history of every organization is the cumulative sum of the decisions that result from these efforts at problem solving. The quality of individual managerial problem solving and decision making is a function of good judgment. This chapter is about how to define and evaluate good judgment. Our central argument is that good judgment is a function of intelligence and personality. The statement seems simple, but it is not. The topic of how to define intelligence is quite vexed, and the largest part of this chapter concerns just that problem. Once that is accomplished, we then try to show how good judgment comes from crossing intelligence (as we define it) with personality.

Peter Drucker, the famous philosopher of management, constantly empha-sized that businesses get in trouble because senior managers exercise bad judgment (cf. Drucker, 2006). Managers are supposed to direct money and energy toward activities that increase profitability. More often, however, they spend time and money solving problems and completing projects that do not matter. It takes clear-minded analysis to determine how to use money and energy appropriately. Clear-mindedness is a function of good judgment.

Drucker also emphasized that virtually every major business crisis results from the fact that the assumptions on which the business was built and is being run no longer fit reality. Drucker called these assumptions the "theory of the business." Constructing a valid theory of the business, and then subse-quently evaluating and revising it, is a function of good judgment.

Menkes (2005) notes that although most people understand the impor-tance of clear-mindedness for the success of business, new managers and executives are rarely hired based on their ability to think clearly. There is a pressing need for sound and defensible methods to evaluate the ability of managers and executives to think clearly and exercise good judgment.

In the mid-1990s, a group of academic researchers issued a statement on scientific evidence on intelligence (see Gottfredson, 1997, p. 13). Their first two conclusions are useful for our discussion:

1. Intelligence is a very general mental capability that, among other things, involves the ability to reason, plan, solve problems, think abstractly, comprehend complex ideas, learn quickly and learn from experience. It is not merely book learning, a narrow academic skill, or test taking smarts. Rather it reflects a broader and deeper capabil-ity for comprehending our surroundings—"catching on," "making sense" of things, or "figuring out" what to do.
2. Intelligence, so defined, can be measured, and intelligence tests mea-sure it well. They are among the most accurate (in technical terms, reliable and valid) of all psychological tests and assessments. They do not measure creativity, character, personality or other important differences among individuals, nor are they intended to.

Standardized intelligence testing was developed to predict academic perfor-mance (Binet and Simon, 1905). In contrast, an effective measure of execu-tive intelligence should predict clear thinking, good judgment, and effective management decision making. Drucker describes the key components of this in very general terms: thinking critically about the theory of the business by reviewing the assumptions on which it was founded and in terms of which it is being operated. We believe this process can be usefully specified in terms of time perspective as follows:

1. *Past perspective*: Are the operating assumptions of the business still valid?
2. *Present perspective*: Given the stated goals of the business, are people currently working on the right problems and tasks?

3. *Future perspective*: Given the stated goal of the business, have people appropriately anticipated the potential future problems and possible future outcomes correctly?

Within each of these perspectives two kinds of thinking will apply. We call them "problem finding" and "problem solving"; we also refer to these two kinds of thinking as "strategic reasoning" and "tactical reasoning." Problem finding involves detecting gaps, errors, or inconsistencies in data, trends, textual materials, existing processes and procedures, or verbal arguments. Problem solving involves finding answers to the problems once they are identified, following arguments to their logical conclusions, and applying well-understood methods to new problem categories.

Our view of competent business reasoning and good judgment starts with the preceding discussion. It assumes that the word *intelligence* refers to clear thinking and is a key component of successful managerial performance. It assumes that intelligence facilitates good judgment. It assumes that two kinds of reasoning are essential to this process—problem finding and problem solving. It assumes that these two kinds of thinking can be measured, and that the results from this measurement process can be used to evaluate good judgment. It assumes that good judgment is a function of both intelligence and personality, that personality can be measured, and that a person's reasoning style can be best estimated using measures of intelligence and personality. And finally, it assumes that the results of this measurement process will predict successful occupational performance.

17.1 The Academic Study of Intelligence

Although psychologists have great faith in the importance of measures of cognitive ability, traditional measures of intelligence suffer from at least five problems. First, the study of intelligence is the most antiintellectual part of American psychology. The Europeans—Piaget (1952), Vygotsky (1978), and Spearman (1923)—have developed some substantive ideas about the nature of intelligence. However, with the exception of Sternberg (1997) and possibly Gardner (1983), there is nothing resembling a plausible theory of intelligence in the American literature. Sternberg and Detterman (1986) asked two dozen experts in the field to define intelligence and each one gave a different answer! In America, conceptual understanding has not greatly advanced past the view that "intelligence is what intelligence tests measure" (cf. Neisser, Boodoo, Bouchard, Boykin, Brody, Ceci, Halpern, Loehlin, Perloff, Sternberg, and Urbina, 1996). In a nutshell, the study of intelligence is a conceptual muddle.

Second, existing measures of cognitive ability may not predict occupational performance as well as the proponents of cognitive ability testing claim. For example, Judge, Colbert, and Ilies (2004) report a fully corrected meta-analytic correlation of .27 between intelligence and leadership (emergence and effectiveness); although they were dismayed by the modest size of the correlation, it is typical. Third, when business decision makers are asked why they want to use a measure of cognitive ability to evaluate job candidates, they seem not to know; this suggests that using measures of cognitive ability to evaluate people is often a faith-based initiative.

Fourth, blacks tend to get lower scores than whites on measures of cognitive ability (cf. Gottfredson, 1997), and we do not believe that blacks are inherently less intelligent than whites (cf. Diamond, 1997); there is something questionable about an assessment methodology that stigmatizes a major segment of the human population. And this stigmatization is very real. The "race realists" Lynn (2006) and Lynn and Vanhanen (2006), for example, state that blacks are inherently less intelligent and less law-abiding than whites, concluding that certain "less intelligent" subgroups threaten the stability of "white" democracies.

Finally, the concept of intelligence has low social penetrance—scores on measures of cognitive ability seem uncorrelated with everyday performance in individuals. For example, we all know people with high cognitive ability scores who are unable to manage their personal affairs; this is an indictment of the validity of the measures. Conversely, anyone who has taught at the college level has met students with modest test scores who have acute powers of analysis.

Perhaps the clearest indication that academic studies of intelligence are unconcerned with everyday performance is that fact the Jensen (1999, p. 48) recommends dropping the word *intelligence* from the language of scientific psychology. He argues researchers should focus on precisely defined abilities such as naming vocabulary, symbol search, arithmetic operations, mental rotation, etc. Jensen also uses the term *psychometric intelligence* to refer to that which is measured by tests of mental abilities. Indeed, in a book on the general factor of intelligence found by analyzing mental ability tests, Jensen (2002) entitled his chapter "Psychometric g."

This change of terminology seems reasonable only if we study cognitive attributes in isolation from everyday human performance. In contrast, Grigorenko (2004) described a substantive body of research on cognitive functioning that has persisted for many years without using the term *intelligence*, and notes that Russian psychology has never used the concept while analyzing cognition in more depth than many Western investigators of psychometric intelligence. The Russian work is theoretically rich and focused on understanding cognitive processes rather than describing features of it as "mental ability," as psychologists typically do. Similarly, researchers in computational intelligence define intelligence explicitly, and build machines that display features of the defined attributes. Consider, for example, this textbook definition (Poole, Mackworth, and Goebel, 1998, p. 1, 2):

> Computational intelligence is the study of the design of intelligent agents. An agent is something that acts in an environment—it does something.

Agents include worms, dogs, thermostats, airplanes, humans, organizations, and society. An intelligent agent is a system that acts intelligently: What it does is appropriate for its circumstances and its goal, it is flexible to changing environments and changing goals, it learns from experience, and it makes appropriate choices given perceptual limitations and finite computation.

The central scientific goal of computational intelligence is to understand the principles that make intelligent behavior possible, in natural or artificial systems. The main hypothesis is that reasoning is computation. The central engineering goal is to specify methods for the design of useful, intelligent artifacts.

The obvious intelligent agent is the human being. Many of us feel that dogs are intelligent, but we wouldn't say that worms, insects, or bacteria are intelligent. There is a class of intelligent agents that may be more intelligent than humans, and that is the class of organizations. Ant colonies are the prototypical example of organizations. Each individual ant may not be very intelligent, but an ant colony can act more intelligently than any individual ant. The colony can discover food and exploit it very effectively as well as adapt to changing circumstances. Similarly, companies can develop, manufacture, and distribute products where the sum of the skills required is much more than any individual could understand. Modern computers, from the low-level hardware to high-level software, are more complicated than can be understood by any human, yet they are manufactured daily by organizations of humans. Human society viewed as an agent is probably the most intelligent agent known. We take inspiration from both biological and organizational examples of intelligence.

This statement contains a perspective that is totally missing from the academic study of intelligence. In computational intelligence, the goal is to design systems that display intelligence, not some abstract statistical model of "psychometric intelligence." It is an attempt to understand what makes organisms, people, and systems "intelligent."

Our work also regards intelligence as a substantive causal component of everyday human performance, in a way that is consistent with the computational, the evolutionary, and the ecological basis of intelligence (Todd and Gigerenzer, 2007).

17.2 An Evolutionary Model of Intelligence: MetaRepresentation

This section outlines our model of intelligence. The word *intelligence* is a recent addition to our language, and it is instructive to note that the ancient

Greeks did not use the word. Rather, they used words like clever, cunning, and wise to describe individual differences in performance. More importantly, all of these words have behavioral referents—people are only called clever if they routinely manifest a certain kind of performance. In our view, intelligence should be defined in terms of certain behaviors, and people refer to these behaviors when they conclude that someone has "good judgment" or, conversely, "poor judgment."

If the word *intelligence* denotes something real, then it must be rooted in biology and promote individual and group survival—there must be adaptive consequences associated with individual differences in intelligent behavior. In a study of self-consciousness, Sedikides and Skowronski (1997) argue that self-awareness—the ability to think about one's impact on one's social environment—is an adaptive outcome of human evolution. They propose that self-awareness gave early humans an advantage relative to their major competitors.

We think that Sedikides and Skowronski (1997) are correct as far as they go—the capacity for self-reflection is a necessary precursor to intelligent behavior. However, we propose that intelligent performance depends on a more general capacity that can be called *metarepresentation*. By metarepresentation, we mean the ability to reflect on our performance (physical, social, and intellectual) across all aspects of experience, to review it, and then evaluate it. The definition of stupidity is to continue doing something that yields a poor outcome but to expect that the outcome will improve if one persists in the same course of action. In contrast, when smart athletes fall behind in a game, they reflect on their performance both on its own terms and relative to the competition, change their tactics, and then improve their performance—and this is why they are called "smart."

Our hominid ancestors evolved (as group living animals) in an environment that was more demanding and less forgiving than ours. In the ancestral environment, survival depended on being able to solve a wide variety of problems, including finding food, water, shelter, and protection from very nasty predators, keeping the peace within the group, and defending oneself and family against attacks by competing human groups. If the group members did not solve these problems correctly, they died; those that solved the entire range of problems prevailed. However, present success is no guarantee of future success. The demands of survival changed constantly; those groups that adapted and improved their survival techniques in the face of constantly shifting environmental pressures became our more recent ancestors—the ultimate winners in the race for survival. Improving one's performance involves correctly anticipating future problems or recalling past performance that yielded better outcomes than those resulting from current performance. In either case, improving performance depends on the capacity for metarepresentation, the ability to reflect on and evaluate one's performance, and then use the results of this reflection to improve subsequent performance.

Intelligent people can detect problems in their performance and then change it. They can also detect problems in other peoples' performance and encourage them to change it. Anthropologists and psychologists have traditionally argued that behavioral flexibility is the most important single human characteristic. *Metarepresentation is the key to behavioral flexibility.* Crocodiles are competent hunters and proficient predators. Over time they have eaten many humans, but because crocodilian behavior is largely wired and inflexible, humans can hunt them to extinction.

17.3   Measuring Intelligence

We believe intelligence is manifested in the ability to solve a wide range of problems correctly, and this includes solving the problem of what to do when the old methods and solutions no longer work. We propose that the capacity for metarepresentation provides the basis for the ability to do this. Spearman (1927) said something similar; namely, he argued that "g," or general intelligence, is the ability to solve a variety of problems correctly. Binet (1903) suggested that the optimal method for measuring intelligence is to give people a large number of qualitatively different problems to solve. Intelligence is whatever underlies the ability to solve the various problems correctly. Such an assessment process, however, would be too unwieldy and time consuming to be useful; in addition, it ignores individual differences in personality. We think that the generic process of problem solving can be broken down into two components, "problem finding" and "problem solving" or "strategic and tactical intelligence." Also, we believe that many previous writers have essentially arrived at the same conclusion.

Early farmers learned to predict when the Nile would flood by watching the stars; certain regular changes in the position of the stars guided the planting process. This model of learning is the essence of science—science emerged and evolved through the process of detecting covariations ("When we do X, Y happens"). In many, if not most, cases that is as far as the analysis goes: neuroscientists still do not know *how* anesthesia works, oenologists still do not know in detail *how* wine fermentation works, every sailor knows the mantra, "Red skies at night, sailors delight; red skies in the morning, sailors take warning," but they do not know why. Based on these considerations, we initially proposed that "intelligence" is formed from two components: The ability (a) to detect covariations (i.e., to identify events that go together reliably), and (b) to recognize when the sequence is recurring or going to recur. More complex combinations come quickly to mind—for example, recognizing when covariations do not occur—as in the case of Sherlock Holmes' dog that did not bark. A more contemporary way of saying the same thing is that intelligence consists of (a) developing schemas to forecast sequences of events in the world; and (b) subsequently applying those schemas appropriately. In any case, it is a two-step process.

17.4   The Structure of Intelligence: The Ubiquitous Two-Component Model

Spearman (1923) suggested that "g" is composed of two components that necessarily work together, but are measured separately, and are only modestly correlated. He called these components *eduction* and *reproduction*. *Eduction* is the ability to abstract meaning from the chaotic information of the world around us; *reproduction* is the ability to recall or describe that information at a later time. Reproduction depends on memory, which can be efficiently assessed using vocabulary measures. Although memory facilitates intelligent performance, it is not the same as intelligence.

Drawing on Spearman's distinction between *eductive* and *reproductive* mental activities, Raven (Raven, Styles, and Raven, 1998) developed the Raven's Progressive Matrices and the Mill Hill Vocabulary test. The matrices are intended to capture *eductive* abilities, and the vocabulary test is intended to capture *reproductive* abilities. These two measures correlate .30, which means that a person can have high scores on one, the other, or both measures. Curiously, although most people regard Raven's matrices as a nearly pure measure of Spearman's "g," it is the vocabulary test that loads most highly on the "g" factor. This is, of course, a problem because memory is not intelligence; rather, it enables intelligent performance.

Spearman's most famous student, R.B. Cattell, proposed that intelligence is composed of what he called fluid and crystallized components. Cattell's most famous student, John Horn, evaluated this model carefully and concluded (Horn, 1994) that Cattell's two factors closely approximated those that Spearman had originally proposed: eduction and reproduction.

Charles Saunders Peirce, along with his friend William James, is the author of the school of American philosophy called pragmatism, a blend of functionalism and evolutionary theory. Peirce, who was a philosopher of science, disagreed with the claim that inductive and deductive logic characterize how scientists reason. He proposed the concept of "abduction" to describe how ordinary people (and scientists) solve problems, as contrasted with how professional logicians claim they solve problems. Abduction is a holistic thinking process that involves (a) forming hypotheses about how things work, and then (b) evaluating those hypotheses. Forming hypotheses resembles eduction; reproduction is essential to evaluating hypotheses.

L.L. Thurstone (1938) also was interested in the distinction between inductive and deductive reasoning. Thinking about the problem from a measurement perspective, Thurstone proposed that inductive reasoning concerned rule finding—inferring rules from a set of particular instances. He then suggested that deductive reasoning concerned applying those rules to new material. The point here is that modern views of the structure of intelligence suggest a hierarchical model with "g" at the top, then verbal, quantitative, and spatial reasoning at the next level below "g" (cf. Lubinski, 2004). In contrast, however, Thurstone suggested (and we agree) that the structure of intelligence is actually composed of rule finding and rule applying, regardless of the stimulus or problem material. In an elegant little paper, Shye (1988) used Guttman's (1968) Smallest Space Analysis and recovered rule-finding and rule-applying factors within the content domains of verbal, quantitative, and spatial reasoning. We think that rule finding and rule applying correspond to strategic and tactical reasoning (or detecting covariations and deciding correctly when they will reoccur).

In the 1930s, Piaget (1952) proposed a highly influential model of the development of intelligence. Piaget had worked with Binet and became interested in the problem of why children made mistakes on Binet's test. He attributed the errors to certain characteristic shortcomings in their reasoning capacity that reflected their level of intellectual development. He proposed a stage theory of cognitive development, and labeled the two final stages "Concrete Operations" and "Formal Operations." In Concrete Operations, children are able to solve problems that are placed in front of them; in Formal Operations, they are able to think about the process of problem solving itself. We believe that Concrete Operations resembles what we have called problem solving or tactical reasoning, and Formal Operations resembles what we have called problem finding or strategic reasoning.

In yet another attempt by philosophers to characterize the nature of scientific thinking, Reichenbach (1951) famously distinguished between "the context of discovery" and "the context of justification." He argued (correctly in our view) that the context of discovery involves finding problems that need to be solved, and the context of justification involves actually solving them. More importantly, he argued that the rules of the scientific method apply only to the justification process, that the process of discovery was not subject to methodological specification. Not surprisingly, we think that Reichenbach's "context of discovery" corresponds to problem finding and "the context of justification" corresponds to problem solving.

Perhaps the most flamboyant effort to describe the components of intelligence is Guilford's (1957) three-dimensional structure of intellect model. Guilford described intelligence in terms of 120 facets. These facets depended on five types of mental operations that seem to fit the two-variable model we are proposing. On the one hand, Guilford hypothesized *Evaluation*—deciding what problems are worth solving—and *Divergent Thinking*—using information to generate a variety of hypotheses regarding solutions to a problem. On the other hand, Guilford hypothesized *Memory*, retention of information; *Cognition*, recognizing patterns and facts; and *Convergent Thinking*, using the information to find a specific right answer. We propose that the mental operations Guilford called Evaluation and Divergent Thinking correspond to problem finding and the operations Guilford called Memory, Cognition, and Convergent Thinking correspond to problem solving.

Following Francis Galton's (1885) efforts to measure the range of individual differences in human performance, James McKeen Cattell developed a series of psychomotor tests (e.g., tapping speed) that he thought should reflect intelligence; if intelligence rests on a neuropsychological foundation, then the speed with which people react to various stimuli should reflect intelligence. A paper by Wissler (1901) brought this interesting line of research to an end. Wissler tested Columbia students using Cattell's measures, correlated their test performance with their grade point average, and found nothing. This measurement effort failed, not because the premise was necessarily wrong, but because of the problem of aggregation: the measures used in the study were unreliable. Modern chronometric research (Luo, Thompson, and Detterman, 2003) has solved the problem of aggregation. Researchers now use what are called elementary cognitive tasks (ECTs) to study intelligence. These tasks fall into two categories: (a) inspection time: the amount of time required to discriminate between a set of stimuli; and (b) response time: the time required to respond to an experimental stimulus. When scores from ECTs are aggregated across *modalities* (auditory, visual), *content* (pictures, numbers, words), and *tasks* (reaction time, inspection time), two general factors emerge: processing speed and working memory. Jensen (2005) suggests that "g" is a function of these two dimensions. This seems plausible; moreover, we believe these factors are related to our notions of problem finding and problem solving.

In a very important paper on the philosophy of science as applied to the behavioral sciences, Haig (2005) critiques the prevailing hypothetico-deductive and inductive theories of science. He argues instead that theory building is about explanation, which he calls *abduction*. He describes the process of theory building in terms of two very general phases, which he calls *phenomena detection* and *theory construction*. Phenomena detection can be characterized as detecting large-scale covariations; moreover, "The importance of phenomena detection in science is underscored by the fact that more Nobel prizes are awarded for the discovery of phenomena than for the construction of explanatory theories" (p. 384). Theory construction involves devising models to explain the phenomena that have been detected.

Finally, it is conventional in the business literature to distinguish between strategic reasoning and tactical reasoning. Strategic reasoning involves anticipating and detecting major movements in customer demand, new technology, and the economy (i.e., problem finding). In contrast, tactical reasoning concerns using established methods to solve problems on a day-to-day basis, so as to promote the larger strategic vision (i.e., problem solving). We think the distinction between strategic and tactical reasoning parallels the two dimensions of cognitive ability that run through the history of cognitive ability measurement. Table 1.1 summarizes the discussion so far.

17.5 An Adaptive Model of Intelligence: Power, Structure, and Style

R. Hogan (1980) critically reviewed the intelligence literature and then, drawing on ideas outlined by Webb (1978), proposed that intelligence is not an

TABLE 17.1

Models of the Structure of Intelligence

Source	Dimension: Strategic	Dimension: Tactical
R. Hogan	Problem finding	Problem solving
C. Spearman	Eduction (problem finding/ solving)	Reproduction (describing the solution)
C.S. Peirce	Forming hypotheses	Evaluating hypotheses
L.L. Thurstone	Rule finding	Rule applying
R.B. Cattell	Fluid intelligence	Crystallized intelligence
J.P. Guilford	Divergent thinking/evaluation	Memory/cognition/convergent thinking
J. Piaget	Formal operations	Concrete operations
H. Reichenback	Context of discovery	Context of justification
B. Haig	Phenomena detection	Theory construction
Business Speak	Strategic reasoning	Tactical reasoning

entity. Rather, it is an evaluation that we place on another person's behavior based on its adaptive adequacy. What constitutes intelligent behavior depends on a person's life circumstances—the demands to which he or she is adapting. The next question concerns the factors underlying or generating intelligent behavior. Webb (1978) called these factors "power, structure, and style."

Power refers to the general efficiency of a person's neurological machinery or his or her capacity for metarepresentation. It is much like Spearman's "g," and it is largely heritable (Deary, Spinath, and Bates, 2006; Petrill, 2002; Plomin, 2003). Power is seen in the speed with which a person acquires new information or "catches on" to ideas, trends, and procedures. It makes sense that academic achievement should be a rough index of power in modern societies—the speed with which a child acquired the technology of its culture once had survival value. This line of reasoning also suggests that power will be less relevant to intelligent behavior in adulthood because the problem for adults is less about acquiring new information and more about integrating and applying the information already acquired. In any case, power is a "metaconcept" and is hard to measure per se.

Structure refers to individual differences in problem finding and problem solving—or strategic and tactical reasoning. Some people excel at problem finding, some at problem solving, some at both, and some at neither. These two domains of reasoning seem important for understanding good judgment; they can also be used to specify a measurement model for assessment. This brings us to the issue of style.

Style refers to how a person uses his or her intellectual capabilities. Some people solve problems in a reflective and deliberative manner, some are cautious and defensive problem solvers, and others are brash and impetuous. We conceptualize style as a combination of (a) strategic/tactical reasoning and (b) personality. It is the combination of these characteristics that forms the components of good judgment; that is, good judgment is a function of intelligence and personality.

17.6   Defining Personality

The foregoing discussion concerns defining (and measuring) intelligence; now we need to define personality. Personality is best defined from two perspectives: that of the actor and that of the observer. Personality from the actor's view is a person's identity—a person's idealized self-view, including his or her hopes, dreams, aspirations, and core values. Personality from the observers' view is a person's reputation, and it is defined in terms of how others describe that person; i.e., as conforming, helpful, talkative, competitive, calm, curious, and so forth; reputation reflects a person's characteristic ways of behaving in public. Reputation describes a person's behavior; identity explains it.

Identity is hard to study and we do not know much about it in a scientific way. Reputation is easy to study, and the science of personality largely rests on the study of reputation. For example, Goldberg (1981) notes that the well-known Five-Factor Model (FFM; Wiggins, 1996) represents the structure of observers' ratings based on 75 years of factor analytic research. These factors are a taxonomy of reputation (cf. Digman, 1990; John, 1990; Saucier and Goldberg, 1996), and are labeled as follows: Factor I, Extraversion or Surgency; Factor II, Agreeableness; Factor III, Conscientiousness; Factor IV, Emotional Stability; and Factor V, Intellect/Openness to Experience (John, 1990).

The FFM reflects the "bright side" of personality and there are well-validated measures for assessing it (cf. R. Hogan and Hogan, 1995, 2007; J. Hogan and Holland, 2003). We believe that we can cross intelligence with bright-side personality to evaluate good judgment.

Reputations are also coded in such terms as bratty, arrogant, and self-centered. These terms point to the "dark side" of personality, an aspect of reputation that coexists with the bright side. The dark side appears when people are stressed, tired, or simply not paying sufficient attention to their social performance. Dark-side tendencies are often masked by good impression management, and typically, they are seen only after prolonged exposure.

Perhaps the first taxonomy of dark-side tendencies can be attributed to Horney (1950). She identified ten "neurotic needs" that she later summarized in terms of three themes: (a) moving toward people: i.e., managing one's insecurities by building alliances; (2) moving away from people: i.e., managing one's inadequacy by avoiding contact with others; and (c) moving against people: i.e., managing one's self-doubts by dominating and intimidating others. Horney's taxonomy provides a useful first step in classifying the dysfunctional dispositions that comprise the dark side of personality.

Following Horney, R. Hogan and Hogan (1997) developed the Hogan Development Survey (HDS). The inventory, which has been validated on several thousand managers and executives, contains scales measuring 11 dark-side dimensions. These dimensions reduce rather neatly into three factors that parallel those originally described by Horney.

The FFM largely concerns competence. The extremes of each FFM dimension reflect interpersonal excesses and deficiencies. For example, an excess of conscientiousness turns into compulsive behavior and a deficit of conscientiousness turns into delinquency. In our view, then, the dark side (as evaluated by the HDS) represents extremes of normal personality factors.

Because the domain of cognitive ability is relatively independent of personality, it is possible to develop a taxonomy linking strategic and tactical reasoning with bright-side/dark-side personality dimensions. Three generalizations seem apparent. First, good judgment is the absence of poor judgment. Low strategic and tactical reasoning ability will lead to problems in judgment for which personality cannot compensate, although a shiny bright-side personality may mask poor reasoning. Second, high strategic and tactical reasoning ability coupled with elevated bright-side personality dimensions should lead to good judgment. Third, virtually everyone has some dark-side tendencies. To evaluate a person's reasoning style, therefore, it is probably most efficient to cross strategic and tactical reasoning with the dark side, because dark-side tendencies degrade judgment. Although good judgment is a critical competency in any management model, it is bad judgment that people want to identify when they ask candidates to complete an assessment. Poor judgment from otherwise good strategic and tactical reasoning is influenced by three dark-side personality configurations.

17.7 Reasoning Style

Table 17.2 is a two-by-two table defined by high and low scores for strategic and tactical reasoning. Consider first persons who score in the Low/Low quadrant. As a result of their limited cognitive abilities across the board, such people should have trouble identifying and solving problems in a timely and efficient manner, regardless of how pleasant or attractive they might seem.

Now consider persons who score in the High/High quadrant. As a result of their superior cognitive abilities across the board, they should be able to identify and solve problems in a timely and efficient manner; other things being equal, they should seem to be intelligent.

Next, consider people with high scores for tactical reasoning and low scores for strategic reasoning. Such people will be "overfocused"; they will do a good job solving whatever problem is placed before them without asking if the problem has been framed correctly or if it is worth their while to work on it. It parallels thinking in the professional military, whose members

TABLE 17.2

Interpreting Tactical and Strategic Reasoning

HI	Focused Problem Solving	Good Problem Solving
TACTICAL	Poor Problem Solving	Unfocused Problem Solving
LO		
	LO **STRATEGIC** **HI**	

state that "Our job is to go where the President says to go and fight whom the President says to fight." There are many circumstances in which such an attitude is not only helpful but even essential.

Finally, consider persons with high scores for strategic reasoning and low scores for tactical reasoning. Such people will be "underfocused." They are blessed with a highly developed critical intelligence that allows them to see problems where others do not. However, they are usually content with pointing out the problems and leaving the solution to others. These people are often professional critics and gadflies; examples in academic psychology would include Paul Meehl or Hans Eysenck—people who are adept at pointing out the problems with other people's research, but unconcerned with finding solutions.

Consider next Table 17.3, which is the heart of our argument. To understand the table, we need to define the variables; specifically, we need to define the three factors of the HDS. Factor I of the HDS is a complex syndrome that can be labeled "negative affectivity" (Tellegen, 1985). High scorers see the world as a dangerous place; as a result, they are alert for signs of criticism, rejection, betrayal, or hostile intent; they are easily upset and hard to soothe. When high scorers think they have detected a threat, they react vigorously in a variety of ways to remove the threat. Low scorers are mellow, calm, and even placid.

Factor II is a complex syndrome that can be labeled "positive affectivity" (Tellegen, 1985). High scorers expect to be liked, admired, and respected; they are self-confident, self-centered, charming, attractive, and driven by their personal agendas. They expect to succeed at every undertaking, resist acknowledging their mistakes and failures (which they blame on others), and are often unable to learn from experience. Low scorers are typically modest, restrained, and even humble.

Factor III is a complex syndrome that can be labeled "restraint" (Tellegen, 1985). High scorers want to please figures of authority; as a result, they have high standards of performance for themselves and others, they work hard, pay attention to details, follow the rules, worry about making mistakes, and

TABLE 17.3

Combining Dark-Side Tendencies with
Tactical and Strategic Reasoning

HDS Factor	Tactical Reasoning	Strategic Reasoning
I: Negative Affectivity	Uneven	Defensive
II: Positive Affectivity	Careless	Overreaching
III: Restraint	Slow	Complicated

Note: HDS refers to Hogan Development Survey.
(From Hogan, R. and Hogan, J. (1997) *Hogan Development Survey Manual.* Tulsa, OK: Hogan Assessment Systems.)

are easy to supervise and popular with their bosses. Their respect for authority seems inversely related to their concern for the welfare of their subordinates. Low scorers are typically independent, skeptical of authority, and considerate of subordinates.

Now consider the first column of Table 17.3 (Tactical Reasoning). Persons with high scores on the first HDS factor and high scores for tactical reasoning will display uneven performance on routine problem-solving tasks. When they are calm, they will be able to focus and use their superior problem-solving abilities, but when they feel threatened—which happens often—they will be distracted, worried, and preoccupied, which will result in uneven performance over time.

Persons with high scores on the second HDS factor and high scores for tactical reasoning will perform routine problem-solving tasks carelessly. Their work will tend to be hasty, they will resist checking their results, and they will not be concerned about making mistakes; giving them feedback will not necessarily help the situation.

Persons with high scores on the third HDS factor and high scores for tactical reasoning will be very accurate problem solvers, but they will also tend to be slow, methodical, meticulous, and inefficient. They will often spend more time solving tactical problems than the problems may actually need or deserve.

Now consider the second column of Table 17.3 (Strategic Reasoning). Persons with high scores on the first factor of the HDS and high scores for strategic reasoning will have a bunker mentality that controls the way they scan the strategic horizon. Their cognitive style will be primarily defensive; they will be preoccupied with identifying ways in which they can be attacked, challenged, and threatened. They will approach the future in a wary and suspicious manner, expecting to be confronted with unpleasant surprises, and they will solve problems in ways that are primarily intended to minimize threats, injuries, and losses, rather than opening up future opportunities for growth.

Persons with high scores on the second factor of the HDS and high scores for strategic reasoning will engage in more or less constant overreach. Their cognitive style will be bold, confident, and boundary-spanning. They will be preoccupied with identifying opportunities for dramatic accomplishment and visible success. For example, CEOs with high scores on the second HDS factor search for acquisitions, make more acquisitions than their peers, pay more for their acquisitions than they should, and relatively more of their acquisitions ultimately fail. They approach the future in a positive and optimistic manner, expect to succeed, ignore failures, and pursue opportunities that others find lofty, grand, and risky. They solve problems in ways that open up opportunities for future success.

Persons with high scores on the third factor of the HDS and with high scores for strategic reasoning will be preoccupied with anticipating possible financial, legal, or procedural errors, and thereby avoiding external challenges, criticism, and litigation. Their cognitive style will be critical, skeptical, incisive, focused, and precise. Their general problem solving style will be to build solutions that are comprehensive, detailed, complex, and often unworkable as a result. They approach the future with a watchful eye toward minimizing exposure rather than maximizing opportunities.

The cognitive styles associated with high scores on each HDS factor tend to be effective in the short term; they persist because they yield results. For organizations using 3-month reporting cycles, this may not be a problem. Over the long term, however, making fear-based (Factor I), ego-based (Factor II), or compliance-based (Factor III) decisions will alienate external partners, internal team members, and subordinates. Ultimately, then, good judgment is a function of (a) a balance between tactical and strategic reasoning capabilities and (b) scores on the HDS that are only moderately elevated.

17.8   Conclusion

Although many businesses routinely evaluate managerial candidates using measures of cognitive ability, they tend not to be very clear about why they do this. We think they are interested in identifying good judgment. In apparent agreement, Menkes (2005) notes that most people understand that clear thinking and good judgment are important for management and business operations.

We suggest that good judgment is a function of intelligence and personality, but that the two terms have not been well defined in much of the academic literature. We propose *conceptualizing* intelligence in terms of the capacity for metarepresentation: the ability to reflect on one's performance and identify problems and performance inhibitors. We propose *measuring* intelligence in terms of two components: "problem finding" and "problem solving," where the former refers to detecting gaps, errors, and inconsistencies in assumptions, arguments, or information, and the latter refers to solving problems correctly once they are identified. Then, we refer to these kinds of thinking as strategic reasoning and tactical reasoning, respectively.

We prefer to define intelligence in terms of a typical kind of performance: performance that is adapted and appropriate to a particular context. We proposed an adaptive model composed of power, structure, and style. Power is the capacity for metarepresentation and reflects the speed with which people acquire new information. Structure refers to individual differences in problem finding and problem solving (or strategic and tactical reasoning). Style refers to individual differences in how people identify and solve problems; i.e., style refers to "cognitive style."

We proposed three factors of cognitive style that degrade or muddle thinking. The first factor involves seeing more danger in the world than actually exists. The second factor involves overestimating one's talents and capabilities and being unable to acknowledge (and learn from) one's mistakes. The third factor concerns being too conscientious, perfectionistic, and eager to please. Finally, we defined good judgment in terms of (a) a balance between problem finding and problem solving and (b) normal elevations on the three factors of cognitive style. Research is currently in progress that will allow us to evaluate these claims.

References

Binet, A. (1903). *L'etude experimentale de l'intelligence*. Paris: Schleicher Freres and Cie.

Binet, A. and Simon, T. (1905). Methodes nouvelles pour le diagnostic du niveau intellectual des anormaux. *L'Année Psychologique*, 11, 191–244.

Chein, I. (1945). On the nature of intelligence. *Journal of General Psychology*, 32, 111–126.

Deary, I., Spinath, F. M., and Bates, T. C. (2006). Genetics of intelligence. *European Journal of Human Genetics*, 14, 690–700.

Diamond, J. (1997). *Guns, Germs, and Steel*. NY: Norton.

Digman, J. M. (1990). Personality structure: Emergence of the Five-Factor Model. *Annual Review of Psychology*, 41, 417–440.

Drucker, P. F. (2006). What executives should remember. *Harvard Business Review*, 84, 145–152.

Galton, F. (1885). On the anthropometric laboratory at the late International Health Exhibition. *Journal of the Anthropological Institute*, 14, 205–219.

Gardner, H. (1983). *Frames of Mind: The Theory of Multiple Intelligences*. New York: Basic.

Goldberg, L. R. (1981). Language and individual differences: The search for universals in personality lexicons. In Wheeler, L. W. (Ed.), *Review of Personality and Social Psychology* (Vol. 2, pp. 141–165). Beverly Hills, CA: Sage.

Gottfredson, L. S. (1997). Mainstream science on intelligence: An editorial with 52 signatories, history, and bibliography. *Intelligence*, 24, 13–23.

Grigorenko, E. (2004). Is it possible to study intelligence without using the concept of intelligence? In Sternberg, R. J. (Ed.), *International Handbook of Intelligence* (pp. 170–211). New York: Cambridge University Press.

Guilford, J. P. (1957). A Revised Structure of Intellect. Reports from the psychological laboratory of the University of Southern California, No. 19.

Guttman, L. (1968). A general nonmetric technique for finding the smallest coordinate space for a configuration of points. *Psychometrika, 33*, 469–506.

Haig, B. D. (2005). An abductive theory of scientific method. *Psychological Methods, 10*, 371–388.

Hogan, J. and Holland, B. (2003). Using theory to evaluate personality and job-performance relations: A socioanalytic perspective. *Journal of Applied Psychology, 88*, 100–112.

Hogan, R. (1980). The gifted adolescent. In Adelson, J. (Ed.), *Handbook of Adolescent Psychology* (pp. 536–559). NY: Wiley.

Hogan, R. and Hogan, J. (1997) *Hogan Development Survey Manual*. Tulsa, OK: Hogan Assessment Systems.

Hogan, R. and Hogan, J. (1995). *Hogan Personality Inventory Manual* (2nd. ed.). Tulsa, OK: Hogan Assessment Systems.

Hogan, R. and Hogan, J. (2007). *Hogan Personality Inventory Manual* (3rd. ed.). Tulsa, OK: Hogan Assessment Systems.

Horn, J. L. (1994). Theory of fluid and crystallized intelligence. In Sternberg, R. J. (Ed.), *Encyclopedia of Human Intelligence* (pp. 443–451). New York: Macmillan.

Horney, K. (1950). *Neurosis and Human Growth*. New York: Norton.

Jensen, A. R. (1999). *The G Factor: The Science of Mental Ability*. Westport, CT: Praeger.

Jensen A. R. (2002). Psychometric g: Definition and substantiation. In Sternberg, R. J. and Grigorenko, E. L. (Eds.), *The General Factor of Intelligence: How General Is It?* (pp. 39–54). Mahwah, NJ: Lawrence Erlbaum.

Jensen, A. R. (2005). Mental chronometry and the unification of differential psychology. In Sternberg, R. J. and Pretz, J. (Eds.), *Cognition and Intelligence*, (pp. 26–50). Cambridge, UK: Cambridge University Press.

John, O. P. (1990). The Big-Five factor taxonomy: Dimensions of personality in the natural language and in questionnaires. In Pervin, L. A. (Ed.), *Handbook of Personality and Research* (pp. 66–100). NY: Guilford.

Judge, T. A., Colbert, A. E., and Ilies, R. (2004). Intelligence and leadership. *Journal of Applied Psychology, 89*, 542–552.

Lubinsky, D. (2004). Introduction to the special section on cognitive abilities: 100 years after Spearman's (1904) General Intelligence, objectively determined and measured. *Journal of Personality and Social Psychology, 86*, 96–111.

Luo, D., Thompson, L. A., and Detterman, D. K. (2003). The causal factor underlying the correlation between psychometric g and scholastic performance. *Intelligence, 31*, 67–83.

Lynn, R. (2006). *Race Differences in Intelligence: An Evolutionary Analysis*. Augusta, GA: Washington Summit Publishers.

Lynn, R. and Vanhanen, T. (2006). *IQ and Global Inequality: A Sequel to IQ and the Wealth of Nations*. Augusta, GA: Washington Summit Publishers.

Menkes, J. (2005). *Executive Intelligence*. New York: HarperCollins.

Neisser, U., Boodoo, G., Bouchard, T. J., Boykin, A. W., Brody, N., Ceci, S. J., Halpern, D. F., Loehlin, J. C., Perloff, R., Sternberg, R. J., and Urbina, S. (1996). Intelligence: Knowns and unknowns. *American Psychologist, 51*, 77–101.

Petrill, S. A. (2002). The case for general intelligence: A behavioral genetic perspective. In Sternberg, R. J. and Grigorenko, E. L. (Eds.). *The General Factor of Intelligence: How General Is It?* (pp. 281–298). Mahwah, NJ: Lawrence Erlbaum.

Piaget, J. (1952). *The Origins of Intelligence in Children*. Oxford, England: International Universities Press.

Plomin, R. (2003). General cognitive ability. In Plomin, R., Defries, J. C., Craig, I. W., and McGuffin, P. (Eds.), *Behavioral Genetics in the Postgenomic Era* (pp. 183–201). Washington, DC: American Psychological Association.

Poole, D., Mackworth, A., and Goebel, R. (1998). *Computational Intelligence: A Logical Approach*. New York: Oxford University Press.

Raven, J. C., Styles, I., and Raven, M. A. (1998). *Raven's Progressive Matrices*. Oxford: Oxford Psychologists Press.

Reichenbach, H. (1951). *The Rise of Scientific Philosophy*. Berkeley, CA: University of California Press.

Saucier, G. and Goldberg, L. R. (1996). The language of personality: Lexical perspectives on the Five-Factor model. In Wiggins, J. S. (Ed.), *The Five-Factor Model of Personality* (pp. 21–50). New York: Guilford.

Sedikides, C. and Skowronski, J. J. (1997). The symbolic self in evolutionary context. *Personality and Social Psychology Review, 1,* 80–102.

Shye, S. (1988). Inductive and deductive reasoning: A structural reanalysis of ability tests. *Journal of Applied Psychology, 73,* 308–311.

Spearman, C. (1923). *The Nature of Intelligence and the Principles of Cognition* (2nd ed.). London: Macmillan.

Spearman, C. (1927). *The Abilities of Man*. New York: Macmillan.

Sternberg, R. J. (1997). *Successful Intelligence*. New York: Plume.

Sternberg, R. J. and Detterman, D. K. (1986). *What is Intelligence?* Norwood, NJ: Ablex Publishing Corporation.

Tellegen, A. (1985). Structures of mood and personality and their relevance to assessing anxiety, with emphasis on self-reports. In Tuma, A. H. and Masser, D. J. (Eds.), *Anxiety and Anxiety Disorders,* (pp. 681–706). Hillsdale, NJ: Erlbaum.

Todd, P. M. and Gigerenzer, G. (2007). Environments that make us smart: ecological rationality. *Current Directions in Psychological Science, 16,* 167–171.

Thurstone, L. L. (1938). Primary mental abilities. *Psychometric Monographs,* (Serial No. 1). Chicago: University of Chicago Press.

Vygotsky, L. S. (1978). *Mind in society*. Cambridge, MA: Harvard University Press.

Webb, R. A. (1978). *Social Development in Childhood*. Baltimore, MD: Johns Hopkins University Press.

Wiggins, J. S. (1996). *The Five-Factor Model of Personality*. New York: Guilford.

Wissler, C. (1901). *The Correlation of Mental and Physical Tests*. New York: Columbia University.

18

Relational Self in Action: Relationships and Behavior

Susan E. Cross and Kari A. Terzino

CONTENTS

18.1 Gender and Relational Self-Construal..378
18.2 Measuring Relational Self-Construal 380
18.3 Relational Self-Construal and Information Processing 380
18.4 Relational Self-Construal and New Relationships 381
18.5 Relational Self-Construal, Relationships, and Well-being..................384
18.6 Culture and Relational Self-Construal 386
18.7 Associations of Similarity Scores with Relational Self-construal 389
18.8 Associations of Similarity Scores with Life Satisfaction....................390
18.9 Relationships and Multifaceted Selves..391
References 392

> Whenever two people meet there are really six people present. There is each man as he sees himself, each man as the other person sees him, and each man as he really is.

William James, 1842–1910

James's quote nicely illustrates the phenomenon of multiple selves: Each person has many facets to the self, depending on his or her social connections. Implicit in James's observation is an assumption that there is a "real self" that is somehow independent of others and social interaction. This assumption pervaded much of the research on self and identity in the twentieth century, but researchers have more recently begun to take seriously the ways that other people—especially close others and important in-groups—shape and define the self.

In this chapter, we review our program of research on this topic, which has focused on the notion of the relational self-construal. We begin with a description of the origins of this program of research, then we provide an overview

of some of the findings to derive from this work, and we close with findings
from a cross-cultural study conducted in Ukraine and the United States.

18.1   Gender and Relational Self-Construal

In our work, we have conceptualized the relational self-construal as the
inclusion of representations of significant others and relationships in the self-
concept. People who define themselves largely in terms of close, dyadic relationships may be said to have developed a highly relational self-construal.
Initially, this construct was developed as a way to explain gender differences
in behavior among men and women in North America (Cross and Madson,
1997). In a review of a wide range of gender differences in cognition, emotion, and social behavior, Cross and Madson appropriated the culture and
self framework developed by Markus and Kitayama (1991) to suggest that
differences in the nature of the self can explain observed gender differences
in behavior. Markus and Kitayama's (1991) articulation of the differences in
interdependent and independent self-construals were the starting point for
this work. Cross and Madson built upon Markus and Kitayama's distinction between independent self-construals, in which the person is viewed as
separate, unique, and autonomous, and interdependent self-construals, in
which the person is understood to be connected to others, defined by group
memberships, and intrinsically embedded in relationships.

Cross and Madson (1997) suggested that the collective or group-oriented
interdependent self-construal that describes members of East Asian cultures
does not adequately describe North American women's self-constructions.
Instead, they theorized that the type of interdependence that characterizes
women in these cultural contexts is based on close, dyadic relationships,
rather than on group memberships. At the time, they termed this the *relational-interdependent self-construal,* but typically it is referred to simply as *relational self-construal.* Both the interdependent self-construal and the relational
self-construal share the process of including others in the self, but they differ in the types of relationships that are included—in-groups versus dyadic
relationships.

How does the construct of the relational self-construal help explain
observed gender differences in behavior? Some of the earliest research on
the self demonstrated its role in information processing (Markus, 1977;
Markus, Smith, and Moreland, 1985). Based on these findings, Cross and
Madson (1997) argued that if women are more likely than men to define
themselves in terms of relationships, they should be more likely than men to
attend to close relationship partners, take their perspective in interactions,
and remember information about them. In fact, a variety of studies in the
domain of cognition and information processing support these hypotheses
(see Cross and Madson, 1997, for a description). Similarly, research on culture

and self-enhancement suggests that people who define themselves interdependently are less likely to engage in self-enhancing processes that focus on independence, uniqueness, and separation from others. Cross and Madson's survey of the gender-related research revealed that very often, women are more modest about their abilities, less likely to boast or show off to others when they have succeeded, and more likely than men to base their self-esteem on their close relationships.

Given that relationships are a central component of self-definition for persons who construct a highly relational self-construal, the development and maintenance of relationships should be a high priority for these individuals. Consequently, the cognitive and emotional processes that affect relationships should be very salient and perhaps highly developed among people with a highly relational self-construal. For example, expression of anger toward a close other may be particularly threatening to a relationship and therefore less likely to be expressed by persons with highly relational self-construals. Indeed, women are less likely to express anger than are men and are more likely to recognize the potential negative consequences of anger for close relationships. The roots of these differences can be found in early socialization; for example, mothers interact with sons and daughters differently with respect to their expressions of anger (see Cross and Madson's 1997 review for more details).

Men and women in North American contexts also differ on other relationship-related processes. Women are more likely than men to disclose sensitive or personal information about themselves, spend more time with close friends talking about personal topics than men, and are more likely than men to signal interest in a conversation partner through nonverbal means, such as smiling or showing involvement in the conversation. Cross and Madson (1997) linked these and other gender differences in interpersonal behavior to different ways that men and women construct and define the self with respect to close relationships with others.

This examination of gender differences in behavior and their hypothesized relations to the self proved fruitful in the articulation of the role of the relational self-construal in psychological processes. At the time of this 1997 review piece, however, there was little recognition of this way of defining the self and almost no method available for assessing it directly. Researchers were left with few options beyond using gender as a proxy for relational self-construal. Some researchers attempted to measure the construct with tasks such as the Twenty Statements Test, in which participants respond to the question "Who am I?" 20 times. These statements can then be coded into categories that reflect an independent self (e.g., statements about one's personality, preferences, hopes, goals, or other individual characteristics), collective self (e.g., statements about one's group memberships), and relational self (e.g., statements about one's close relationships; see Gabriel and Gardner, 1999). In general, studies using this method reveal that women are more likely than men to describe themselves in terms of close relationships (Gabriel and Gardner, 1999; McGuire and McGuire, 1982), but it is a cumbersome and unwieldy method of tapping relational self-construal. For research to move beyond examination of gender differences, it was necessary to have a new tool for measuring relational self-construal.

18.2 Measuring Relational Self-Construal

Recognizing the need to assess relational self-construal independently of gender, Cross and her colleagues (Cross, Bacon, and Morris, 2000) set about developing a measure of this construct. This measure was developed with the forms of relatedness that characterize North Americans in mind—close, dyadic relationships. This contrasts with the collective interdependence that is thought to characterize members of East Asian cultures (as described by Markus and Kitayama, 1991, and Triandis, 1989), in which people define themselves in terms of group memberships and important roles (see Singelis, 1994, for a measure of this *interdependent self-construal*). The *Relational-Interdependent Self-Construal Scale* (termed *RISC* Scale; Cross et al., 2000) assesses an individual's sense of the self as defined by close relationships, in contrast to other measures that focus on other dimensions of a relational orientation, such as the affective consequences of investment in close relationships (e.g., Kashima, Yamaguchi, Kim, Choi, Gelfand, and Yuki, 1995) or the expectation of equity or reciprocity in close relationships (e.g., communal orientation, Clark, Ouellette, Powell, and Milberg, 1987). Research with more than 4000 participants examined the reliability and validity of this new measure and its association with a variety of other personality measures, measures of well-being, and relationship-oriented measures. These studies revealed that the Relational-Interdependent Self-Construal Scale had good internal and test-retest reliability, correlated moderately positively with other measures of relatedness or communal orientation, and was uncorrelated with measures of independence or individualism. Women tended to score higher on the scale than did men; effect sizes (Cohen's *d*) across eight samples ranged from −.17 to −.57, indicating small-to-moderate gender differences in the scores. As expected, individuals who scored high on the RISC scale tended to evaluate their relationships more positively than did others, reported that they were more likely to disclose to relationship partners, were more committed to their relationships, and perceived more social support from others, compared to participants who scored low on this measure (Cross et al., 2000).

18.3 Relational Self-Construal and Information Processing

With the availability of this measure of the relational self-construal, research on the topic has begun to blossom. Initially, our focus was on the cognitive, motivational, and relational processes influenced by relational self-construal. For example, Cross, Morris, and Gore (2002) built on previous research linking the self to social cognition to investigate hypothesized links between relational self-construal and relational cognition. Classic work has shown that when individuals define themselves in terms of a domain, they tend to have positive views of themselves with respect to that domain, they pay close attention to domain relevant stimuli, and they develop elaborate memory structures for that domain. As a result, they will have better memory for information pertaining to that domain (Baumeister, 1998; Markus and Wurf, 1987). Using the Implicit Association Task (Greenwald, McGhee, and Schwartz, 1998), Cross and her colleagues found that the Relational-Interdependent Self-Construal Scale predicted positive associations among relationship terms (such as "together," "us," and "commitment"; Cross et al., 2002, Study 1). Other studies revealed that participants with highly relational self-construals were more likely than others to remember relational information about a target person and to organize information about a target person in terms of their close relationships (Cross et al., 2002; Studies 2 and 3). Thus, as earlier research on the self and social cognition suggests, highly relational self-construal serves to orient and direct the processing of relationship-relevant information.

Classic research also found that self-views influence perceptions of others; people often use their own self-schemas in thinking about other people (Markus, Smith, and Moreland, 1985). In most respects, Western psychologists have taken the view that people seek to individuate themselves from others and see themselves as distinct and different from other people (Tesser, 1988). This differentiation from others serves to affirm the independent self-construal. However, if the self is defined in terms of close others and representations of close relationships are included in the self, then characteristics of relationship partners may overlap with characteristics of the self (as Aron, Aron, Tudor, and Nelson, 1991, describe). For individuals with highly relational self-construals, perceptions of similarity with close others may therefore contribute to a sense of closeness in the relationship.

Cross et al. (2002; studies 5 and 6) examined this hypothesis in two studies of college students, who were asked to rate the descriptiveness of a variety of trait terms, ability dimensions, and values statements for themselves and for a close friend. As hypothesized, participants with highly relational self-construals were more likely to describe themselves and their friend similarly than were low relationals. Cross and her colleagues suggested that this sense of similarity promotes positive affect and a sense of belonging in the relationship for highly relational persons, and so is preferred to perceptions of the self as distinctive or very different from close others.

18.4 Relational Self-Construal and New Relationships

These information-processing functions of relational self-construal are also evidenced in interactions with relationship partners. People who define themselves in terms of close relationships should pay attention to and remember information about relationships to build and enhance their own relationships. Other behaviors and skills that facilitate the development and maintenance of satisfying relationships, such as self-disclosure and responsiveness to others' disclosures (Reis and Shaver, 1988; Reis and Patrick, 1996), should also be associated with high levels of relational self-construal. Sharing personal information with a new partner communicates that one trusts and likes the partner, and indicates openness to further closeness in the relationship (Collins and Miller, 1994). Sensitive responding to the partner's disclosures also signals caring, understanding, and interest (Reis and Patrick, 1996). Thus, we examined these relationship-enhancing processes in several studies. In one of our earliest studies using the Relational-Interdependent Self-Construal Scale, we examined the degree to which individual differences in relational self-construal were associated with self-disclosure and responsiveness to another person's disclosure in a laboratory acquaintanceship task (Cross et al., 2000; Study 3). In this study, we asked previously unacquainted pairs of women to engage in a 15-min task designed to help them become acquainted with one another (Aron, Melinat, Aron, Vallone, and Bator, 1997). After this period of asking and answering questions that revealed increasingly sensitive information about themselves, the partners were separated and completed several measures assessing the interaction and their impressions of their partner's behavior. We found that the highly relational participants tended to self-disclose more than the low relationals and that their partners described them as more sensitive and responsive to their own disclosures, compared to the partners of women with low relational self-construals. Partners of the highly relational women were also more satisfied with the interaction than were the partners of the women who reported a low relational self-construal. Thus, even in situations in which they were required to interact with a stranger, highly relational persons tended to behave in a way that affirmed a relational self: they were open and disclosing about their own experiences, and they responded sensitively and with understanding to their partner's experiences.

Laboratory studies are, admittedly, artificial environments in which to study human relationships. Thus, we followed this study with studies of actual developing relationships. Cross and her colleagues (Cross and Morris, 2003; Gore, Cross, and Morris, 2006) hypothesized that the early stages of a new relationship may be most influenced by the self-construals of the persons involved in the relationship; highly relational individuals should be more likely than those with low relational self-construals to self-disclose, to attend closely and respond sensitively to the partner's disclosures, and to remember the other's disclosures. It is relatively difficult, however, to capture the early stages of a developing relationship outside the laboratory. One context in which relative strangers are most likely to develop a relationship is in university residence halls. Very often, strangers are assigned to live together in a room or an apartment. Some of these new relationships will blossom into supportive and caring friendships; others will be barely tolerated until the end of the term, when the contract can be dissolved. Because of these external contractual constraints, and unlike other new relationships, these roommate relationships are not easily dissolved. Thus, Cross and her colleagues hypothesized that persons with a highly relational self-construal will seek to have a positive, harmonious relationship with their roommate, even if they are not destined to be close friends (Cross and Morris, 2003; Gore, Cross, and Morris, 2006). If a positive roommate relationship is self-affirming for the highly relational person, he or she will be open about his or her own concerns and listen closely to the roommate's concerns to build a positive and perhaps close relationship with the roommate.

In two studies of new college roommates, Gore et al. (2006) found that persons with highly relational self-construals were more likely to self-disclose than were lows, even when the initial closeness of the relationship was controlled. This suggests that even when a new relationship is not viewed as close or supportive, highly relational persons will tend to share personal information about themselves, perhaps to improve the relationship. Persons with highly relational self-construals were also perceived as more open and responsive by their roommates, who in turn reported higher levels of closeness in the relationship, compared to the roommates of persons with low relational self-construals. Using the same samples of new roommates, Cross and Morris (2003) discovered that highly relational participants were more likely than lows to accurately predict a new roommate's responses to a measure of their values and beliefs, indicating that they had paid close attention to and remembered their roommate's disclosures. This effect was moderated, however, by the closeness of the relationship. The association between relational self-construal and accurate prediction was strongest when the relationship was not very close; this association was weaker when the relationship was very close and satisfying. Cross and Morris (2003) suggested that in already close relationships, people listen and remember information shared by a partner because that communicates that one cares and is invested in the relationship. In contrast, when a relationship is not very close, highly relational persons will be more likely than lows to pay attention and remember the partner's disclosures because this may help them to prevent later problems, anticipate the roommate's behavior, or find insight into ways to enhance the relationship.

In summary, these studies revealed that persons with highly relational self-construals engage in a variety of behaviors that facilitate the development of new relationships, and that they tend to do this even when the relationship is evaluated as distant or not very satisfying. Cross and Morris (2003, Study 2) also found that people with highly relational self-construals seemed to have an optimistic view of their relationship with their roommate: Even when the relationship was not very close, highly relational participants were more positive in their prediction of their roommate's sense of closeness in the relationship than were those with low relational self-construals. Taken together, these studies show that relational self-construal is associated with a variety of cognitive and behavioral processes that communicate openness and interest in new relationships, which in turn contribute to partners' satisfaction with the relationship. Such relationships should affirm the highly relational person's self-views, verifying their perceptions of themselves as connected to and interdependent with close others.

18.5 Relational Self-Construal, Relationships, and Well-being

Researchers in Western cultural contexts have repeatedly found that close, satisfying relationships are an important, if not the most important, contributor to happiness and well-being (Baumeister and Leary, 1995; Myers and Diener, 1995). Given that highly relational persons are more likely than those with low relational self-construals to tend to and nurture intimate relationships with others, do they also experience higher levels of well-being as a result? The evidence thus far is mixed. Simple correlations between the Relational-Interdependent Self-Construal Scale and measures of well-being (such as life satisfaction, depression, and self-esteem) range from almost zero for depression and life satisfaction (Cross et al., 2000) to small positive associations (Cross and Morris, 2003).

If close relationships are more important for people with highly relational self-construals, perhaps the association between relationship closeness and well-being is stronger for them than for people with low relational self-construals. We examined this hypothesis in the studies of new college roommates mentioned earlier (Cross and Morris, 2003; Study 1). Cross-sectional data collected during the first half of the semester revealed that the association between closeness to one's roommate and life satisfaction was positive for persons with highly relational self-construals, but negative for those with low relational self-construals. Study 2 assessed change in life satisfaction over a 1-month period. For highly relational persons, variation in closeness to the roommate was not associated with changes in well-being, but for low relationals, high levels of closeness were associated with declines in life satisfaction. Cross and Morris (2003) hypothesized that the low relationals may perceive a very close relationship as infringing on their independence or separateness; they may also have fewer relationship skills, resulting in more problems dealing with the inevitable conflicts that occur in close living arrangements. These results point to the importance of investigating involuntary relationships (i.e., those that have not been self-selected) and their influence on well-being. In addition, these results highlight the importance of consideration of the self-construal in understanding the part close relationships play in fostering well-being.

The interactive effects of self-construal, relationships, and well-being were investigated in another series of studies that examined the importance of cross-situational consistency for persons with a highly relational self-construal (Cross, Gore, and Morris, 2003). Consistency theories play a central role in contemporary social psychological theories, including theories of attribution (Kelley, 1967), cognitive dissonance (Aronson, 1969; Festinger, 1957), and self-verification theory (Swann, Stein-Seroussi, and Giesler, 1992). Consistency plays an even larger role in personality theory. For example, Lecky (1945) and Allport (1973) argued that the integrity of the self is based on personal consistency. Others have viewed consistency as the foundation of good mental health and successful adaptation in the world (Rogers, 1959). Multiple studies reveal a positive association between self-described consistency across situations and well-being (Block, 1961; Donahue, Robins, Roberts, and John, 1993; Sheldon, Ryan, Rawsthorne, and Ilardi, 1997).

The universality of the consistency motive has been challenged by cross-cultural research, however. In particular, studies conducted in East Asian cultures reveal that consistency is not as highly valued there as in Western cultures (Kitayama and Markus, 1998). In many East Asian cultures, people are expected to adapt their behavior to fit the situation; thus, a person may behave inconsistently across situations. As a result, attempts to replicate Western studies based on consistency theories have often failed. For example, Japanese participants are much less likely than North Americans to engage in the dissonance reduction processes demonstrated in traditional cognitive dissonance paradigms (see Heine and Lehman, 1997, for a review). Suh (2002) found that Koreans described themselves less consistently across relationships than did North Americans. Furthermore, self-described consistency was less strongly related to well-being for the Koreans than for North Americans in Suh's (2002) study.

We drew upon this cross-cultural pattern in our studies, hypothesizing that consistency across relationships would be less strongly related to well-being for persons with highly relational self-construals compared to those with low relational self-construals. Behaving consistently with one's self-views, attitudes and beliefs likely affirms the internal, private, or individual self; thus, it should be strongly related to well-being for very independent, individualistic persons. However, persons with highly relational self-construals may be less likely to derive satisfaction or well-being from cross-situational consistency; instead, sources of well-being for them may be more interpersonal or relationship-oriented in nature.

Consistency in these studies was assessed by asking participants to rate themselves on a set of approximately 30 attributes (e.g., caring, selfish, sincere) in general, and then to rate the extent to which these attributes described them in each of five different self-selected close relationships (such as with close friends, parents, or romantic partners). Consistency scores were computed for each participant by conducting factor analyses of the ratings across the six types of ratings (five relationships and the self-in-general rating; see Donahue et al., 1993 and Block, 1961, for a description of this method). Three studies of college students using different measures of well-being supported our hypothesis. Across all three studies, self-reported consistency was more strongly related to well-being for persons with low scores on the relational self-construal measure than for high scorers.

In short, these studies suggest that being the same person across different relationships is not as important for the well-being of highly relational persons as has been traditionally assumed. Instead, for highly relational persons, a sense of well-being is based on close and supportive relationships with others. It is in the context of close relationships that persons with highly relational self-construals will be well, will understand and interpret their world, and will accomplish their goals (see Gore and Cross, 2006, for studies of relational self-construal and goal attainment).

18.6   Culture and Relational Self-Construal

The relational self-construal was originally conceptualized as a form of interdependence, American-style; that is, it focused on dyadic, close relationships rather than interdependence within important in-groups (which was the collectivist or interdependent self-construal that Markus and Kitayama, 1991 and Triandis, 1989, described). Others outside North American and Western European contexts, however, have argued for the importance of the concept of relational self-construal for understanding behavior in their societies. For example, Kagiticibasi (1996) has maintained that the contrast between independence and interdependence is not sufficient for understanding the self in Turkish settings. Instead, she suggests that Turkish people are more appropriately characterized by an *autonomous-relational self*, in which personal agency is understood in the context of close interpersonal relationships (rather than as separate from relationships).

Others have suggested that these three representations of the self—as independent, relational, and collective—are available to people in all societies, but that cultures vary in the extent to which they emphasize them or provide opportunities to develop each dimension (see Sedikides and Brewer, 2001). In North American society, the independent, individual self is emphasized in many social practices and traditions, whereas the relational or collective selves are encouraged and emphasized in fewer situations (see Markus and Kitayama 1994 for a description of how cultural practices and institutions create opportunities for the development of different self-construals in the United States and Japan). In Japan, educational, employment, and socialization practices often focus on the person's place in important in-groups and so train attention on the collective self.

Therefore, we have begun to examine relational self-construal and its association with psychological processes in other cultural contexts. In one study, Cross collaborated with a Ukrainian researcher, Dr. Larissa Ponomarinko, in an investigation of American and Ukrainian women's attitudes toward themselves and others. Although the purpose of the study was not to examine the relational self-construal cross-culturally, the data allowed us to examine several hypotheses regarding the relational self-construal and well-being. In specific, we investigated associations among relational self-construal, life satisfaction, and three types of self-discrepancy. Previous research has shown that discrepancies between one's ratings of one's current self and one's ideal self are negatively related to well-being (Higgins, 1987). However, other types of discrepancy—or its opposite, similarity—may also be related to self-construal and well-being. For example, just as one can use an ideal as a standard for rating the self, one can also use a negative, to-be-avoided standard of comparison (as in a feared possible self; Markus and Nurius, 1986). Thus, in this study we examined similarities between women's ratings of their actual selves and their ratings of two other representations: their conceptions of how an ideal woman would respond to the questions and their conceptions of how a woman who is to be avoided would respond to the items.

Finally, we also examined the similarities between these women's self-ratings and their ratings of how they thought their mother would respond to the items. In our previous research, we found that relational self-construal was positively related to perceptions of similarity with a close other (Cross et al., 2002). At the time, we hypothesized that for persons with a highly relational self-construal, viewing the self as very similar to a close relationship partner could serve to create close bonds or a greater sense of interdependence in the relationship. Yet our previous research was conducted in the United States, where relationships, even family relationships, may be construed as volitional and discretionary. For example, Bellah and his colleagues (Bellah, et al., 1985), in an insightful study of American individualism and commitment, described the struggle for identity this way, "...[T]he meaning of one's life for most Americans is to become one's own person, almost to give birth to oneself …. It involves breaking free from family, community, and inherited ideas." (p. 82). In short, Americans are expected to be independent of their families, and continued relatedness is based on conscious choices and cost–benefit analyses (Fiske, 1992). Thus, in the United States, relationship closeness, even between mothers and daughters, is very often viewed as an active choice that must be worked on and consciously affirmed.

In collectivist cultural contexts, relationships, especially family relationships, may not be viewed as volitional or discretionary. Instead, the individual is assumed to be embedded in relationships, which are obligatory and communal (Adams, Anderson, and Adonu, 2004; Fiske, 1992). Thus, in these cultural contexts, women may be less likely to actively seek similarity with a close relationship partner, such as their mother. Because people are ontologically embedded into their relationships with family members, which are perceived as an "inescapable fact of human existence" (Adams et al., 2004, p. 323), perceived similarity between oneself and a close relationship partner may be less strongly related to well-being among people in more collectivist, embedded cultures. Collectivism scores are not generally available for Ukraine; it was not studied in other multi-country studies of individualism and collectivism (Hofstede, 1980). It shares a cultural history with Russia, however, which has been described as having collectivist values and ideals

(Triandis, 1995). Thus, we assume that Ukraine represents a more collectivist society than the United States.

We examined American and Ukrainian young women's perceptions of themselves, an ideal woman, a woman who should be avoided, and their mother. In brief, the women were asked to respond to items that assessed attitudes toward a variety of elements of life, including attitudes toward professional life, personal development, family and private life, and civic engagement. This questionnaire was developed by Ponomarinko and her colleagues based on the Psychosemantic approach of Petrenko (1993), which seeks to identify individuals' personal meaning systems, as seen through their stereotypes, attitudes, and perceptions of others. (More information about this measure is available in Ponomarinko, 2004.) Participants were asked to estimate the likelihood of engaging in a variety of behaviors themselves (or the likelihood that their ideal woman, a woman to be avoided, and their mother would engage in these behaviors). Sample items included "To start your own business" and "To work hard at an enjoyable job, even for little pay" (professional life); "To smoke" and "To feel alone and not do anything about it" (personal development); "To share housekeeping equally with one's husband" and "To preserve old friendships after marriage, even with people one's husband doesn't like" (family and private life); and "To raise money for a charity cause" and "To participate in a strike or official protest against a governmental policy decision with which you disagree" (civic engagement). After rating their own attitudes on these items, the participants were asked to rate the items from three other perspectives: how their ideal woman would rate them; how a woman they would seek to avoid would rate them; and how they thought their mothers would rate them. We created within-person correlations of the self-ratings with each of the other ratings across all the items, resulting in indices of the similarity of self-ratings with the ideal woman ratings, the woman-to-avoid ratings, and the mother ratings. In addition, participants completed our measure of the Relational-Interdependent Self-Construal (RISC, Cross, et al., 2000) and a widely used measure of well-being, the Satisfaction with Life scale (Diener, Emmons, Larson, and Griffen, 1985). Reliability of these measures was acceptable in both countries (alphas ranged from .63 to .87).

First, we examined mean scores of the U.S. and Ukrainian participants on these measures. We found significant differences between the groups on the RISC and life satisfaction scales, with U.S. participants scoring higher on relational self-construal and reporting greater life satisfaction than Ukrainian participants (see Table 18.1). We also found cultural differences on the similarity ratings. U.S. participants' self-ratings were marginally more similar to their ratings of their ideal self than were the self-ratings of the Ukrainian participants. U.S. participants evaluated themselves as less similar to the self they would like to avoid than Ukrainian participants. Finally, U.S. participants evaluated themselves and their mothers significantly more similarly than did Ukrainian participants.

We suspect that the cultural difference in mean scores on the RISC scale may be attributable to differences in the extent to which relationships are

TABLE 18.1

Means and Standard Deviations of the Variables Used in the Study of U.S. and Ukrainian College Women

Variable	U.S. Participants (N = 228)	Ukraine Participants (N = 197)
Relational-Interdependent Self-Construal Scale	4.08a(0.59)	3.63b(0.51)
Life Satisfaction Scale	3.64a(0.81)	3.15b(0.87)
Myself-Ideal Similarity	0.54a+(0.23)	0.50a+(0.22)
Myself-Avoid Similarity	−0.43a(0.25)	−0.12b(0.37)
Myself-Mother Similarity	0.40a(0.30)	0.34b(0.22)

Note: Similarity scores derived from within-person intraclass correlations. Values with different subscripts are significantly different, $ps \leq .03$. Subscript a+ represents marrginal significance ($p = 0.06$). Standard deviations are in parentheses.

viewed as inescapable, de facto aspects of social life or as volitional bonds that are to be negotiated or created by the individual. In the U.S., an individual is assumed to be independent of social ties and to define the self primarily on the basis of individual attributes, choices, abilities, and preferences. In this cultural context, individuals who have defined the self in terms of close relationships will likely view themselves as relatively distinctive compared to their social groups and the cultural norm, and so may rate the self fairly extremely on a measure of relational self-construal (when a typical Likert scale with anchors such as *strongly disagree* and *strongly agree* are used). In other cultural contexts, to be connected to others and to define oneself in terms of relationships with others may be taken for granted or assumed. When asked to evaluate the self on items that tap this construct, individuals in these cultural contexts will likely compare themselves to others within their group and rate themselves relatively moderately on this dimension. (See Heine, Lehman, Peng, and Greenholtz, 2002, for further description of the influence of reference group effects in cross-cultural research.) Thus, in this and other research (e.g., Oyserman, Coon, and Kemmelmeier, 2003), expected cultural differences in constructs such as relational self-construal or collectivism may not be obtained when people rate themselves using ordinary Likert-scaled items. In contrast, the similarity scores in this study, which are derived from intraclass correlations, should be less influenced by reference group effects or other such biases.

18.7 Associations of Similarity Scores with Relational Self-construal

We examined the relation of the relational self-construal measure with the similarity measures. Interestingly, the myself-ideal similarity score was

TABLE 18.2

Correlations Among Variables in the Study of Ukrainian and American Women's Self-Evaluations and Evaluations of Others

Variable	1.	2.	3.	4.	5.
1. RISC	—	.26***	.03	−.17*	.27***
2. Life Satisfaction	.16*	—	.24***	−.15*	.18**
3. Myself-Ideal Similarity	.08	.16*	—	−.31***	.16*
4. Myself-Avoid Similarity	−.06	-.11	−.14*	—	-.16*
5. Myself-Mother Similarity	.07	.00	.14	-.05	—

Note: Correlations above the diagonal represent U.S pparticipants; correlations below the diagonal represent Ukrainian participants. RISC = Relational-Interdependent Self-Construal Scale.

* $p < .05$; ** $p < .01$; *** $p < .001$.

not associated with relational self-construal for either group of women (see Table 18.2). There was a small negative association between the RISC scale and the myself-avoid ICC for U.S. participants ($r = -.17$, $p < .05$), but no significant association for Ukrainian participants ($r = -.06$). The only significant cultural difference was in the association between relational self-construal and myself-mother similarity scores. This relation was stronger among the Americans ($r = .27$, $p < .01$) than among the Ukrainian women ($r = -.05$, *ns*). This finding replicates earlier work that showed a correlation between RISC scale scores and self-other similarity for the Americans, but not for the Ukrainian women.

18.8   Associations of Similarity Scores with Life Satisfaction

Research in North American cultural contexts indicates that similarity between a person's current self-views and their ideals for themselves is positively related to well-being (Higgins, 1987). In this study, myself-ideal similarity was positively related to life satisfaction for both groups, although this association was somewhat stronger for U.S. participants. There was not a significant cultural difference in the correlation between myself-avoid similarity and life satisfaction. In contrast, we found a cultural difference in the association between life satisfaction and myself-mother similarity scores; this relation was significantly stronger for U.S. participants ($r = .18$, $p < .01$) than for Ukrainian participants ($r = .00$).

Why would perceived similarity between the self and one's mother be more strongly related to life satisfaction for the U.S. sample than the Ukraine sample? It is possible that Americans affirm the mother–daughter relationship by viewing themselves as much like their mothers, which in turn creates a sense of acceptance and belonging. Other studies using North American populations show that perceived similarity is related to closeness in relation-ships (e.g., Tesser et al., 1998). In contrast, a study by Heine and Renshaw (2002) demonstrated that the relation between similarity and liking or close-ness found in the West is not as strong among Japanese participants. Perhaps similarity with another individual is not as important in the nature of the relationship in collectivistic cultures as it is in individualistic cultures. In individualistic cultures, similarity may serve to enhance the self, or form the foundation for a bond, whereas in collectivistic cultures, the bond exists outside the two people (in the *nature* of the relationship). In these cases, the individuals are already connected by the relationship itself, and perceived similarity is not necessary to create or strengthen relational bonds.

In summary, this study revealed both similarities and differences in the self-evaluations and evaluations of others among Ukrainian and American college women. Most notably, the similarity of the self-evaluations and the evaluations of one's mother were more strongly correlated with relational self-construal and life satisfaction for the American students than for the Ukrainian students. This suggests important cultural differences in the understanding of close relationships, such as in the degree to which relation-ships are volitional, negotiated, and discretionary (as in the United States) versus embedded, involuntary, and existing independently of the person's construction of them (as in many collectivist cultures; Adams et al., 2004). There may be other distinctions in the ways that close family relationships are construed in different cultural contexts that are importantly related to self-construals and well-being. Research on cultural viewpoints on close relationships has unfortunately lagged behind the research on the self, but given the central role of close relationships in the construction of the self and in the maintenance of well-being, culturally informed research on close relationships should be given high priority.

18.9   Relationships and Multifaceted Selves

In closing, we return to James's quote about the self and social interaction. He posits that in each interaction, there is a "real self" that somehow is separate from the selves that are presented or perceived in social interactions. What James does not consider is the self that is defined and constructed *within* social interactions, especially social interactions with stable and close relationship partners. Our research has sought to take this *relational* self seriously, and to examine the ways that relational self-definition influences behavior.

Most of this work has focused on relational self-construal among North American participants, so the value of this concept for understanding the selves of members of other cultures is still unknown. Unfortunately, explicit measures of self-definition, such as the Relational-Interdependent Self-Con-strual scale (Cross et al., 2000), may not be effective instruments in contexts in which the relational self is the de facto, taken for granted, or assumed norm for most persons. Instead, in these cultural situations, researchers may need to develop more indirect or implicit measures of self-definition. In addition, there may be other important dimensions of self-definition that have largely gone unnoticed by researchers, but that markedly influence behavior. Collaborations among researchers from different cultures can provide important insight from within and observations from outside cultures, resulting in new theories and concepts that illuminate the varieties of selves in social situations.

References

Adams, G., Anderson, S. L., and Adonu, J. K. (2004). The cultural grounding of closeness and intimacy. In Mashek, D and Aron, A. (Eds.), *The Handbook of Closeness and Intimacy*, (pp. 321–339). Mahwah, NJ: Lawrence Erlbaum Associates.

Allport, G. W. (1937). *Personality: A Psychological Interpretation*. New York: Henry Holt.

Aron, A., Aron, E. N., Tudor, M., and Nelson, G. (1991). Close relationships as including other in the self. *Journal of Personality and Social Psychology*, 60, 241–253.

Aron, A., Melinat, E., Aron, E. N., Vallone, R. D., and Bator, R. J. (1997). The experimental generation of interpersonal closeness: A procedure and some preliminary findings. *Personality and Social Psychology Bulletin*, 23, 363–377.

Aronson, E. (1969). A theory of cognitive dissonance: A current perspective. In Berkowitz, L. (Ed.), *Advances in Experimental Social Psychology* (Vol. 4, pp. 1–34). New York: Academic Press.

Baumeister, R. F. (1998). The self. In Gilbert, D., Fiske, S. T., and Lindzey, G. (Eds.), *Handbook of Social Psychology* (4th ed., pp. 680–740). Boston: McGraw-Hill.

Baumeister, R. F. and Leary, M. R. (1995). The need to belong: Desire for interpersonal attachment is a fundamental human motivation. *Psychological Bulletin*, 117, 497–529.

Bellah, R. N., Madsen, R., Sullivan, W. M., Swidler, A., and Tipton, S. M. (1985). *Habits of the Hear: Individualism and Commitment in American Life*. Berkeley, CA: University of California Press.

Block, J. (1961). Ego-identity, role variability, and adjustment. *Journal of Consulting and Clinical Psychology*, 25, 392–397.

Clark, M. S., Ouellette, R., Powell, M. C., and Milberg, S. (1987). Recipient's mood, relationship type, and helping. *Journal of Personality and Social Psychology*, 53, 94–103.

Collins, N. L. and Miller, L. C. (1994). Self-disclosure and liking: A meta-analytic review. *Psychological Bulletin*, 116, 457–475.

Cross, S. E., Bacon, P., and Morris, M. (2000). The relational-interdependent self-construal and relationships. *Journal of Personality and Social Psychology*, 78, 791–808.

Cross, S. E. and Madson, L. (1997). Models of the self: Self-construals and gender. *Psychological Bulletin*, 122, 5–37.

Cross, S. E. and Morris, M. L. (2003). Getting to know you: The relational self-construal, relational cognition, and well-being. *Personality and Social Psychology Bulletin*, 29, 512–523.

Cross, S. E., Morris, M. L, and Gore, J. (2002). Thinking about oneself and others: The relational-interdependent self-construal and social cognition. *Journal of Personality and Social Psychology*, 82, 399–418.

Diener, E., Emmons, R. A., Larsen, R. J., and Griffin, S. (1985). The Satisfaction with Life Scale. *Journal of Personality Assessment*, 49, 71–75.

Donahue, E. M., Roberts, R. W., Roberts, B. W., and John, O. P. (1993). The divided self: Concurrent and longitudinal effects of psychological adjustment and social roles on self-concept differentiation. *Journal of Personality and Social Psychology*, 64, 834–846.

Festinger, L. (1957). *A Theory of Cognitive Dissonance*. Stanford, CA: Stanford University Press.

Fiske, A. (1992). The four elementary forms of sociality: Framework for a unified theory of social relations. *Psychological Review*, 99, 689–723.

Gabriel, S. and Gardner, W. L. (1999). Are there his and her types of interdependence? The implications of gender differences in collective and relational interdependence for affect, behavior, and cognition. *Journal of Personality and Social Psychology*, 75, 642–655.

Gore, J. S. and Cross, S. E. (2006). Pursuing goals for us: Relationally-autonomous reasons in long-term goal pursuit. *Journal of Personality and Social Psychology*, 90, 858–861.

Gore, J. S., Cross, S. E., and Morris, M. L. (2006). Let's be friends: The relational self-construal and the development of intimacy. *Personal Relationships*, 13, 83–102.

Greenwald, A. G., McGhee, D. E., and Schwartz, J. L. K. (1998). Measuring individual differences in implicit cognition: The implicit association test. *Journal of Personality and Social Psychology*, 74, 1464–1480.

Heine, S. J. and Lehman, D. R. (1997). The cultural construction of self-enhancement: An examination of group-serving biases. *Journal of Personality and Social Psychology*, 72, 1268–1283.

Heine, S., Lehman, D., Peng, K., and Greenholtz, J. (2002). What's wrong with cross-cultural comparisons of subjective Likert scales? The reference-group effect. *Journal of Personality and Social Psychology*, 82, 903–918.

Heine, S. J. and Renshaw, K. (2002). Interjudge agreement, self-enhancement, and liking: Cross-cultural divergences. *Personality and Social Psychology Bulletin*, 28, 578–587.

Higgins, E. T. (1987). Self-discrepancy: A theory relating self and affect. *Psychological Review*, 94, 319–340.

Hofsteade, G. (1980). *Culture's Consequences: International Differences in Work-Related Values*. Beverly Hills, CA: Sage.

Kagitcibasi, C. (1996). The autonomous-relational self: A new synthesis. *European Psychologist*, 3, 180–186.

Kashima, Y., Yamaguchi, S., Kim, U, Choi, S. C., Gelfand, M. J., and Yuki, M. (1995). Culture, gender, and self: A perspective from individualism-collectivism research. *Journal of Personality and Social Psychology*, 69, 925–937.

Kelley, H. H. (1967). Attribution theory in social psychology. In Levine, D. (Ed.), *Nebraska Symposium on Motivation*, (Vol. 15, pp. 192–240). Lincoln, NE: University of Nebraska Press.

Kitayama, S. and Markus, H. R. (1998). Yin and yang of the Japanese self: The cultural psychology of personality coherence. In Cervone, D. and Shoda, Y. (Eds.), *The Coherence of Personality: Social Cognitive Bases of Personality Consistency, Variability, and Organization*, (pp. 242–302). New York: Guilford.

Lecky, P. (1945). *Self-Consistency: A Theory of Personality*. New York: Island Press.

Markus, H, (1977). Self-schemata and processing information about the self. *Journal of Personality and Social Psychology*, 35, 63–78.

Markus, H. and Kitayama, S. (1991). Culture and the self: Implications for cognition, emotion, and motivation. *Psychological Review*, 98, 224–253.

Markus, H. R. and Kitayama, S. (1994). A collective fear of the collective: Implications for selves and theories of selves. *Personality and Social Psychology Bulletin*, 20, 568–579.

Markus, H. and Nurius, P. (1986). Possible selves. *American Psychologist*, 41, 954–969.

Markus, H., Smith, J., and Moreland, R. L. (1985). Role of the self-concept in the perception of others. *Journal of Personality and Social Psychology*, 49, 1494–1512.

Markus, H. and Wurf, E. (1987). The dynamic self-concept: A social psychological perspective. *Annual Review of Psychology*, 38, 299–337.

McGuire, W. J. and McGuire, C. V. (1982). Significant others in self space: Sex differences and developmental trends in social self. In Suls, J. (Ed.), *Psychological perspectives of the self*, (Vol. 1, pp. 71–96). Hillsdale, NJ: Erlbaum.

Myers, D. G. and Diener, E. (1995). Who is happy. *Psychological Science*, 6, 10–19.

Oyserman, D., Coon, H. M., and Kemmelmeier, M. (2002). Rethinking individualism and collectivism: Evaluation of theoretical assumptions and meta-analyses. *Psychological Bulletin*, 128, 3–72.

Petrenko, V. (1993) Meaning as a unit of consciousness. *Journal of Russian and East European Psychology*, 31, 5–19.

Ponomarenko, L. (2004). Peculiarities of the Ukrainian and US' women social representations (cross-cultural analysis) (in Ukrainian). *Social Psychology*, 1.

Reis, H. T. and Patrick, B. C. (1996). Attachment and intimacy: Component processes. InHiggins, E. T. and Kruglanski, A. W. (Eds.), *Social Psychology: Handbook of Basic Principles* (pp. 523–563). New York: Guilford Press.

Reis, H. T. and Shaver, P. (1988). Studying social interaction with the Rochester Interaction Record. In Zanna, M. P. (Ed.), *Advances in Experimental Social Psychology* (Vol. 24, pp. 269–318). San Diego, CA: Academic Press.

Rogers, C. R. (1959). A theory of therapy, personality, and interpersonal relationships as developed in the client-centered framework. In Koch, S. (Ed.), *Psychology: A Study of a Science*, (Vol. 3, pp. 184–256). NY: McGraw-Hill.

Sedikides, C. and Brewer, M. B. (Eds.). (2001). *Individual Self, Relational Self, Collective Self*. Philadelphia: Psychology Press.

Singelis, T. M. (1994). The measurement of independent and interdependent self-construals. *Personality and Social Psychology Bulletin*, 20, 580–591.

Sheldon, K. M., Ryan, R. M., Rawsthorne L. J., and Ilardi, B. (1997). Trait self and true self: Cross-role variation in the big-five personality traits and its relations with psychological authenticity and subjective well-being. *Journal of Personality and Social Psychology*, 73, 1380–1393.

Suh, E. M. (2002). Culture, identity consistency, and subjective well-being. *Journal of Personality and Social Psychology*, 83, 1378–1391.

Swann, W. B., Stein-Seroussi, A., and Giesler, R. B. (1992). Why people self-verify. *Journal of Personality and Social Psychology*, 62, 392–401.

Tesser, A. (1988). Toward a self-evaluation maintenance model of social behavior. In Berkowitz, L. (Ed.), *Advances in Experimental Social Psychology* (Vol. 21), 181–227. New York: Academic Press.

Tesser, A., Beach, S. R. H., Mendolia, M., Crepaz, N., Davies, B., and Pennebaker, J. (1998). Similarity and uniqueness focus: A paper tiger and a surprise. *Personality and Social Psychology Bulletin,* 24, 1190–1204.

Triandis, H. C. (1989). The self and social behavior in differing cultural contexts. *Psychological Review,* 96, 506–520.

Triandis, H. C. (1995). *Individualism and Collectivism.* Boulder, CO: Westview.

Index

A

ACC, *see* Anterior cingulate cortex
Activity
  goal, vocational training, 55
  leading, 260
  notion of accepted and unexplained
    variability, 232
  object of, 247
  performance, learning and, 57
  psychic determination of, 303
  self-regulation, vocational training
    and, 46
  suprasituational level of its
    determination, 307
Activity, discourse in, 247–266
  activity theory model, 249, 250, 251
  agencies of cultural reproduction,
    254
  artifacts, 249
  Bernstein, 252–264
    classification, 253
    contradiction and change in
      transmission practices, 259–263
    culture and language, 255–258
    discursive hybridity, 263–264
    framing, 254–255
    organizational practices, 253
    social positioning, 258–259
  change in transmission practices, 259
  classification, 253
  communities of practice approach,
    254
  cultural artifact, 255, 256
  culture, 255
  discourse–subject relation, 252
  discursive hybridity, 263
  division of labor, 253
  ecosocial systems, 259
  Engeström's interpretation of activity
    theory, 248–251
    first generation, 249
    second generation, 249–250
    third generation, 250–251

  framing, 254
  horizontal discourse, 257
  individual experience analysis, 249
  instability and contradiction, 250
  internal contradictions, 250
  internal relationship, 257
  language of description, 248
  leading activity, 260
  mediated action, 249
  microstudies, 260
  networks of activity systems, 262
  object of activity, 247
  object-oriented activity, 248
  order of discourse, 260–261
  pedagogic discourse, 256
  principles of control, 252
  professional identities, 263
  regulative discourse, 256
  rules of cultural historical formation,
    248
  Russian tradition, 248
  social positioning, 252, 253, 258
  stages of mental development, 260
  subject–object relations, 251
  virtual collaboration, 262
Activity theory (AT), 221, 247
  Marxist philosophy and, 221, 222
  model
    first-generation, 249
    second-generation, 250
    third-generation, 251
  operational, 230
Activity theory (comparative analysis
    of Eastern and Western
    approaches), 221–245
  action style, 240
  action theory, 240–242
  activity analysis system of units, 224
  activity composition, 225
  activity decomposition, 234
  activity structure, 225
  algo-heuristic method, 239
  algorithmic analysis, 230–231, 233

approximate theory of action, 241
AT learning, 238–239
behavioral actions, 229
cognitive actions, 229
consciousness, 223
control of behavior, 227
cultural meanings, 222
decision-making actions, 229
ecological psychology, 242
feedback, 241
functional analysis, 236
general activity theory, 222–227
goal-directed activity, 230
goal-oriented behavior, 240
limitations, 242
mediation of cognition, 227
mental activity, 226
mental acts, 231
mental development, 239
motive–goal vector, 228, 241
notion of accepted and unexplained
        variability, 232
object-oriented activity, 223
operational AT approach, 230
personality principle, 226
Piaget's theory, 239
principle of creative activity, 223
production rules, 238
psychological units of analysis, 234
quasi-algorithmic description, 233
self-regulation, 236, 241
sign-mediated nature of activity, 227
signs as mental tools, 222, 223
SSAT learning, 239
structure of learning, 239
subject–object relationship, 229
subject-oriented activity, 224
symbol-oriented-activity, 224
System Structural Activity Theory,
        230–238
technological unit, 235
Western AT, 227–230
ACT-R
    architecture, 33
    hierarchy of movements, 36
    knowledge types, 34
    model of human driving
            performance, 37–38
    module types, 34
    motor module, 35

system for issuing motor commands,
        36
AFM, *see* Method of additive factors
Alertness
    gold standard for assessment of, 170
    quantification, 172
Alertness and Memory Profiling system
        (AMP), 176, 177
Ambiguity intolerance index, 315
AMP, *see* Alertness and Memory
        Profiling system
Anterior cingulate cortex (ACC), 205,
        208
Assembly workers, disability with neck
        injury, 85
AT, *see* Activity theory
Attention, selection for action, error
        processing, and safety, 203–218
    alert attention network, 205
    attention disengagement, 213
    attention networks, 205–206
        disturbance in, 203, 206
        errors, 206–207
    automation cuing, 213
    brain–computer communication, 213
    brain functional–anatomic neural
            networks, 205
    brain mapping studies, 209
    conflict processing, 210
    covered shift of attention, 207
    efficiency of attention, 204
    error positivity, 212
    error-related negativity, 210
    executive attention model, 207–208
    executive network and anterior
            cingulate cortex, 207–214
        EEG studies, 210–212
        human error, ERN, Pe, feedback
            ERN, and neuroadaptive
            interfaces, 213–214
    executive neural networks, 213
    fight–flight conditions, 212
    hidden risk, 204
    knowledge disruption, 198
    medial frontal negativity, 211
    neural indices of cognitive workload,
            213
    neuroadaptive interface states, 213
    neuronal attention system, 207

neuronal substrate for error
   detection, 214
routine actions, 212
saccadic eye movements, 207
scanning process, 207
Stroop-like tasks, 209
task demands, 204
technological progress, attention,
   and errors, 204–205
tunnel vision, 207
Augmented Cognition, 179–180

B

B-Alert® system, 171, 172
Bar-On Emotional Quotient Inventory,
   318
BCPE, *see* Board of Certification in
   Professional Ergonomics
Behavior, *see* Relational self in action
Board of Certification in Professional
   Ergonomics (BCPE), 4
Brain activation maps, neural
   processing and, 144–152
   modular processes in number
      comparison, 145–146
   modular processes for stimulus
      encoding and response
      selection, 146–152
Brain, mind, and machine (new
   interface of), 167–188
   Alertness and Memory Profiling
      system, 176, 177
   alertness quantification, 172
   augmenting human capacity, 179–180
   B-Alert® system, 171, 172
   brain–behavior relationships, 179
   circadian effect, 172
   computer-based models, 180
   detection of drowsiness, 169–173
   differing susceptibility of
      individuals to effects of sleep
      deprivation, 174–175
   EEG-based feedback alarms and
      driving performance, 175
   EEG metrics for engagement, 178
   effectiveness of feedback alarms, 175
   elevated distraction, 181
   engagement levels, 182

future of education, 180–182
   group dynamics, 182
   intersection of brain, mind, and
      machine, 167–169
   learning styles, 180
   Maintenance of Wakefulness Test,
      172
   neural prosthetics, 168
   neuroassays for alertness and
      memory, 176–177
   neuroergonomics, 168, 179
   neuropharmacoassays, 178–179
   NeuroTeam, 182–184
   nicotine withdrawal, 178
   paired-associate learning/memory
      task, 172
   sleep debt, 170
   sleep-technician-observed
      drowsiness, 172
   social neuroscience, 168
   venerable history of EEG, 169
   wavelets eye blink identification
      routine, 172
   workload dynamics, 181
Burnout syndrome, 341, 345, 347, 350

C

Cane, movements of prior to locomotion
   judgments, 267–300
   abstract perception of probes, 294
   actor decisions, 296
   arm–rod system posture, 287
   cane aberrations, 294
   center of tapping, 282, 283, 289, 293
   change of direction, 279, 285
   cognitive impenetrability of
      important perceptual
      processes, 271
   cognitive inaccessibility, 271
   complementary attitudes, 271
   data pitfalls, 272
   electronic devices, 274
   experiment 1, 277–287
      data analysis, 279
      discussion, 286–287
      method, 278–279
      results, 280–286
   experiment 2, 287–290

data analysis, 288
 discussion, 289–290
 method, 287–288
 results, 288–289
experiment 3, 291–297
 discussion, 294–297
 method, 292
 results, 293–294
exploration with cane, 275–297
exploratory taps, 284
exploratory variables, 290
eye movement research, 270
gap-crossing experiments, 277, 280
gap–probe combination, 286
GERIATRIX board game, 273
high-tech travel aids, 268
historical analysis, 268
inclination of cane instructors, 275
information pickup, 276
intellectual impediment, 269
limitations of prior research on
 nonvisual locomotion, 270–275
manner of exploration, 290
noise level, 275
patterns of exploration, 276
perception, 270
perceptual illusions, 276
perceptual manipulations, 294
preferred movement, 283
probe manipulations, 295
probe weighting, 290
psychological curricula, 272
shorelining, 277
stereotype labels, 273
technique of blind equestrienne, 272
text-reading machines, 274
training assumption, 274
T-shaped rod, 291
video analysis, 279
weighted-branch manipulation, 295
Carpal tunnel syndrome (CTS), 73
Chair design
 improvement of, 10
 seat height adjustability, 12
Clear-mindedness, 358
Cognition
 mediation of, activity theory and, 227
 role of anterior cingulate cortex in,
  209
Cognitive emotional ability, 306

Cognitive ergonomics, CWA and, 23
Cognitive style, 373
Cognitive work analysis (CWA), 21
 effectiveness in cognitive
  ergonomics, 23
 first stage of, 22
 social organization and cooperation
  stage of, 22
 strategies analysis, 22
 work domain analysis, 22
Computational intelligence, scientific
 goal of, 361
Computer test performance, 99
Constraints, classes of, 18
Construct operationalization, *see*
 Emotional intelligence
Continuous positive airway pressure
 (CPAP), 177
Control task analysis, 22
CPAP, *see* Continuous positive airway
 pressure
Creative activity, principle of, 223
CTS, *see* Carpal tunnel syndrome
CWA, *see* Cognitive work analysis

D

Database, of movements, 32
Declarative knowledge, 34, 43
Dexterity, 18
DHMs, *see* Digital human models
Digital human models (DHMs), 29, 30,
  see also Virtual product design,
  integrating cognitive and
  digital human models for
Discovery learning, AT and, 239
Donders' subtraction method, 128–129,
 154
Driving performance
 ACT-R model of, 37–38
 EEG-based feedback alarms and, 175
Drowsiness, detection of, 169

E

Ecological ergonomics, 3–28
 aesthetics, 24
 ANSI/HFES Standard, 13
 buttock-popliteal length, 15

cause-effect paradigm, 5
clearance, 22
cognitive work analysis, 21, 22
constraint classes, 18
control task analysis, 22
cubist posture, 9
definition of comfort, 19
dexterity, 7, 18
discomfort, 23
discovery and exchange of
    knowledge, 4
ecological approach to psychology, 7
elbow rest height, 12
emotions associated with ease, 24
ergonomic disorders, 8
ergonomic-related injuries, 4
ergonomics identity crisis, 3–6
ergonomics problem and strength,
    6–8
example (seated work posture), 8–15
    interdependencies in seated
        posture, 9–15
    relationship between working
        conditions and disorders, 8–9
framework for ecological research
    and practice, 20–24
    cognitive work analysis as tool for
        integration, 21–23
    epistemological issues, 23–24
HF/E, 3, 5
implementing fit with multiple
    degrees of freedom, 15–17
means-end abstraction hierarchy, 21
MEPS study, 16
misconceptions, 5
necessity for ecological approach,
    17–20
OCRA, 17
office work with VDTs, 9
part-whole decomposition, 21
perception–action cycle, 18, 22
person–environment
    complementarity, 17
physical form, 21
physical function, 21
popliteal height, 12
posture–workstation fit, 22
purpose-related function, 22
relationship between people and
    technology, 3

RULA, 17
seated eye height, 14
symvatology, 6, 20
system functional purpose, 22
task microstructure, 19
task-related postural requirements,
    15
trading zones, 24
work domain analysis, 21
worker competencies, 22
Ecosocial systems, 259
ECTs, *see* Elementary cognitive tasks
EEG, *see* Electroencephalogram
EI, *see* Emotional intelligence
Elbow rest height, 12
Electroencephalogram (EEG), 168
    -based feedback alarms, driving
        performance and, 175
    future application of in classroom,
        183
    future application of in corporations,
        183
    history of, 169
    quantification of, 171
Electronic assembly workers, 72, *see
    also* Workstation design and
    testing, laws of ergonomics
    applied to
Electronic travel aid (ETA), 274, 275
Elementary cognitive tasks (ECTs), 366
Emotion(s)
    experienced, 331
    indicators of, 331, 332
    regulation, psychological
        components of, 329
    typical for education process, 330
    unidentified, 322
Emotional burnout, 341, 345, 347, 350
Emotional competence, components of,
    326
Emotional intelligence (EI), 303–324
    ambiguity intolerance index, 315
    assessment criteria, 322
    avoidance strategy, 316
    awareness of positive life
        functioning, 318
    Bar-On Emotional Quotient
        Inventory, 318
    choice spatial structure, 312
    cluster-analysis statistics, 315

cognitive emotional ability, 306
coping strategies, 314
deeds, 313
Diaries of Emotions, 319
dynamic unity, 307, 309
ego control, 315
emotional functioning hierarchy, 311
emotional processes, 314
emotional self-efficacy, 306
evaluative attitudes to reality, 313
exteriorization of experience, 310
formation, mechanism of, 310
frequency of positive emotions, 320
future research, 321–323
hierarchy of levels, 322
impulsive acts, 313, 319
information processing, 305, 323
intensity of emotions, 321
interiorization of experience, 310
internal and external components of
      activity, 312–314
internal ontological facets, 315
internal phenomenological facets of,
      315, 317
mechanism of EI formation, 310
model, 305
nature of choices, 321
observation data, 319
optimism, 309
perceptual appraisal, 303
personality dispositional component,
      321
personality traits, 304, 315, 321
positive psychological functioning,
      316
problem–focused strategy, 316
psychic determination of activity, 303
psychic reflection of reality, 308
psychological adaptability, 318
psychological categories, 311
psychological well-being, 314
relevance of approach to
      operationalization of, 314–316
resistance to stress, 318
results and discussion, 316–321
scores, 318
self-confidence, 316
self-driven activity, 307
self-esteem, 314
self-regulation, 313

self-scaling of emotional intensity,
      321
sensitivity to stressful factors, 318
structural units, 310
unidentified emotions, 322
unity of external and internal in
      psychic determination of
      behavior, 307–312
voluntary acts, 313
Emotional process, degree of mediation
      of, 314
Emotional regulation, *see* Learning
      process, emotional regulation
      of
Emotional self-efficacy, 306
Emotional tension, 52
EQ, *see* Emotional intelligence
Ergonomics
   categories, 4
   cognitive, CWA and, 23
   first law of, 74, 76
   focus of, 5
   reductionist approach, 17
   second law of, 75, 76
   third law of, 76
   workstation design and, 74
ERN, *see* Error-related negativity
ERP, *see* Event-related potential
Error
   operator, analysis of, 89
   positivity, 212, *see also* Attention,
         selection for action, error
         processing, and safety
   processing, *see* Attention, selection
         for action, error processing,
         and safety
   -related negativity (ERN), 210, 212
ETA, *see* Electronic travel aid
Event-related potential (ERP), 116, 136,
      210
   amplitude, modular processes, 158
   attention and, 210
   semantic satiation, 158
Exploration, cane, *see* Cane, movements
      of prior to locomotion
      judgments
Eye Gaze Development System, 78, 80
Eye movement research, cane
      movements and, 270

F

Feedback alarms, effectiveness of, 175
FFM, *see* Five-Factor Model
Fitness-for-work, operator, *see* Operator functional state and fitness-for-work, day-to-day monitoring of
Fitts' law, 35, 50
Five-Factor Model (FFM), 368
fMRI, *see* Functional magnetic resonance imaging
FSA, *see* Functional system of activity
Functional magnetic resonance imaging (fMRI), 136, 145, 168, 203
Functional state, operator, *see* Operator functional state and fitness-for-work, day-to-day monitoring of
Functional system of activity (FSA), 98
Functional systems, Anokhin's theory of, 94

G

Gaussian model, equal-variance, 125
GERIATRIX board game, 273
Goal
    -directed activity, SSAT and, 230
    -directed self-regulative process, 43, 68
    formation
        AT focus on, 239
        process, functional model of, 237
    -oriented behavior, activity theory and, 240
    -related motivation, vocational training and, 64
Good judgment, 357–375
    abduction, 364
    academic study of intelligence, 359–361
    adaptive model of intelligence, 366–368
    business reasoning, 359
    clear-mindedness, 358
    cognitive styles, 372, 373
    conceptualizing intelligence, 372
    context of discovery, 365
    convergent thinking, 365
    defining personality, 368–369
    distinction between inductive and deductive reasoning, 364
    divergent thinking, 365
    education, 364
    evolutionary model of intelligence, 361–363
    Hogan Development Survey, 369, 370
    intelligence components, 364
    intelligence structure models, 367
    measures of cognitive ability, 359–360
    measuring intelligence, 363, 372
    mental ability, 360
    metarepresentation, 362, 363
    nature of scientific thinking, 365
    negative affectivity, 370
    neurotic needs, 368
    personality taxonomy, 368
    phenomena detection, 366
    positive affectivity, 370
    problem finding, 359, 363, 372
    problem solving, 359, 363, 372
    psychometric intelligence, 360, 361
    race realists, 360
    reasoning style, 369–372
    reproduction, 364
    restraint, 370
    scientific evidence on intelligence, 358
    scientific goal of computational intelligence, 361
    strategic reasoning, 359, 363, 372
    structure of intelligence, 364–366
    tactical reasoning, 359, 363
    theory construction, 366

H

HDS, *see* Hogan Development Survey
HF/E, *see* Human factors/ergonomics
HFES, *see* Human Factors and Ergonomics Society
Hick–Hyman law, 50
Hogan Development Survey (HDS), 369, 370
Human factors/ergonomics (HF/E), 3, 5, 24

Human Factors and Ergonomics Society (HFES), 4
Human information processing, 112
Human reliability control, complexity of, 94
HUMOSIM, 32–38
  integrating HUMOSIM into cognitive architecture, 35–37
  mind simulation, 33–35
    how it works, 34
    motor system in detail, 35
  rules of thumb for interface design, 37
  specific task, 37–38
  Stretch Pivot method, 32

I

IEA, *see* International Ergonomics Association
IK algorithms, *see* Inverse kinematics algorithms
ILR, *see* Image Learning and Recognition
ILR-I, *see* ILR with Interference
ILR with Interference (ILR-I), 178
Image Learning and Recognition (ILR), 178
Implicit Association Task, 381
Index of difficulty, 51
Information
  processing
    EI, 305
    oscillatory structure of, 95
    relational self-construal and, 380–381
  system, automated, 97
Informer fallacy, *see* Cane, movements of prior to locomotion judgments
Institute for Occupational Medicine, 90
Intelligence, *see* Good judgment
Interactions and transformations, law of, 76
International Association of Activity Theory, 221
International Ergonomics Association (IEA), 3
Inverse kinematics (IK) algorithms, 30

K

Knowledge
  acquisition, AT and, 239
  declarative, 34, 43
  disruption, learning and, 198
  emotion understanding, 326
  level, operator, 90
  procedural, 34, 43
  professional, 93
  spontaneous, 44
  thinking and, 44
  transfer, 44
  vocational training, 43, 45

L

Lateralized-readiness potential (LRP), 138
Learning
  activity
    character of, 65
    performance and, 57
  AT structure of, 239
  discovery, AT and, 239
  knowledge disruption and, 198
  self-regulative concept of, 239
Learning process, emotional regulation of, 325–339
  attention switching, 334
  components and skills of emotional competence, 326
  domination of inner motive, 330
  experienced emotions, 331
  formative experiments, 335
  knowledge, 326
  levels of emotional–motivational regulation, 330
  meaning of experience, 326
  method of emotion regulation, 331–333, 337
  psychological components of emotion regulation, 329
  purposes and methods of study, 328–331
  rehabilitation of emotional conditions, 334
  student age differences, 333–334
  student emotional sphere regulation, 332

systematic development of emotional stability, 336
system of emotional regulation, 332
teacher emotional maturity, 334
theory of multiple intelligences, 325
understanding of emotion, 327
verification of emotion regulation method, 333–336
LRP, *see* Lateralized-readiness potential
Luminance ratio, 125

M

Machine, *see* Brain, mind, and machine (new interface of)
Magnetoencephalography (MEG), 168
Maintenance of Wakefulness Test (MWT), 172
Maximum volunteer contraction (MVC), 79
Means-end abstraction hierarchy, 21
Medial frontal negativity (MFN), 211
MEG, *see* Magnetoencephalography
Memory, interaction of sleep and, 189–202
    AB–AC interference paradigm, 193
    accident influences, 190
    active consolidation hypothesis, 194, 195
    behavioral drive, 189
    consolidation, 189
    conventional sleep terminology, 198–199
    IQ, 197
    long-term potentiation of memory traces, 195
    memory consolidation
        definition of, 189
        enhanced, 196
        neurobiological processes and, 190, 198
        sleep and, 191, 193
        SWS and, 195
        ultradian cycles and, 197
    nondeclarative processes, 190
    passive-protection perspective, 192
    permissive consolidation hypothesis, 194
    REM sleep, 191

repair of damaged memories, 197–198
    reverse replay, 195
    sleep facilitating memory, 193–198
    sleep not facilitating memory, 191–193
    sleep as temporary shelter, 192
    stability-plasticity problem, 198
    test of transitive inference, 196
    ultradian cycles, 196, 197
    verbal memories, 193, 94
Mental modules, identification of, 111–134
    additive factors, 117
    combination rule, 116
    composite measure, 116
    correct rejection probability, 124
    decision-specific factors, 124
    discussion, 131–132
    Donders' subtraction method, 128
    effect of sleepiness on cognitive functioning, 128
    empty trials, 118
    food deprivation, 131
    food trials, 118
    functional distinctness, 114
    human information processing, 112
    inferential logic, 124, 130, 131
    isolation of timing module in rat, 118–119
    luminance ratio, 125
    mapping familiarity, 120
    measure of discriminability, 126
    mental processing stages inferred from reaction times, 119–123
        method of additive factors, 119–120
        selectivity of effect of sleep deprivation in process with stages, 120–123
        support for stages from analysis of neural process, 123
    mental-process module, 113
    modular analyzers, 129
    modular decomposition, 113
    modules and modularity, 112–113
    multiplicative combination rule, 130
    neural-process module, 113
    partial modularity, 124
    payoff matrix, 127

process decomposition method,
   114–118
   overview of examples and issues,
      117–118
   processes and their measures,
      pure and composite, and
      combination rules, 114–117
   separate modifiability, process-
      specific factors, selective
      influence, and functional
      distinctness, 114
process decomposition versus task
   comparison, 127–129
   Donders' subtraction method,
      128–129
   finding which process is influenced
      by manipulation, 128
   two methods compared, 127
process-specific factors, 114
pure measures, 114, 115
reinforcement ratio, 125
response bias, 124
response process, 119
response-rate function, 119
selective influence, 114, 119
sensory-specific factors, 124
separate modifiability, 112
signal detection theory, 123–127
   finding of only partial modularity,
      123–125
   modularity of sensation and
      decision, 125–127
sleep state, 120
spatial-frequency analyzers, 130
stimulus quality, 120
summaries of other examples,
      129–131
   modular processes for learning
      and motivation, 131
   modular spatial-frequency
      analyzers from detectability of
      compound gratings, 129–131
   modular spatial-frequency
      analyzers from selective
      adaptation, 129
task comparison method, 118
task process modules, 122
MEPS study, 16
Method of additive factors (AFM), 118,
   120

MFN, *see* Medial frontal negativity
Mind, *see* Brain, mind, and machine
   (new interface of)
Mind simulation, HUMOSIM, 33
Model(s)
   activity theory
      first-generation, 249
      second-generation, 250
      third-generation, 251
   digital human, 29, 30
   emotional intelligence, 305
   executive attention, 208
   Five-Factor, 368
   Gaussian, equal-variance, 125
   goal formation process, 237
   human driving performance, 37–38
   intelligence
      adaptive, 366
      evolutionary, 361
      structure, 367
   multilevel regulation of social
      behavior, 311
   personality, Big Five, 304, 315, 321
   sleep, REM-like, 198
Motion database search, 33
Motivation
   AT focus on, 239
   learning and, 328
   preconscious stage of, 64
   process-related stage of, 65
   time standards and, 66
Motor-cortex asymmetry, index of, 138
MSDs, *see* Musculoskeletal diseases
Musculoskeletal diseases (MSDs), 71
   disabilities caused by, 73
   epidemiology, electronic assembly
      work, 72
   ergonomic improvements reducing, 9
   workstation lowering risk of, 86
Mutual adaptation, law of, 74
MVC, *see* Maximum volunteer
   contraction
MWT, *see* Maintenance of Wakefulness
   Test

N

Neural modules, identification of, 135–164
   category discrimination, 142

combination rule, 154
comments and questions, 159–161
    implications of brain metabolism
        constraints, 161
    quantitative versus qualitative
        task changes, 159160
    relation between mental and
        neural modules, 160
    separate modifiability as criterion
        for modularity, 160–161
    specialized processors and
        modular processes, 160
    task-general processing modules,
        159
decomposing neural processes with
    lateralized readiness potential,
    138–144
    analysis of reaction-time data in
        example, 139–141
    parallel modules for
        discriminating two stimulus
        features, 141–144
    serial modules for preparing two
        response features, 138–139
Donders' subtraction method, 154
evidence from ERP amplitude for
    modular processes in semantic
    classification, 158–159
Go-NoGo Discriminability, 142, 143
lateralized-readiness potential, 138
localized neural processor, 160
location discrimination, 142
LRP-based pure measures, 141
mapping familiarity, 146–148
measure of relative sensitivity, 153
modular mental processes, 145
motor-cortex asymmetry, 138
neural processing modules inferred
    from brain activation maps,
    144–152
    modular processes in number
        comparison, 145–146
    modular processes for stimulus
        encoding and response
        selection, 146–152
numerical proximity, 146
overadditive interaction, 151
overview of examples and issues,
    136–138
partial information, 142

patch-to-key mapping, 148
process decomposition, 160
process decomposition versus task
    comparison, 152–155
    analog of Donders' subtraction
        method applied to fMRI data,
        154–155
    tactile perception tasks, 152–154
semantic satiation, 158
sleep deprivation, 155
sparse coding, 161
spatial mapping compatibility,
    146–148
stimulus-response mapping, 142, 146
use of TMS to associate mental
    processes with brain regions,
    155–158
    number comparison and rTMS,
        157–158
    visual search and TMS, 156–157
Neuroergonomics, 168, 179
Neuropharmacoassays, 178–179
Neurotechnology, 167
Nonvisual locomotion, 269, *see also*
        Cane, movements of prior to
        locomotion judgments

O

Object-oriented activity, 223, 248
Obstructive sleep apnea (OSA), 177
On-the-job training, 41
Operational tension, 52
Operator(s)
    errors, analysis of, 89
    knowledge level of, 90
    performance efficiency, 92
    professional selection of, 90
    psychophysiological condition, 90
    reliability control, functional
        structure of, 93
    task performance with new
        ergonomic equipment, 16
Operator functional state and
    fitness-for-work, day-to-day
    monitoring of, 89–107
    Anokhin's theory of functional
        systems, 94
    approach methodology, 93–98

selection of informative psychophysiological parameters, 94–97
system engineering, 97–98
background, 89–91
chronic diseases, 104
computer test performance, 99
fluctuating structure of information processing, 95
human–machine system control, 93
human reliability control, 94
Institute for Occupational Medicine research, 90
interindividual distinctions, 96
levels of activation, 95
measurement of physiological parameters, 91
mental activity dynamics, 104
methodological attitude, 94
methods, 98–100
nonoptimal effort, 91
operator errors, 89
operator performance efficiency, 92
operator psychophysiological condition, 90
operator reliability control, 93
oscillatory structure of information processing, 95, 102
physiological cost of mental workload, 105
power unit operator, 90
preshift control system, 98
professional selection of operators, 90
regulating influence, 97
results and discussion, 100–104
    correlation between task performance and physiological indices, 101–102
    fluctuations of task performance time, 103–104
starting intrashift, 98
system design principles, 97–98
system hypothalamus-hypophysis-adrenal glands, 96
target setting, 91–93
task performance time, 95
time pressure modeling, 100
Traube–Gering waves, 103

variations in operator performance, 104
OSA, *see* Obstructive sleep apnea
OSHA, *see* U.S. Occupational Safety and Health Administration
OSHA Web site, 4

P

Paired-associate learning/memory task (PAL), 172
PAL, *see* Paired-associate learning/memory task
Payoff matrix, 127
PCB, *see* Printed circuit board
PCC, *see* Posterior cingulate cortex
Perception–action cycle, 18
Personality, *see also* Good judgment
    defining, 368–369
    intelligence and, 368
    principle, activity theory, 226
    taxonomy, 368
    traits, Big Five, 304, 315, 321
PET, *see* Positron emission tomography
Phenomena detection, 366
Piaget's theory, AT and, 239
Positron emission tomography (PET), 168, 203
Posterior cingulate cortex (PCC), 208
Power spectral density (PSD), 171, 172
Printed circuit board (PCB), 72, 77
    image angle, 81
    study of optimal angle of, 77–81
        conclusion, 80
        equipment, 77–78
        method, 77
        participants, 78
        procedure, 78–81
Procedural knowledge, 34, 43, 44
Process decomposition method, 118
Professional trainer, emotional resources of, 341–354
    behavioral activity components, 347
    cognitive-resource deficiency, 347
    definition of objectives, 352
    emotional burnout syndrome, 345, 347, 350
    emotional stability, 344
    formulation of general problem, 341–348

global criteria, 343
individual separation, 351
maintenance of well-being, 352
mental burnout, 345
mental tension, 344
methods and procedures, 348
narrative, 342–343
negative influences, 352
predisposition to conflict, 346
professional burnout, 346
psychological potential, 342
psychological skills, 353
qualities, 343–344
results, 349–353
self-estimation, 346
significant individual quality, 344
social activity, 344
tension response forms, 345
triggering mechanisms of emotional
    experience, 344
ways to transmit experience, 342
PSD, *see* Power spectral density
Psychology, ecological approach to, 7
Psychometric intelligence, 360, 361

R

Race realists, 360
Reasoning style, 369–372
Relational self in action, 377–392
autonomous-relational self, 386
close relationships, 384, 391
consistency motive, 385
gender differences in behavior, 379
Implicit Association Task, 381
inescapable fact of human existence,
    387
influence on behavior, 391
involuntary relationships, 384
life satisfaction, 384, 388, 390, 391
measures of self-definition, 392
myself-ideal similarity score, 389–390
obligatory relationships, 387
personal meaning systems, 388
phenomenon of multiple selves, 377
Psychosemantic approach, 388
real self, 377
Relational-Interdependent Self-
    Construal Scale, 380, 382, 384
    relational self-construal
        culture and, 386–389
        gender and, 378–380
        information processing and,
            380–381
        measuring, 380
        new relationships and, 381–384
        relationships, well-being, and,
            384–386
        similarity scores and, 389–390
    relationship-related processes, 379
    relationships and multifaceted
        selves, 391–392
    self-disclosure, 382, 383
    self-enhancing processes, 379
    women's self-ratings, 387, 388, 391
Relationships, *see* Relational self in
    action
Repetitive transcranial magnetic
    stimulation (rTMS), 152, 157
rTMS, *see* Repetitive transcranial
    magnetic stimulation

S

Safety, *see* Attention, selection for
    action, error processing, and
    safety
SDT, *see* Signal detection theory
Seated work posture, 8–15
    backward-leaning posture, 11
    buttock-popliteal length, 15
    cubist posture, 9
    effectiveness of interventions, 9
    elbow rest height, 12
    forward-leaning posture, 11
    interdependencies in seated posture,
        9–15
    keyboard heights and, 14
    popliteal height, 12
    relationship between working
        conditions and disorders, 8–9
    seated eye height, 14
    seat height adjustability, 12
    task-related postural requirements,
        15
Selection for action, *see* Attention,
    selection for action, error
    processing, and safety

Self-regulation, mechanism of, 43, 68
Signal detection theory (SDT), 123
Skills acquisition process, student
        attention during, 45, 47
Sleep, *see also* Memory, interaction of
        sleep and
    debt, 170
    deprivation
        effect of in process with stages,
            120–123
        error correction ability and, 212
        reaction time and, 155
        susceptibility to effects of, 174
    models, REM-like, 198
Sleepiness, effect of on cognitive
        functioning, 128
Slow wave sleep (SWS), 195
Social neuroscience, 168
Social positioning, activity and, 252,
        253, 258
Spontaneous knowledge, 44
SSAT, *see* Systemic-structural activity
        theory
Stretch Pivot method, 32
Student(s)
    differences in performance time, 58
    elements of work, 61
    emotional sphere regulation, 332
    emotional state of, 328
    fatigue, 68, 183
    learning, models of, 181
    life satisfaction, 391
    problem-solving task, discovery
        learning and, 239
    productivity dynamics, 60
    skills acquisition process, 45, 47
    task performance, 62
SWS, *see* Slow wave sleep
Symvatology, 6, 20
System engineering, 97
Systemic-structural activity theory
        (SSAT), 46, 230
    activity analysis, 232
    activity decomposition, 231, 235
    algorithmic analysis, 231
    analysis activity within, 232
    cognitive analysis, 230
    learning, 238, 239
    morphological analysis, 231
    units of analysis, 234, 242

vocational training and, 42

T

Task(s)
    changes, quantitative versus
        qualitative, 159
    cognitively demanding, 209
    comparison method, mental modules
        and, 118
    complication of, 56
    -general processing modules, 159
    microstructure, 19
    performance
        activity during, 232
        average time of, 54
        time (TPT), 99, 101, 102
    performance strategy
        externally given instructions and,
            57
        motor action and, 55
        spatial position in, 56
Technology, operator performance and,
        92
Theory construction, 366
Thinking, *see* Good judgment
TMS, *see* Transcranial magnetic
        stimulation
TPT, *see* Task performance time
Trading zones, 24
Trainer, professional, *see* Professional
        trainer, emotional resources of
Training fallacy, *see* Cane, movements of
        prior to locomotion judgments
Transcranial magnetic stimulation
        (TMS), 136, 137
    mental processes and, 155
    visual search and, 156
Tunnel vision, 207

U

Ultradian cycles, 196, 197
U.S. Occupational Safety and Health
        Administration (OSHA), 8

V

VDTs, office work with, 9

Verbal Memory Scan (VMS), 178
Virtual product design, integrating
    cognitive and digital human
    models for, 29–40
    ACT-R cognitive architecture, 34
    database of movements, 32
    declarative knowledge, 34
    declarative memory module, 34
    digital human models, 29, 30–32
    HUMOSIM, 32–38
        integrating HUMOSIM into
        cognitive architecture, 35–37
        mind simulation, 33–35
        specific task, 37–38
    interactive virtual design
        assessment, 31
    inverse kinematics algorithms, 30
    motion database search, 33
    procedural knowledge, 34
    Stretch Pivot method, 32
    summary, 38
    virtual prototypes, 29
Visual impairment, *see* Cane,
    movements of prior to
    locomotion judgments
VMS, *see* Verbal Memory Scan
Vocational training, time study during,
    41–70
    above real-time training, 50
    acquisition of professional
        knowledge and skills, 45–48
    activity goal, 55
    average time of task performance, 54
    cognitive functions, 56
    declarative knowledge, 43
    elements of work, 61
    emotional tension, 52
    energy expenditure, 49
    Fitts' law, 50
    functional analysis of pace formation
        process, 53–57
    goal-directed self-regulative process,
        43
    goal-related motivation, 64
    Hick–Hyman law, 50
    index of difficulty, 51
    influence of time standard on
        vocational school student
        performance, 57–63

knowledge and skills classification,
    43–45
knowledge transfer, 44
learning activity, 65
main purpose of teaching, 44
monotonous work, 65
motivation, 64, 65
on-the-job training, 41
operational tension, 52
pace evaluation, 48
pace of performance, 48–53
performance speed, 53
performance strategies, 46
procedural knowledge, 43, 44
productivity dynamics, 66
purpose, 42
purpose of craft classes, 60
reducing monotony, 59
repetition without repetition, 47
skill acquisition learning curves, 47
spontaneous knowledge, 44
State Committee of Vocational
    Education of Ukraine, 58
student behavior, 59
student productivity dynamics, 60
students raising the bar, 67
systemic–structural activity theory,
    42
task performance time, 62
theoretical classes, 44
thinking, 44
time standard, 67
time standards and trainee
    performance, 63
time study and work motivation,
    64–68
transformation, 45
work pace, 48

W

Work
    domain analysis, 21
    pace
        definition of, 48
        evaluation, 48, 49
    process, main source of errors in,
        203, 206
    productivity, measurement of, 83

Worker competencies, 22
Workplace design, application of
        ergonomic principles to, 9
Workstation(s)
    assembly, comparison of
            productivity, 84
    data comparison, 85
    electronic assembly, 74, 75
    existing, work comfort compromised
            by, 81
    postural criteria, 13
Workstation design and testing, laws of
        ergonomics applied to, 71–87
    carpal tunnel syndrome, 73
    disabilities caused by MSD, 73–74
    electronic assembly workers, 72
    experimental study of PCB optimal
            angle, 77–81
        equipment, 77–78
        method, 77
        participants, 78
        procedure, 78–81
    Eye Gaze Development System, 78,
            80
    industrial comparative testing of
            assembly workstations, 83–85

invention of assembly workstations
        with indirect observation of
        operations and negative tilt of
        work surface, 81–82
laboratory testing of new
        workstations, 82–83
law of interactions and
        transformations, 76
law of mutual adaptation, 74
laws of ergonomics, 74–77
law of work structures plurality, 75
musculoskeletal diseases, 73
optimized work surface height and
        angle, 82
repetitive wrist flexion and
        extension, 74
work productivity measurement,
        83
Work structures plurality, law of, 75

Z

Zone of proximal development (ZPD),
        327
ZPD, *see* Zone of proximal development